NEW ENERGIES

Copenhagen

Land Art Generator Initiative

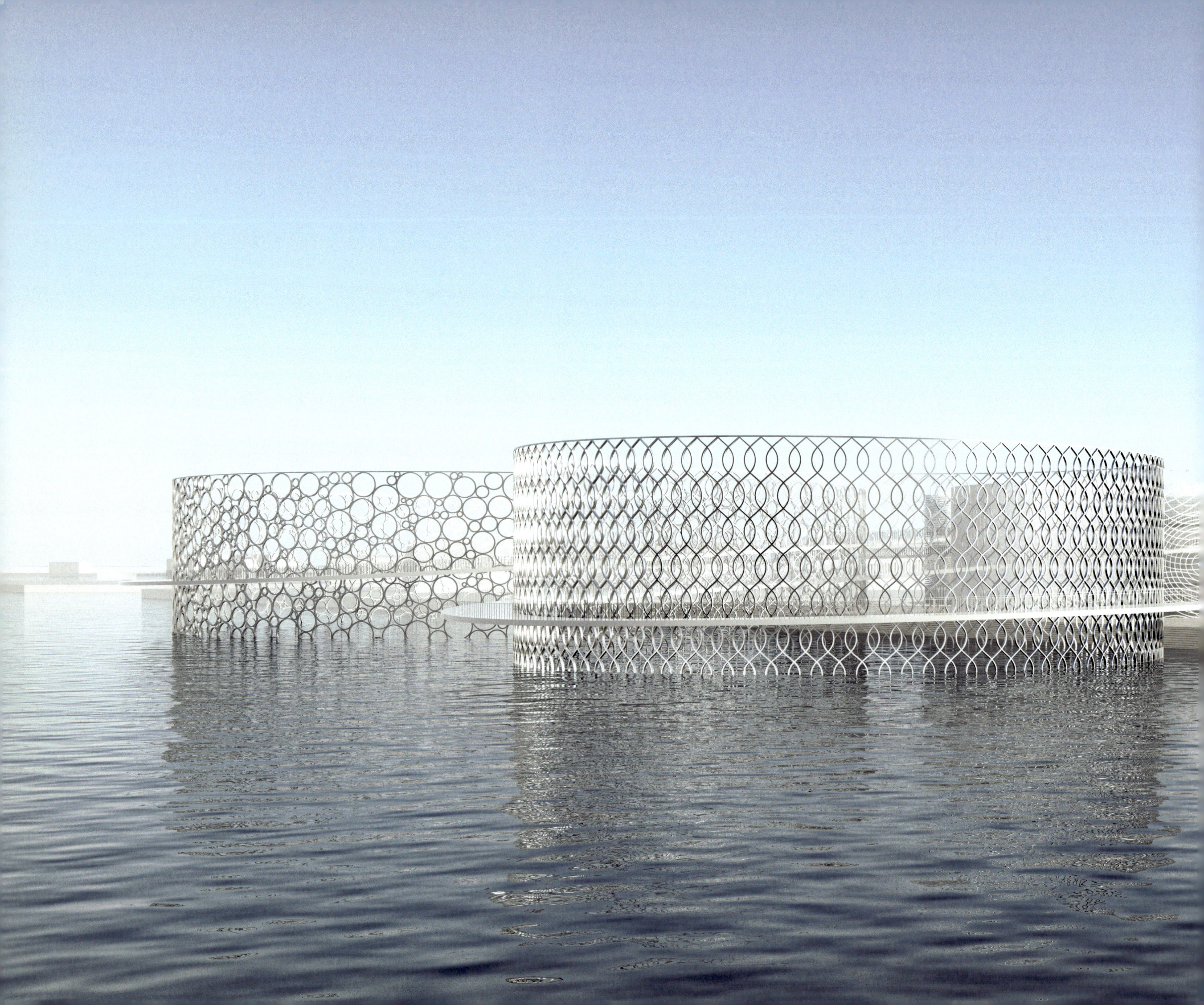

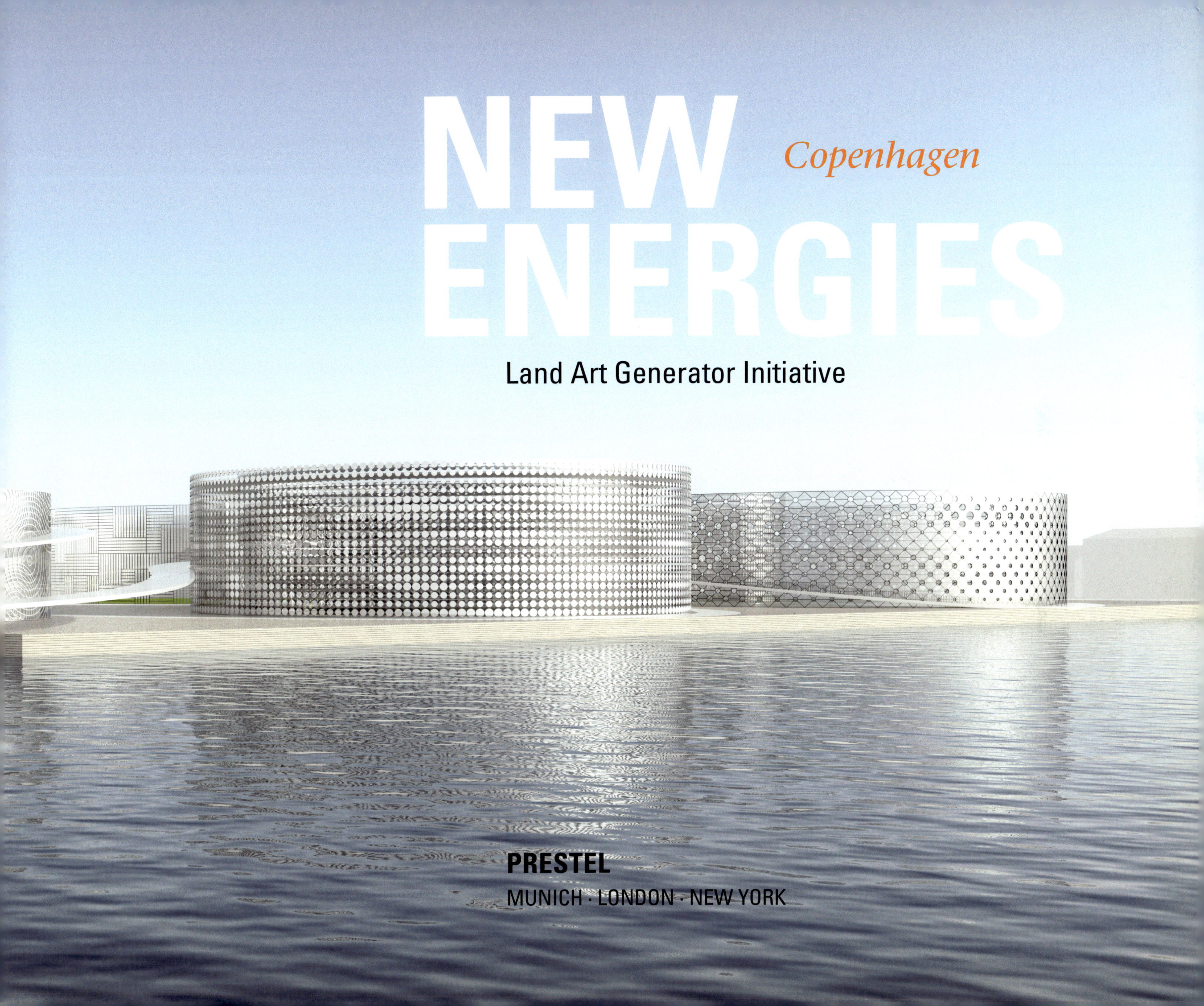

NEW ENERGIES

Copenhagen

Land Art Generator Initiative

PRESTEL

MUNICH · LONDON · NEW YORK

CONTENTS

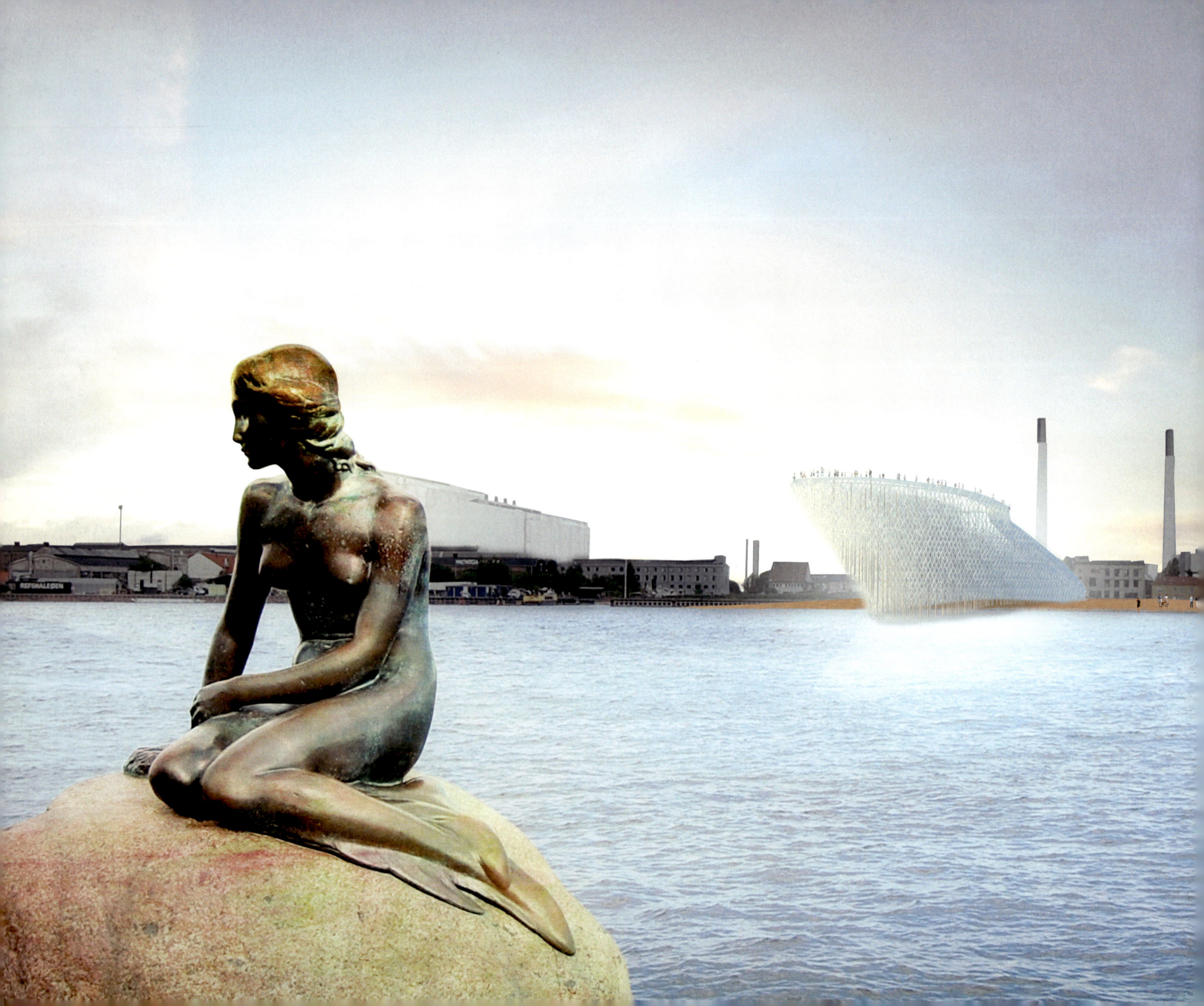

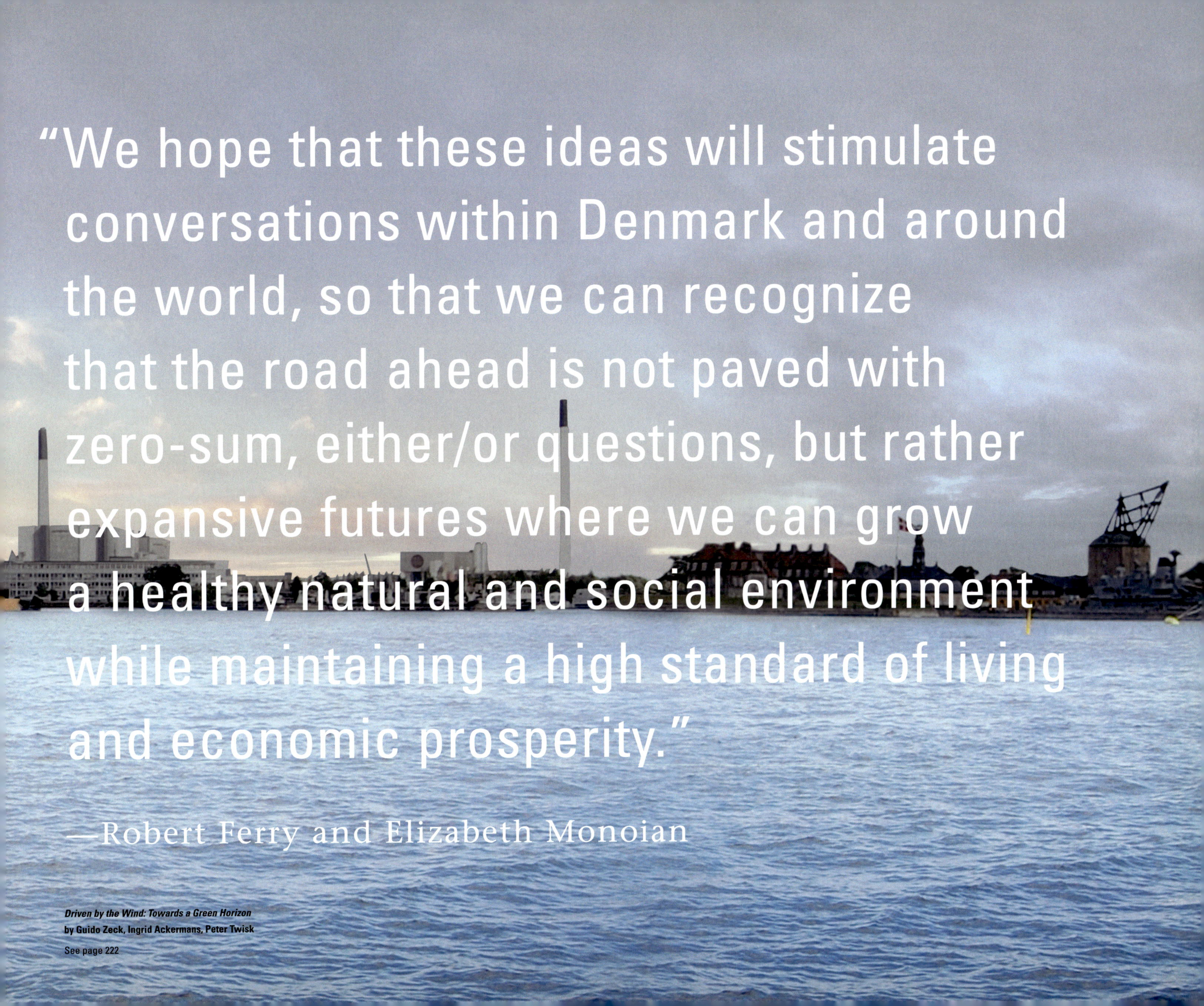

"We hope that these ideas will stimulate conversations within Denmark and around the world, so that we can recognize that the road ahead is not paved with zero-sum, either/or questions, but rather expansive futures where we can grow a healthy natural and social environment while maintaining a high standard of living and economic prosperity."

—Robert Ferry and Elizabeth Monoian

Driven by the Wind: Towards a Green Horizon
by Guido Zeck, Ingrid Ackermans, Peter Twisk
See page 222

INTRODUCTION

New Energies

Robert Ferry and Elizabeth Monoian

In 2012, the Land Art Generator Initiative design site at Freshkills Park brought forth hundreds of ideas about how we can creatively adapt our reclaimed landfills for renewable energy infrastructures, seamlessly integrated into a regenerated natural environment—beautiful public energy parks to power hundreds of homes while providing safe places for people to recreate and learn. The juxtaposition of reclaimed landfill and energy infrastructure design provided a fertile conceptual terrain onto which artist teams imagined their work.

In 2014, LAGI has explored new territory with new conceptual frameworks. The 2014 site at Refshaleøen is an industrial brownfield site. Its history is unique (as you will read in the essay by Annette Skov on page 16), but every city has a site similar to this. These are places where residential development is not yet possible due to high levels of environmental pollutants deposited year after year during a long industrial past. In places such as these, there is a great opportunity to conceive of iconic public art installations that can stand watch over history until the ground and air are clean enough to be sustainably developed into the fold of the growing city. In the meantime, these sentinels can serve to reduce our dependence on fossil fuels by generating clean, renewable electricity for the city.

The entries to this year's competition have once again pointed the way to a future in which our sustainable energy systems can exist in complete harmony with the living city around us.

Will humanity transition to 100% carbon-free energy? Yes, we will because we must, and the solutions already exist.[1] Fossil fuels are a finite resource and their unrestrained extraction/combustion is damaging to human health and the environment.[2] The question is: how difficult will the change be, and how long can we put off the inevitable before the effects of climate change bring suffering to millions and ruin to the global economy?

If we are to make a fairly seamless transition, then we must possess an appreciation for history and acknowledge that individuals and corporations acting without adequate government coordination and leadership will not be able to find solutions quickly enough.

Consumer movements, activism, and corporate responsibility have brought us to a point in time where a clear majority of people are in favor of public investment in renewable energy. If we are to continue to rely mostly on the marketplace to dictate the terms of the transition, all progress will be threaded through the financing needle of net present value determinations using time horizons that are too short-sighted for our long-term collective interest.[3]

Instead, we desperately need a coordinated effort organized by governments, and policies that can override the short-term economic models. According to the International Energy Agency, in order to meet the global carbon reduction challenge, the world will need to invest $6 trillion over the next 25 years.[4]

When we are facing a collective challenge such as this, we would be wise to transcend narrow commercial interests and recognize the value of economy-wide returns like those associated with renewable energy investment. We have done this before during times of critical mass determination, in order to address social injustices or to expand the boundaries of scientific progress.[5]

1 Mark Z. Jacobson, Guillaume Bazouin, Zack Bauer, et al., "100% Wind, Water, Sunlight (WWS) All-Sector Energy Plans for the 50 United States," (Atmosphere/Energy Program, Dept. of Civil and Environmental Engineering, Stanford University, July 17, 2014). **http://web.stanford.edu/group/efmh/jacobson/Articles/I/USStatesWWS.pdf (http://thesolutionsproject.org)**

2 If we are to avert the most devastating effects of global climate change by staying at or below 2° Celsius of warming, we must abandon (leave in the ground and not use) 80% of the proven reserves of fossil fuels that the industry is now claiming on their balance sheets as assets. For more information, refer to the Carbon Tracker Initiative's report, "Unburnable Carbon" at **http://carbontracker.org**

3 A rate of return of 12–15% (IRR) and investment horizons of 7–10 years are a barrier to some renewable energy investments, which may require more than a decade to achieve strong returns on investment. Technological innovation to reduce the EROI (energy return on energy invested) of renewable energy systems (to make them more competitive with fossil fuels in the marketplace) requires expensive research and development that is not easily borne by the private sector. Feed-in-tariffs, portfolio standards for utilities, and tax incentives can make renewable projects more commercially attractive to private investors, but the fact that global GHG emissions continue to increase year-on-year shows that these mechanisms alone are not sufficient.

It is time to come together again in the name of renewable energy investment, and the way forward is to look to the best models that are already in place.

The success of wind power in Denmark is an example of what can happen when people come together as a collective and consider energy and climate change from the point of view of the common good.[6]

In our recent conversation with Natalie Mossin (February 2014), the head of the Danish Architects' Association (Arkitektforeningen), she noted:

> The overall objective of a sustainable society is, in the end, a collective objective. Taking over ownership of energy production is not an end in itself. It is the taking ownership of the collective solution that is interesting. The initiatives of individuals should be a part of that. The Andelsbevægelsen, or cooperative movement, in Denmark is a really strong part of our culture and it shows beautifully how each individual can take ownership in the collective solution. Each person has a role and influence, but all are acting for the common good.

This strong history of the economic cooperative has always influenced Denmark's energy sector[7] and has played a pivotal role in defining the strategy for the implementation of wind power in Denmark in the later decades of the 20th century. During that time, most wind farm developments were financed collectively by cooperatives of 20 to 40 households with some government assistance and the requirement that all those living within three kilometers of a turbine were invited to invest in its installation. A strong feed-in tariff[8] helped to ensure these installations were profitable with the earnings shared among the cooperative stakeholders.[9]

In the early years of the 21st century, changes were made to these cooperative-friendly policies. The three-kilometer rule was dropped and feed-in-tariffs were removed in favor of market incentives. The result was that investment in wind power declined rapidly between 2004 and 2006, and instances of public resistance to new installations increased. In 2008, some of the incentives were brought back, but many fear that resistance to new installations will clash with Denmark's ambitious renewable energy goals.

In order to meet the goal of 50% electricity from wind by 2020, Denmark will need to install an additional 500 MW of wind power onshore and 1,500 MW offshore.[10] Given decreasing land resources, these installations will inevitably need to come into closer proximity with residential neighborhoods and populated shorelines. Recent popular resistance to new wind turbine installations has led to research efforts to study the causes, such as the Danish Strategic Research Council's 23.6 million DKK (4.5 million USD) grant to study the dynamics of local acceptance.[11]

Denmark is setting the highest standard for the public embrace of renewable energy infrastructure, but as we have seen, even in this relatively idyllic political environment there are still a great many challenges to overcome.

[4] Brendan Pierpont, Uday Varadarajan, David Nelson, and Anne Schopp, "Renewable Energy Financing and Climate Policy Effectiveness CPI Analysis Framework (Working Paper)," (The Climate Policy Initiative, July, 2011). **http://climatepolicyinitiative.org**

[5] A dedicated investment on a par with that provided to the Apollo program ($160 billion in 2014 dollars) could be exactly what is needed to lay the infrastructural foundation for a shift to 100% renewable energy.

[6] Denmark has set a realistic goal of a complete transition to 100% renewable energy by 2050. A great overview of the strategy and history can be found in the 2012 Danish Energy Agency publication, "Energy Policy in Denmark." (Danish Energy Agency, 2012). **http://www.ens.dk/sites/ens.dk/files/dokumenter/publikationer/downloads/energy_policy_in_denmark_-_web.pdf**

[7] Denmark's first district heating plant (waste-to-energy) was running in 1903 in Frederiksberg municipality (just outside of Copenhagen). **http://dbdh.dk/district-heating-history**

[8] A statutorily guaranteed buy-back rate from the utility company for kilowatt-hours produced in excess of those consumed at a given site, such as a residence with solar panels on the rooftop.

[9] Information from Professor Frede Hvelplund of Aalborg University, featured in a 2012 article by Aedan Kernan. **http://www.leonardo-energy.org/danish-community-and-wind-power-feature**

[10] Ibid.

[11] "Wind2050 – Multidisciplinary study on local acceptance and development of wind power projects" (2014–2017) led by Kristian Borch with the Technical University of Denmark in Lyngby. From the abstract:

> The overall analytical perspective considers wind power facilities as socio-technical systems, which allows the work packages to use different scientific perspectives and methods in order to understand why and how different institutions, regulations, actors and perceptions induce or block deployment of wind power. The analyses are finally integrated in strategic scenarios, which provide recommendations and decision support for future deployment of wind power and other renewable energy sources.

The Land Art Generator Initiative sees these challenges as a welcome opportunity. In this book, you will read about how we can make energy visible by recognizing the inherent mystery of electrical power generation (p. 26). You will learn about the ways in which ecological art movements can model how to intervene in our environments and public spaces (p. 46), and how interdisciplinary collaborations between artists and scientists are opening up new avenues of design exploration (p. 42). You'll be asked to think about how science education can be advanced through the imagining of potential futures (p. 50).

You will learn about how research in Denmark is casting a light on the way that humans adjust to sustainability-centered cultural shifts (p. 28) and how actor-network theory can be applied to the complex history of the emergence of wind power as a dominant force in Danish energy production (p. 30). We find a particular relevance in Peter Karnøe's essay—a story of city-integrated and design-oriented renewable energy infrastructure, which is also a story of interacting actor networks and an evolution from establishment to emerging social systems.

Our sincere thanks are due to all those who participated in LAGI 2014 Copenhagen: as artists and designers, as project partners, as jurors and committee members, and as messengers of the information that is contained within this volume.

We hope that these ideas will stimulate conversations within Denmark and around the world, so that we can recognize that the road ahead is not paved with zero-sum, either/or questions, but rather with expansive futures where we can grow a healthy natural and social environment while maintaining a high standard of living and economic prosperity.

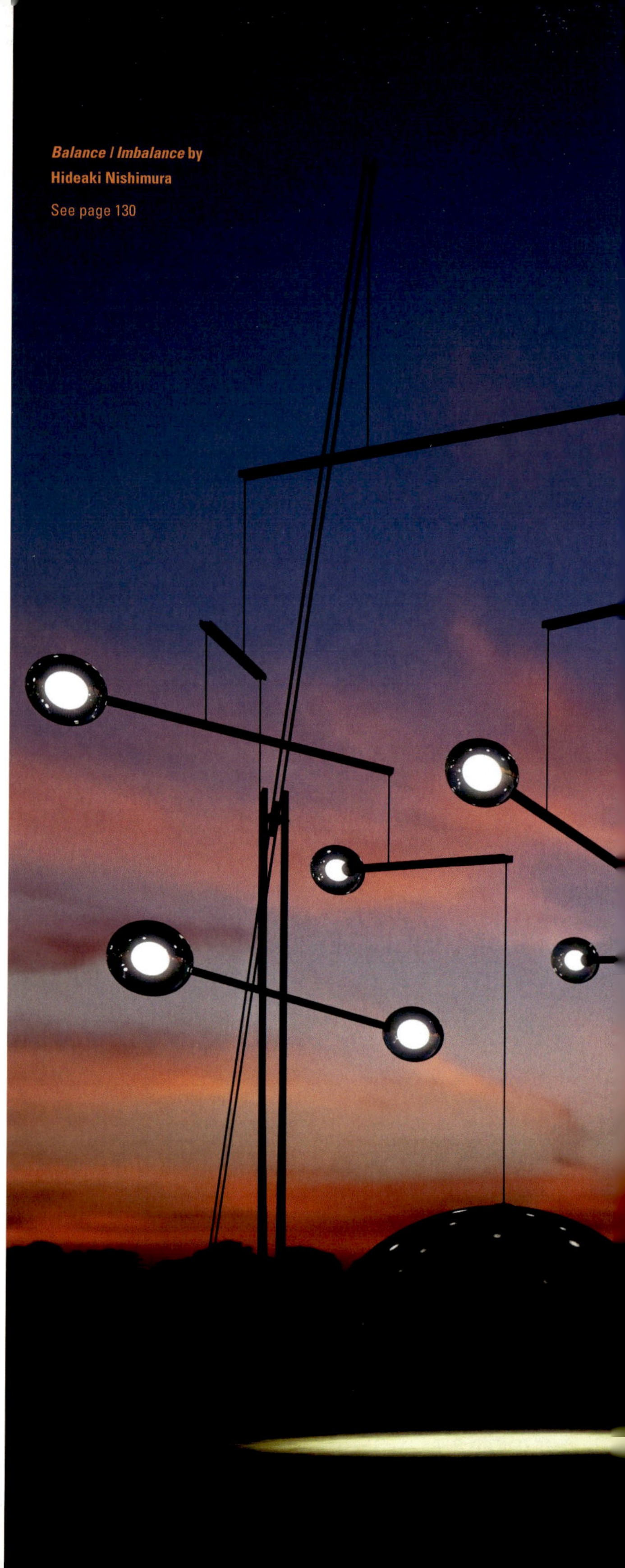

***Balance / Imbalance* by Hideaki Nishimura**

See page 130

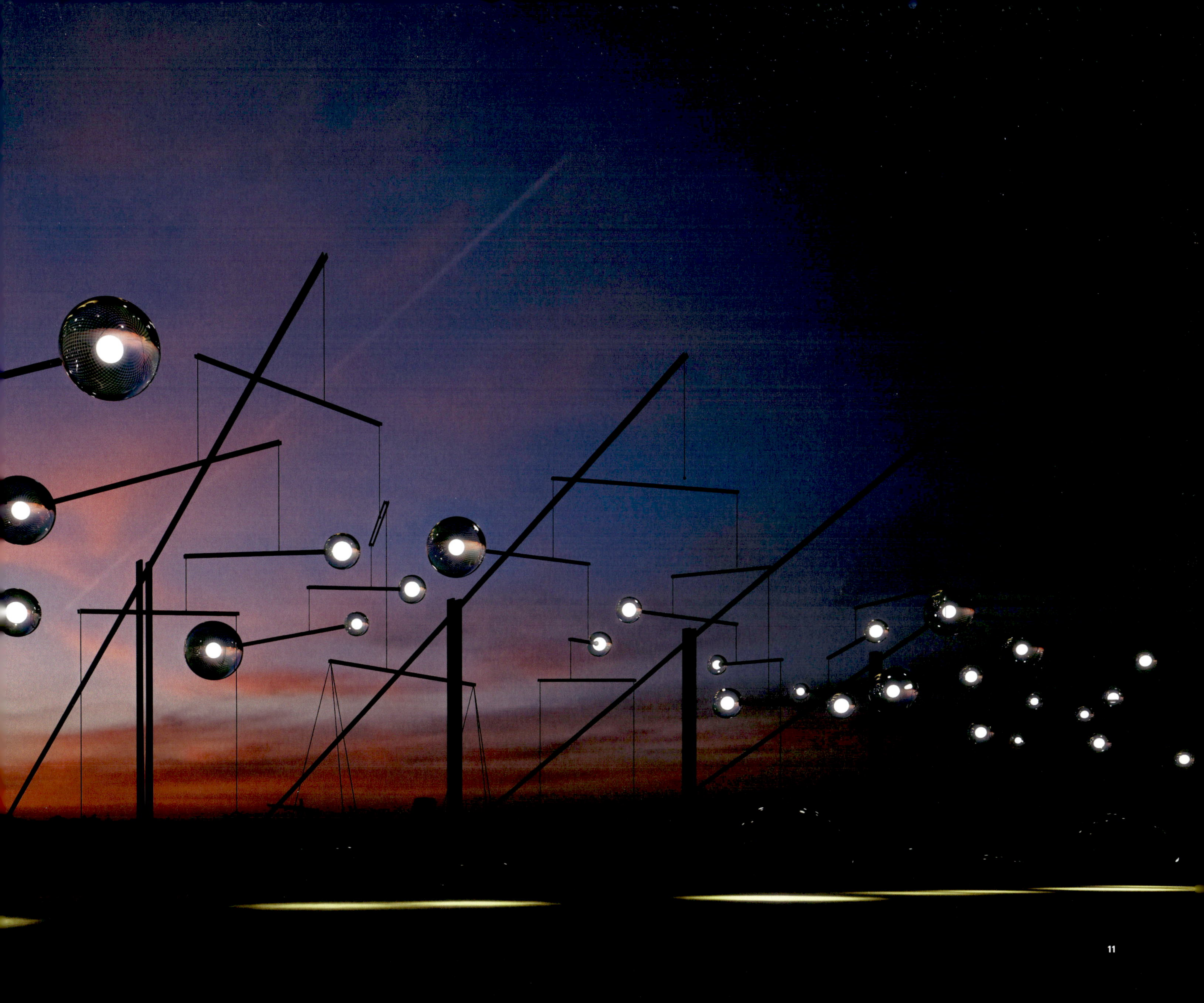

LAGI 2014 DESIGN GUIDELINES

The LAGI 2014 design guidelines were developed closely with Refshaleøen Holding and the City of Copenhagen.

The guidelines constituted the rubric for determining qualified entries to the LAGI 2014 competition.

PROJECTS MUST:

Consist of a three-dimensional sculptural form that has the ability to stimulate and challenge the mind of visitors to the site. The work should aim to solicit contemplation from viewers on such broad ideas as ecological systems, human habitation and development, energy and resource generation and consumption, and/or other concepts at the discretion of the design team;

Capture energy from nature, convert it into electricity, and have the ability to store, and/or transform and transmit the electrical power to a grid connection point to be designed by others. Consideration should be given to the artful housing of the required transformer and electrical equipment within the project boundary and restricting access to those areas for the safety of visitors to the site;

Not create greenhouse gas emissions and not pollute their surroundings. The work must not impact the natural surroundings negatively. Each entry must provide a brief (approx. 300 words) environmental impact assessment as a part of the written description, in order to determine the effects of the project on the natural ecosystem and give reference to a mitigation strategy addressing any foreseeable issues;

Be pragmatic and constructible, and employ technology that can be scalable and tested. There is no limit on the type of technology or the proprietary nature of the technology that is specified. It is recommended that the design team make an effort to engage the owners of proprietary technology in preliminary dialogue, as a part of their own research and development of the design entry. The more pragmatic the proposals are, the greater the likelihood will be that one of them may get built;

Be well informed by a thorough understanding of the history, geography, details of the design site, and the broader contexts of Refshaleøen, Copenhagen, and Denmark;

Be safe to people who would view it. Consideration must be made for viewing platform areas and boundaries between public and restricted areas;

Be designed specifically to the constraints of the design site at Refshaleøen as shown in the Location Plan;

Designs must not exceed 125 meters in height.

LAGI 2014 JUDGING CRITERIA

The LAGI 2014 jury made their decisions based on the following criteria:

- Adherence to the Design Brief;
- The integration of the work into the surrounding environment and landscape;
- The sensitivity of the work to the environment, and to local and regional ecosystems;
- The estimated amount of clean energy that can be produced by the work;
- The way in which the work addresses the public;
- The embodied energy required to construct the work;
- The perceived return on capital investment of the work, judged by the complexity of the design in relation to the energy it produces each year;
- And the originality and social relevance of the concept.

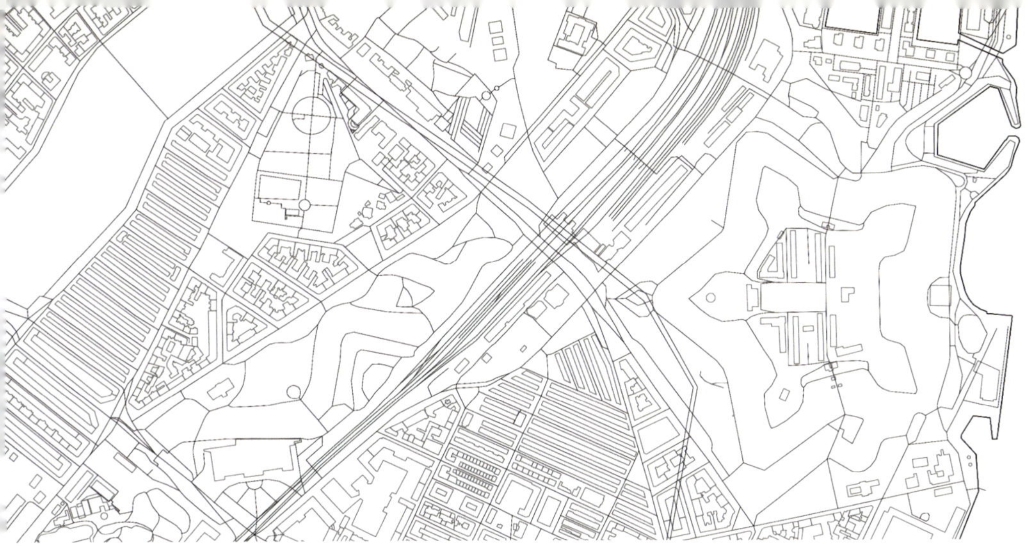

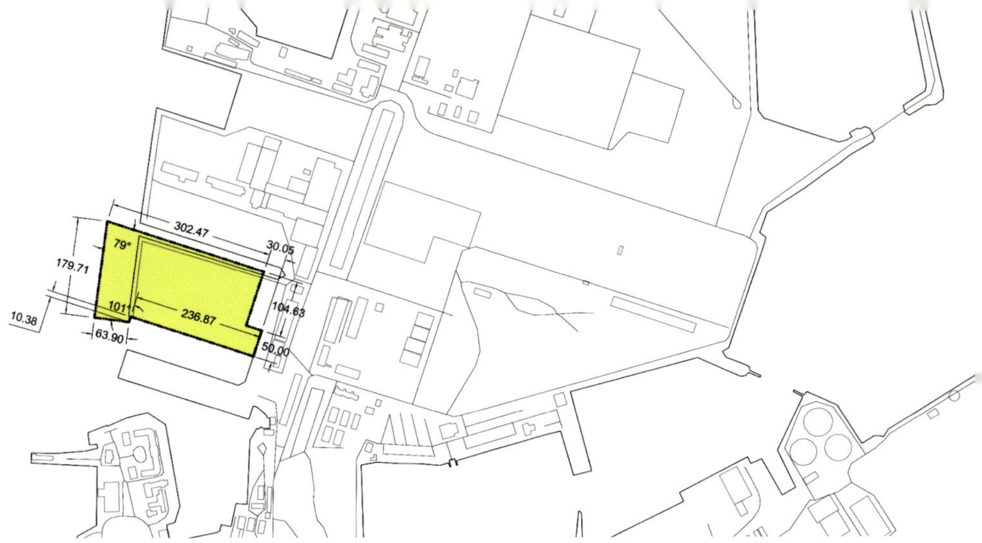

LAGI 2014 Design Competition Site

The jury and voting process took place through an online format over the month of July 2014.

OFFICIAL LAGI 2014 JURORS

Connie Hedegaard
European Commissioner for Climate Action

***Martin Lidegaard**
Danish Minister for Climate, Energy, and Building

Tina Saaby
City Architect of Copenhagen

Lars Aagaard
Director, Danish Energy Association

Jason F. McLennan
CEO, International Living Future Institute

Nille Juul-Sørensen
CEO, Danish Design Centre

Kent Martinussen
CEO, Danish Architecture Centre

Christian Herskind
CEO, Refshaleøen Holding A/S

Steen Christiansen
Mayor, Albertslund Municipality

Maria Hørmann
Change Maker—Design & Innovation; Blog Editor, Hello Materials Blog, Danish Design Centre

Else Marie Bukdahl, Dr. Phil.
Danish Art Historian, Ålborg University

Agnete Fog
Chairman of Green Cities 2014–2015

Sharon Chang
CEO and Founder, Yoxi; NYU Trustee

Chris Fremantle
Producer, ecoartscotland

Matthew Rosenberg
Founding Partner, M-Rad; 2nd Place Winner LAGI 2012

Michael Singer
Principal, Artist and Designer, Michael Singer Studio

Stig L. Andersson
Founding Partner, SLA; Creative Director, Professor

*Unable to participate due to change in position.

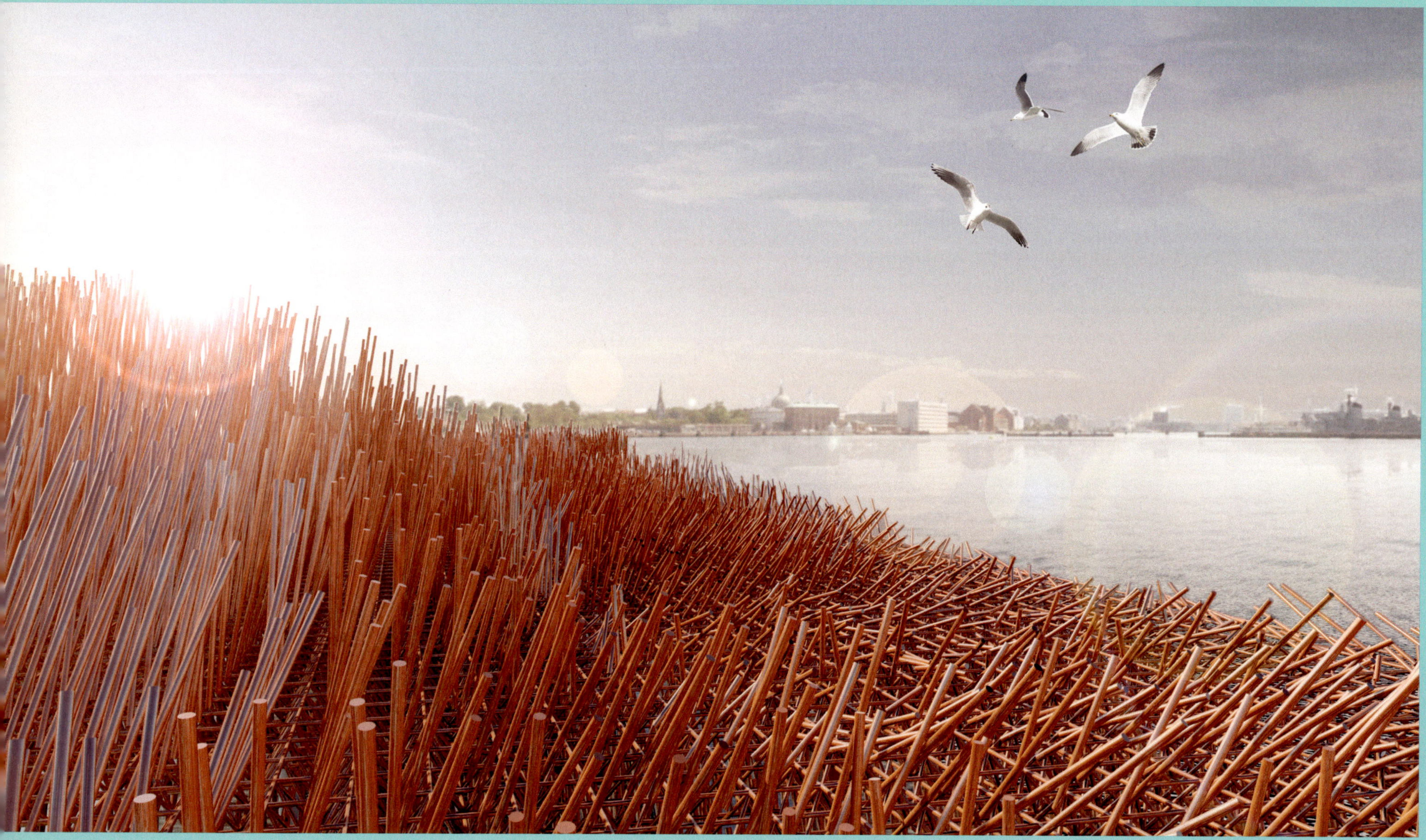

Super Cloud by Lucas Jarry, Rita Serra e Silva, Lucas Guyon, Marianne Ullmann

Piezoelectric generators at the base of each orange tube turn breezes into electricity.

See page 146

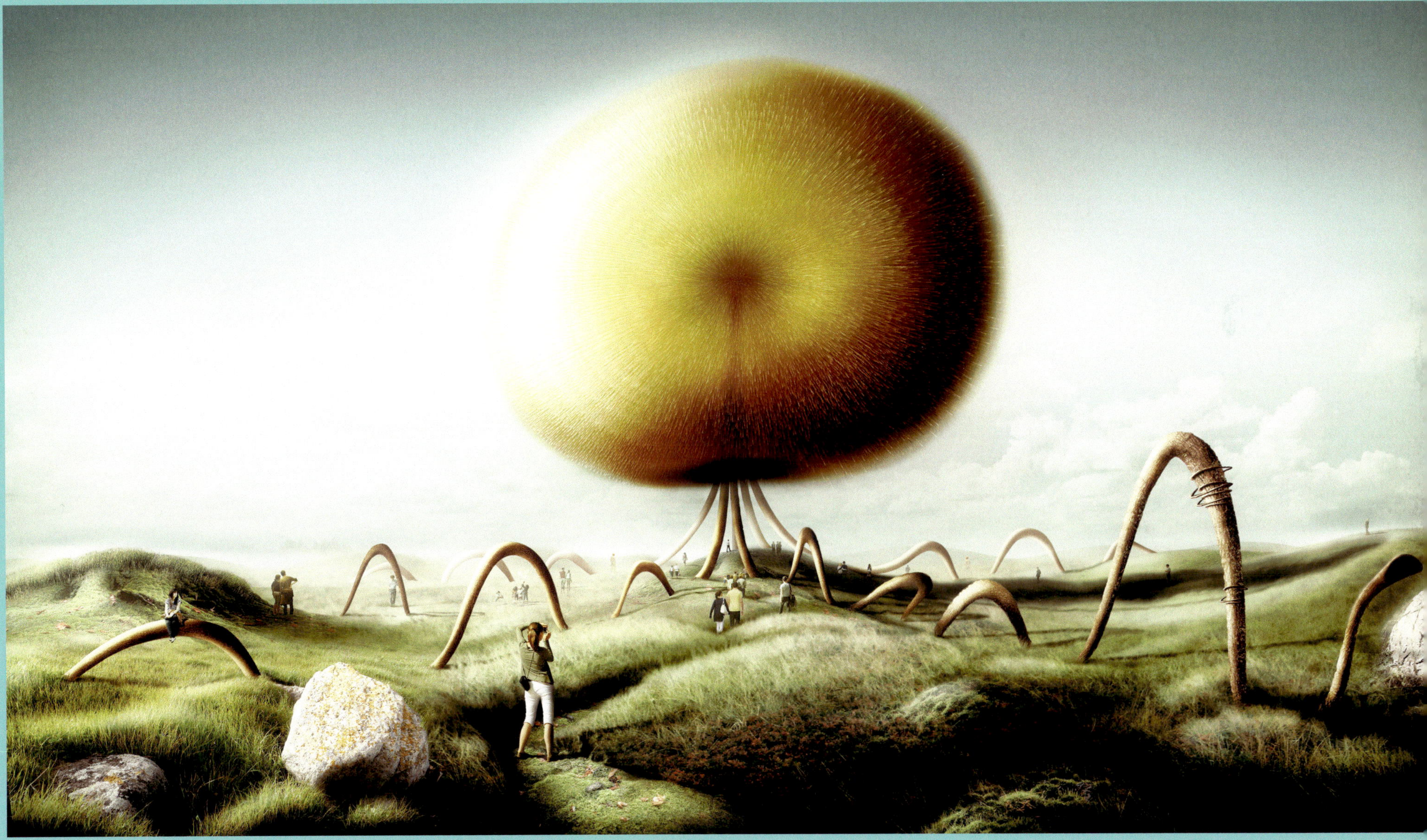

Tree of Knowledge by Boguslaw Barnas, Rafal Barnas, Omar Gharaibeh, Natalia Jejer

The roots warm the air at the ground plane. By way of the chimney effect, the air then rises naturally up to pass through a microturbine.

See page 206

ESSAY

The Glowing Island

Annette Skov

Images courtesy of Dieselhouse (www.dieselhouse.dk)

Rusty nails and bolts. Round-arched rigging sheds, crane tracks, and halls with ceiling heights of 68 meters. Worn out steel hawsers, a wrinkled schedule of guard shifts, and a dark wooden shed.

Refshaleøen is roughly 500,000 square meters of artificial land and until 1996 was the home of the Burmeister & Wain (B&W) shipyard. A peninsula in the Copenhagen harbor, it still carries the traces of 150 years of legendary industrial history.

For the workers whose days were spent at Refshaleøen, the area contains a chain of memories. For the rest of us, it is a mysterious, coded landscape for which we don't have a manual. But we can create a channel back in time through historical images, objects, and stories.

In 1956, Henry Tullin started as an apprentice at B&W. He was 15 years old. Very early each morning, he would put on his shabbiest clothes—layers upon layers—and cycle out to the workplace with his lunchbox on the carrier. Sixteen slices of rye bread with paté, sausage, and cheese. And then he would slog away with iron and tools.

There were a lot of men working in the marking hall; many different things were happening all the time, so there was a lot of noise. The magnetic cranes could drop the plates—it was rough. Unfortunately we also had a few accidents where the plates got pulled in the wrong direction and people had their legs chopped off. It's always been a dangerous place to work.

Henry continued his training; he was an active member of the sports club, and he had his friends at the shipyard. But one night in 1996 the boss called, *Don't come into work tomorrow, Henry. There's no shipyard anymore. We're bankrupt.*

That night, Denmark lost a workplace where upwards of 10,000 people had built nearly 1,000 ships. But when the shipyard closed its doors a new era opened: the transformation from heavy industry to a new part of the capital city.

THE PLACE AND THE NAME

The first part of the name *Refshaleøen* refers to the old Danish word *rævshale* or foxtail. That was because someone thought that seen from above, the island looked like a foxtail.

The area is surrounded by clean water, like the rest of the Copenhagen harbor. A few minutes' ride on the yellow harbor ferry and you are on the other side, very close to the Little Mermaid. The sculpture was placed on her rock by the waterfront in 1913 and has since become a favorite tourist attraction. Few are aware that Refshaleøen is the backdrop of the photos they take of this small bronze sculpture.

Refshaleøen neighbors the Holmen naval station, which today is the home of many educational art institutions, as well as the world famous Christiania Freetown established in 1971.

EARLY TIMES

The story of B&W begins in 1846 with the Baumgarten & Burmeister machine factory, located in the city district of Christianshavn, a few kilometers from Refshaleøen. The factory made engines, boilers, and centrifuges among other items, and in 1858 it built the first Danish steam-powered iron ship.

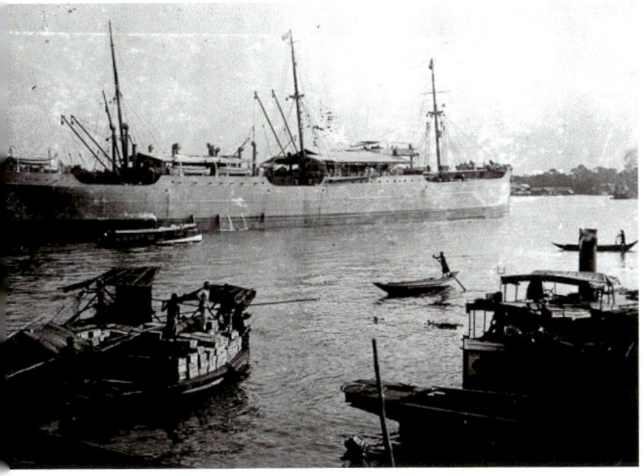

Selandia, the world's first oceangoing, diesel-powered ship

The factory was expanding rapidly and needed more space, which it found on Refshaleøen. At the time, this part of Copenhagen was merely sand banks, but the Copenhagen port authorities filled the flat area with tons of soil and gravel, and thus created new land. The first buildings stood ready in 1872.

The expansion of the island and the shipyard continued over many years. The island had ample space for everything—shipbuilding, administration, and leisure activities. Massive halls, office buildings, commissaries and sporting facilities. Beddings, floating docks, marking halls, a sanatorium, studios, a fire station, a tennis court, welding halls, and wooden sheds for the men who removed the snow from the area.

Most Copenhageners had some connection to the shipyard—a son, a father, an uncle who worked there. The story goes that the island was lit day and night. The citizens of Copenhagen could see the lights of the yard when they woke up and when they went to bed. And with their eyes, they could follow the low wooden boats that transported the workers back and forth across the harbor entrance.

THE AREA TODAY

When B&W left the area in 1996, a secret life took over for a few years. Rumor has it that bikers with pony-sized dogs guarded a stockpile in one of the halls, illegal workers lived in campers stacked on top of each other, and containers housed porn producers.

Today the island is sprawling with an uneven hoard of settlers—artists, creative entrepreneurs, chefs, musicians, athletes, designers, architects and antique dealers, welders, climbers, carpenters, rocket builders, paintball players, set designers, yachtsmen and women, mechanics, fashion models, and music producers. And once in a while, festivals of art, food, theatre, music, and sports momentarily seize the island: Copenhell, Reggae Festival, Uhørt, Asteroiden, Move Copenhagen, and Distortion—a block party that transforms Copenhagen every year in June.

The owners of the area—four pension funds—have a set goal that this new part of the city must maintain a connection to the old industrial history. Already, different experimental combinations of shipyard traditions and new temporary initiatives can be found there.

At Copenhagen Yacht Service, modern pleasure boats are stored in the former repair hall. The original welding halls, erected in 1942, are now home to The Royal Theatre's designers, who build sets for the capital's opera and theatre productions. The marking hall, constructed in 1951, contains northern Europe's largest climbing center, where more than 150 routes draw both the inexperienced and some of the world's best climbers. The ceiling remains the same as when Henry stood beneath it in 1957. And the Optical Tower, that holds the enormous projector which magnified and transferred ship blueprints on to the steel plates in the hall below, is now a meeting room for the climbing club—fittingly, with 85 steep steps to a floor 25 meters up, where 16 windows provide an absorbing view of the city's skyline.

Every summer on Sønder Hoved, the oldest part of Refshaleøen, one can spot a towering multi-colored circus tent. One kilometer south of Refshaleøen, the world-renowned Norma restaurant, along with other representatives of New Nordic Cuisine, has initiated an annual summit where the culinary elite discuss new challenges. Chefs and farmers from all over the world gather under canvas tents while waiters in white carry silver dishes over the rough field of weeds and are received with applause.

NATURE AND ENERGY

Refshaleøen is surrounded by a humming warmth and energy. At night, the smokestacks glow at Amagerværket, a coal and oil-fired power and heating station. A wind farm surrounds the northern edges of Refshaleøen, and the seagulls soar over the basins of the Lynetten water treatment plant. Its ovens and silos also glow in the dark—red, yellow, and blue—so beautifully that the plant received a lighting award in 2012.

In 2017, Refshaleøen will have a new neighbor when the construction of the Amager Bakke energy station is completed. Bjarke Ingels Group (BIG) is creating the world's most advanced and green waste-to-energy station, which will also contain recreational spaces. The station will be constructed as a hill, 85 meters tall, that invites skiing, hiking, running, and other "mountain activities."

On Refshaleøen, the yard buildings and paved routes take up most of the space, but the area also contains open horizons, fields of grass, bushes, and other low vegetation. Sea-buckthorn, berries, pines, and willows mark the view. And the weeds spread from every crack at the first sign of the spring sun.

THE FUTURE

Refshaleøen is designated as a so-called Perspective Area in the 2011 Municipal Plan. This means that the area cannot be developed until after the 12-year planning period that runs from 2011 to 2023. Until then the capital is happy to have this laboratory for creative and temporary initiatives just a 15-minute bike ride from the City Hall Square. When the time is right the former shipyard area will be developed, with mixed housing and commercial real estate on one of city's prime locations, between the Copenhagen harbor and the Oresund strait.

Refshaleøen Historic Timeline

Annette Skov

1624

A block house is built on the original island (then much smaller) to guard the entrance to the Copenhagen harbor and St. Ann's Fortress (now Kastellet) on the coast just north of the city.

1766

As a measure to defend the city, the Lynetten sea-fort is built north of the existing Refshaleøen. The half-moon shaped entrenchment is not armed with heavy artillery until 1885. The fort is decommissioned in 1922.

1787

Construction begins on Trekroner Fort. This fortification at the entrance to the Copenhagen harbor is a vital part of the Danish defence line until 1920. During World War II, it is used by the German occupiers. Today it is a designated protected area and a favorite place for an outing near Refshaleøen. The original Trekoner Fort was actually situated a few hundred meters to the north and was constructed in 1713, when three old naval warships were run ashore and filled with rocks. One of the ships carried the name "Tre Kronor," from which the fort got its name.

1872

Reclamation increases the size of the island considerably as the port's waterways are deepened. Spoil is dumped on the shallow sand banks of the Copenhagen harbor and B&W starts building ships on Refshaleøen.

An ordinary day at the B&W shipyard in 1872 is described in the regulations for workers: "Normal daily working hours are from 6am until 6pm with a lunch break from 8 to 8:30am and dinner break from 12 to 12:30pm. On Saturdays, however, work ends at 4pm."

Rules regarding piecework and the use of tools and aquavit are also described: "It is strictly forbidden to drink aquavit in the factory, and anyone proven to have done so will immediately be dismissed from the factory, and his wages will not be paid to him but instead collected to support the welfare fund."

1877

B&W is now Denmark's largest company, with more than 1,100 employees.

The ships are built on bedding, meaning outdoors on a foundation of wooden scaffolding that carries the hull. Work is organized in gangs—small groups that work in all sorts of weather conditions.

The riveting gang fastens and assembles the plates that form the hulls of the massive ships. A gang consists of three men: one who heats up the bolt in a portable forge, one who holds the bolt and the riveter who flattens it. In the early days, this was done with a hammer and later with an air hammer. Since the workers' hearing is often impaired, the gang communicates using special codes. Four knocks on the plate if the bolt is too long, two if it is too short, and three if it is too cold.

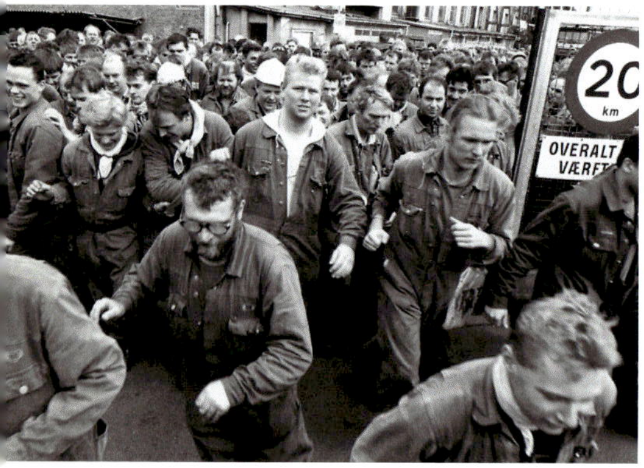

Workers at Burmeister & Wain

1879

His Danish Majesty's Ship Dannebrog, a paddle steamer, is built at B&W's shipyard in 1879. The 196-foot ship had a crew of 56 and was used for royal voyages until 1934.

1895

B&W builds many types of ships over the years: Paddle steamers, luxurious passenger ships and transatlantic liners, royal ships, training ships, warships, tugs and lifeboats, customs and police craft, cargo ships, a whaler, the world's first gas diesel ship, a giant crane, a drilling rig. Many of them are revolutionary, for example the Standard, a yacht ordered by Princess Dagmar of Denmark, now the Tsarina Maria Feodorovna, for her husband Alexander III and launched in March 1895,—a 367-foot floating palace staffed by 275 servants, decorated with plush furniture, 1,000 electric lamps, gold tableware, and icons. A new, 145-meter dry dock has to be constructed to build the luxury ship.

1900

The shipyard now employs 1,600 people.

1906

A floating dock is constructed. The company is now the largest in the Nordic region, with 3,200 employees.

1912

B&W creates a sensation with the Selandia, the world's first ocean-going, diesel-powered ship that emits the exhaust through the rear mast instead of a funnel. The ship's maiden voyage takes her from Copenhagen to Bangkok. Selandia is ordered by East Asiatic Company and is fundamental to the development of the company's trading with the Far East. This ship marks the beginning of a new era, when coal-fired paddle steamers are replaced by diesel-powered ships, which have a greater radius of movement and are less polluting than the steamers.

1919

More beddings, new halls, and workshops are added.

1940

Nazi troops occupy Denmark and take over the shipyard.

1942

The welding halls are constructed. Today these are home to The Royal Theatre's design team and are where the sets for the capital's opera and theatre stages are made.

1943

January 27. B&W's engine factory on Christianshavn is the target of an air raid by Britain's Royal Air Force. Seven people are killed. The attack underscores the importance the British ascribed to the shipyard due to its capability to produce German submarines. Four bombs had already been dropped over the area in a smaller air raid in 1942; however, the attack was never widely publicized because of the German censorship of the press.

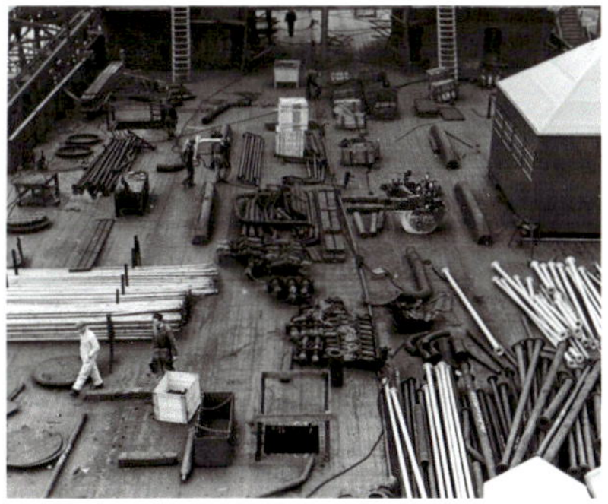

Burmeister & Wain slips

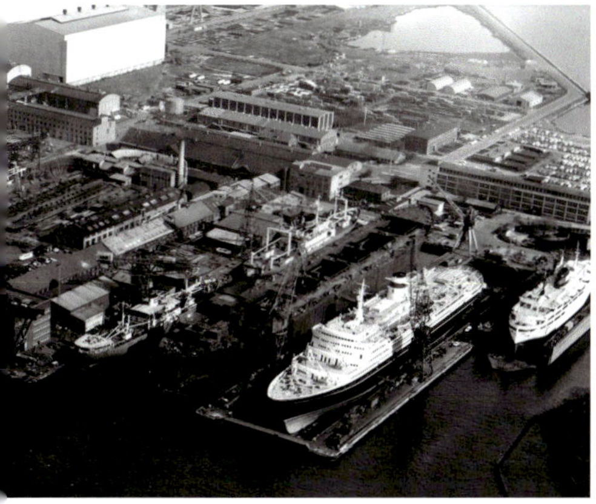

Burmeister & Wain shipyard

1951

Modernization of the shipyard begins. More buildings are erected and sectional welding is introduced.

The production process at B&W is further developed over the years, but shipbuilding still consists of two main stages: 1. the steelwork to build the hull, and 2. the process of filling the ship with its "innards."

During the 1950s, bolting is replaced by welding—a revolution in the working process, as bolt work was hard and cumbersome.

When the ship's hull is finished, it is transferred by tugboat from the building dock to the equipping pier. Here, different craftsmen take over: welders join the larger sections, metalworkers deal with the smaller ones. Boilermakers construct the boilers and funnels, and plumbers are in charge of installing the internal piping. Mechanics work in the engine room, the joiners make cabins, carpenters build scaffolding and wooden decks, and the electricians insert the wiring. When all of these processes are complete, the ship receives its name and can then leave the pier.

A member of B&W's staff recounts:

> I was responsible for sending out invitations to the naming ceremony. We, the ordinary employees, did not take part, but the naming ceremonies were always a huge thing for the yard. A godmother had to be found—preferably a woman who was married to a man of influence, often wives of politicians. Famous people were good publicity.

The guests arrived at the administration building in the afternoon, then walked out to the ship here on the pier. The ship was christened with a champagne bottle, and this was followed by a fancy dinner and party at night in Langeliniepavillonen. It's right over on the other side, so they could sit there and look at the ship and the yard.

1956

Modernization is completed, with the largest repair dock in the Nordic region. The yard employs 7,900 people.

1959

Section halls S1 and S2 are built: the newest part of the yard.

The two section halls are slightly staggered and connected along their sides. The skylight is 126 by 20 meters and fastened with 12,000 screws.

Until recently, S2 was Denmark's largest single interior space: 150 meters long, 35 meters wide, and 65 meters tall. The dock gate measures 2,000 square meters and weighs 120 tons.

During the 1950s, B&W changes its shipbuilding methods. Instead of building the ship on a bedding—the outdoor construction site—the company switches to constructing the ship from smaller parts called sections. These are built in workshops and assembled inside the assembly halls before the final construction in the ship dock. This transition is completed in 1960.

1961

B&W's building dock to complete the ships is laid in 1961. It is 240 meters long, 38 meters wide, and has a free passage of 6.5 meters.

The launch of the completed ships becomes much easier with the building dock. Previously the ship would have to roll into the water from the bedding on a kind of wooden rest. But now the ship can just float out of the dock once it is filled with water—64 million liters of water!

Once the work in the assembly hall is complete, there can be upwards of 50 sections to be assembled in the dock. The deck is installed last of all, giving the workers access to the engine room for as long as possible.

A former administrative employee at B&W recounts:

> When the ship was ready it had to be out-docked—meaning to leave the dock where the workers had assembled the plates and prepared the hull. Everyone left his or her post! We just ran out and saw the out-docking by the dock gate. We clambered up onto things and stood there watching. There were lots of us. Water gushed into the dock through the dock gate, and then the ship could glide out and be tugged over to the equipping pier where it was finished. There was a very special atmosphere. And we all felt that we had each helped build the ship—that we all had a part in it.

The dock has pumps, a warning system, a gate, and drainage. It works as well now as in B&W's time; however, it hasn't been used for shipbuilding since 1992.

In 2000, it was used to build the foundations for the Middelgrunden offshore wind farm. And again recently, a construction company used these same facilities to form elements for wind turbines.

1972

After World War II, B&W becomes the world's leading manufacturer of large container vessels. The first large container vessel, Selandia III, is built in 1972 and measures 900 feet long—the maximum allowed to pass through the Panama Canal.

Selandia III is equipped with three B&W engines totalling 75,000 h.p., with a top speed of an impressive 31.5 knots recorded on its sea trials in the Baltic Sea. As the world's fastest freighter, it sets a new world record with a voyage around the world in 40 days, 3 hours, and 5 minutes, revolutionizing the way cargo is handled.

Selandia III is sold to the United States in 1993 and is renamed USNS Gilliland. To this day it operates as a supply vessel in the U.S. Navy.

1979

The company faces liquidity problems and has to sell half of B&W Diesel to MAN, the German industrial group. In 1980 it is renamed MAN B&W Diesel A/S and is still in business at the time of writing.

1980

The Lynetten water treatment plant opens in 1980. It is Denmark's largest facility for wastewater treatment. The plant is located in a dense area on the north side of Refshaleøen, and has a treatment capacity for three quarters of a million people.

In 1997, the plant is reconstructed and expanded. Lynetten can now remove nitrogen by nitrification and denitrification, and phosphorus by a combination of biological phosphorus removal and chemical absorption with iron salts.

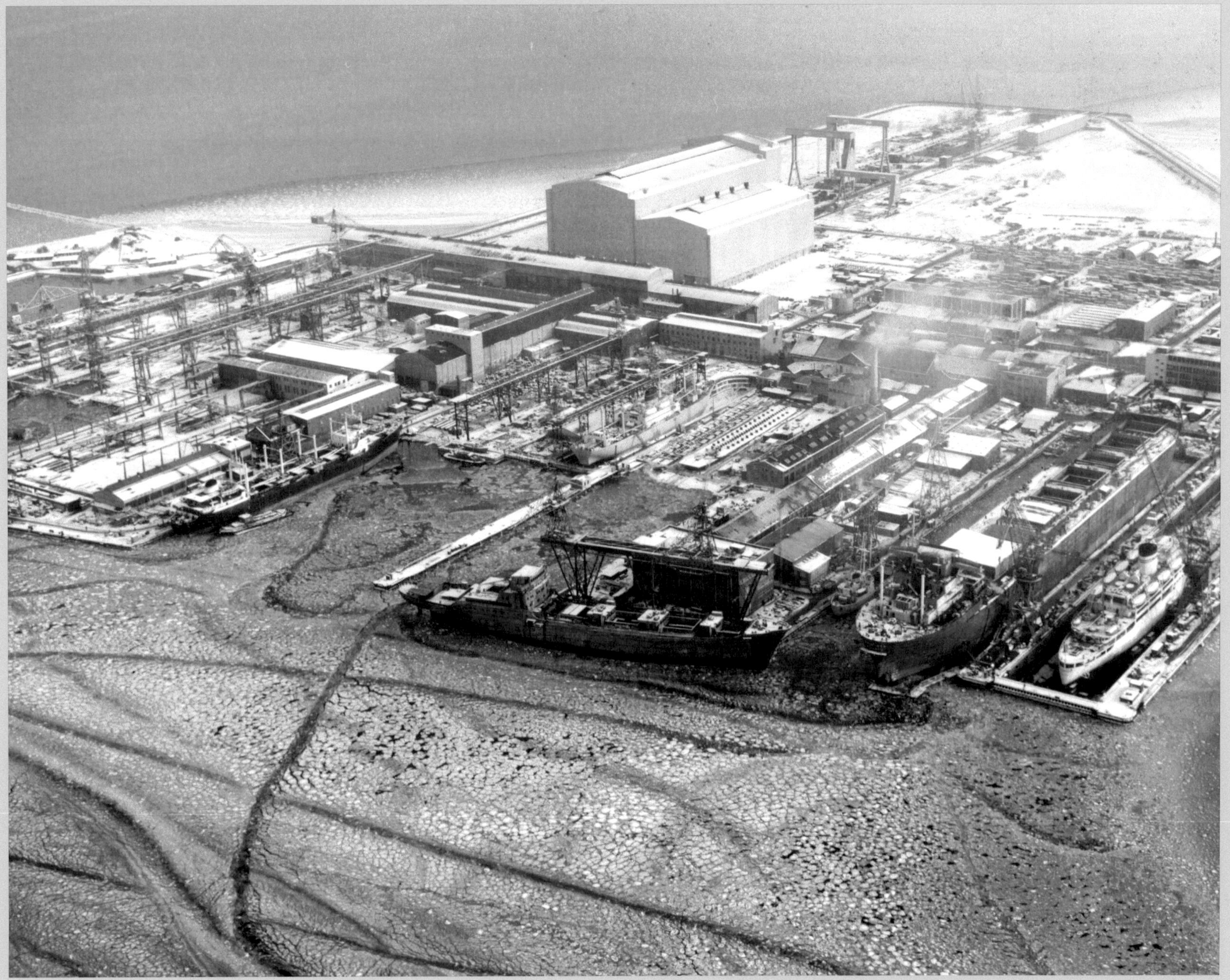

Burmeister & Wain shipyard

1993

Ship number 942, Siboelf, is built at B&W. It is made of 13,000 tons of steel, 13 kilometers of pipes, 76 kilometers of electrical wiring, and 200,000 liters of paint. 1,650 employees work on it for 8 months. The ship measures 751 feet and can transport oil, grain, coal, and ore.

1996

B&W goes bankrupt. The shipyard and the port authorities sell off their properties to Refshaleøens Ejendomsselskab A/S, which is owned by four pension funds and administers the properties and the development of the area.

2011

In the new 2011 Municipal Plan, Refshaleøen becomes a designated Perspective Area. This means that the area cannot be developed until after the 12-year planning period that runs from 2011 to 2023.

The American architect and artist Michael Singer, who works holistically with sustainable urban development, visits Refshaleøen. In the course of a one-week workshop, he and the participants develop a plan for The Edge, a pier area on Refshaleøen.

2013

The Land Art Generator Initiative announces that LAGI 2014 will take place at Refshaleøen. An interdisciplinary and international jury is appointed to judge the submissions.

2014

The Eurovision Song Contest takes place on Refshaleøen, which is renamed Eurovision Island. The two large section halls dating from 1959 are reconstructed to hold the 11,000 guests, who will be exposed to the latest European pop music against a backdrop of past industrial history.

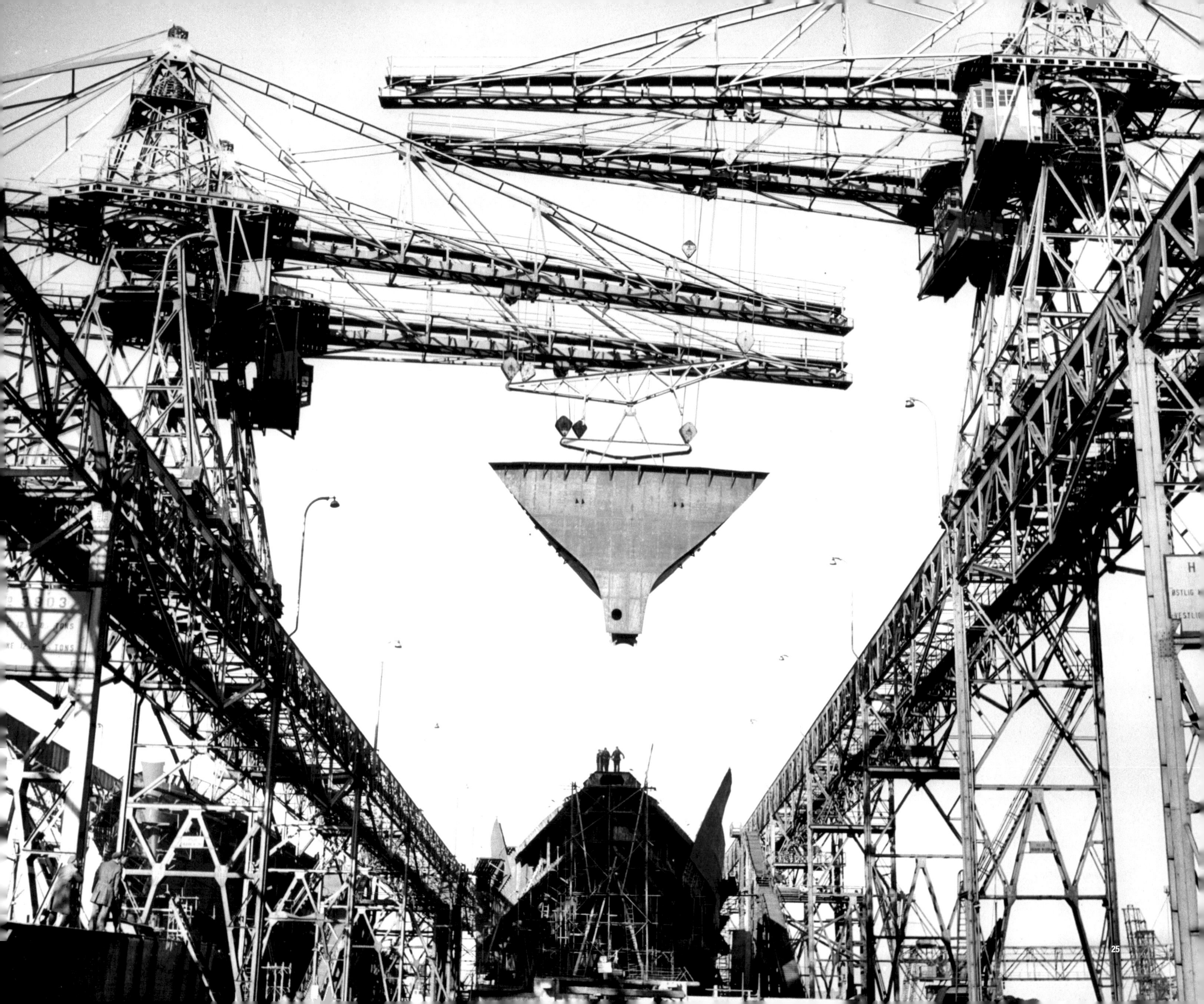

ESSAY

Energy as Alien

Laura Watts and Brit Ross Winthereik

IT University of Copenhagen

Energy is so familiar to us that it is almost banal to think about electricity sockets and plugs, batteries and bicycle lights. Yet so much of our energy infrastructure is invisible: the cables buried in the walls and earth, the power stations that transform fossil fuels into electricity, the standards and metrics, all of which ensure our phone charges instantly when we plug it in. When our energy infrastructure is working well, it is invisible to us, and so it feels banal to think about it.

Yet, something is changing. In our landscape there are growing numbers of wind turbines, solar panels, and geothermal pipelines, and in our seas there are a growing number of wave energy devices. These renewable energy technologies are often unfamiliar, disconcerting, contested and therefore more visible. Along with electricity pylons (which were unfamiliar a few generations ago), these new and unfamiliar energy entities are "breeding" on our land and in our seas. Their unfamiliarity troubles many people. They do not seem to fit. So how can we form a less troubled relationship with these entities and make them part of our energy infrastructure?

Aliens in science fiction stories may be different from us, but they often have something that intrigues us—a quality that humans in these stories learn to care about. For example, Frankenstein's monster is perhaps one of the oldest imagined aliens. The monster is comprised of bits and pieces of people and technology mixed together. Its relationship with its maker is ambivalent, and it has an uncertain future. Through Mary Shelley's novel, we as readers learn to care for this lonely monster. Renewable energy technology is also uncertain and contested, comprised of people and technology. Its place in society is still being determined and its long term costs and benefits are still to be fully known. It is a monster and alien in the sense that we are not even sure what "it" is.

In our university research project, "Marine Renewable Energy as Alien," we use the figure of the "alien" to think about the difference between familiar and invisible, and unfamiliar and visible, energy infrastructures. We are concerned with how renewable energy technology is being integrated into our existing electricity infrastructure, and how it might be otherwise. We interviewed a senior manager at a Danish electricity company. He said to us, "You ask about what makes renewable energy alien. Well, all energy is alien." He gave a reason for this sense of alienness, "We have burned coal for more than 100 years, and not even people who had physics in high school can explain what a kilowatt-hour is. But as the cost of energy increases, one must begin thinking about it."

We agree with him that, as citizens and consumers, we need to think more about the costs of energy. Today, the cost of energy is multiple: both economic and environmental. It is climate-changing. Renewable energy is a move to reduce those costs. But it is not simply a matter of swapping coal for wind or waves—it's more complicated. The uncertainty we have spoken of means there is an opportunity for us to respond, to participate, and to be moved by energy as consumers, citizens, and human beings. We should not be alienated by our energy nor see it as only a matter for physicists, as in the example with the kilowatt-hour.

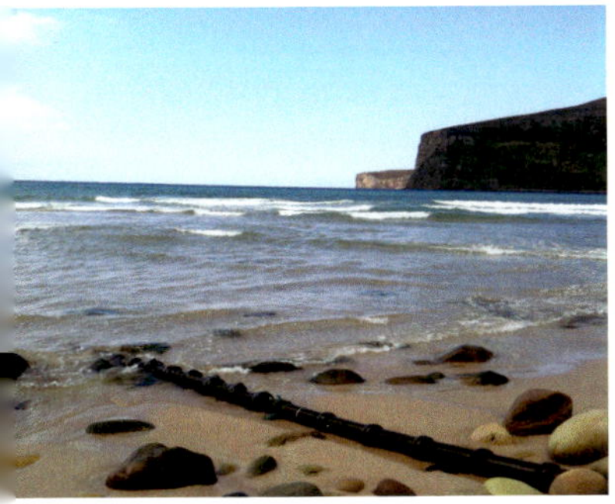

The invisible infrastructure of electricity: an undersea electricity cable

Copyright: Laura Watts

In our research project, we see the alienness of renewable energy as a good thing. If renewable energy is alien, it is no longer invisible, and the uncertainty it creates—its fuzziness—is something positive. As with Frankenstein's monster, we can learn to care for renewable energy because of its fragility and uncertainty, not despite it. Uncertainty can lead to an openness and to new ideas—a situation in which energy infrastructure can be imagined otherwise. It means that there is room for exploring new possibilities for a relationship with renewable energy, perhaps even a caring one. The Land Art Generator Initiative can help us all notice energy and its infrastructures. LAGI is a site for exploring how to grow new relationships between humans and energy technologies.

By exploring renewable energy as alien we can explore its possible domestication. Knowing how we relate to energy infrastructure, as alien or as family, is a way forward for sustainable infrastructure re-design, and public participation in that process. Read more and tell us about your own "alien" energy experiences at *www.alienenergy.dk* or via the "Alien Energy" facebook page.

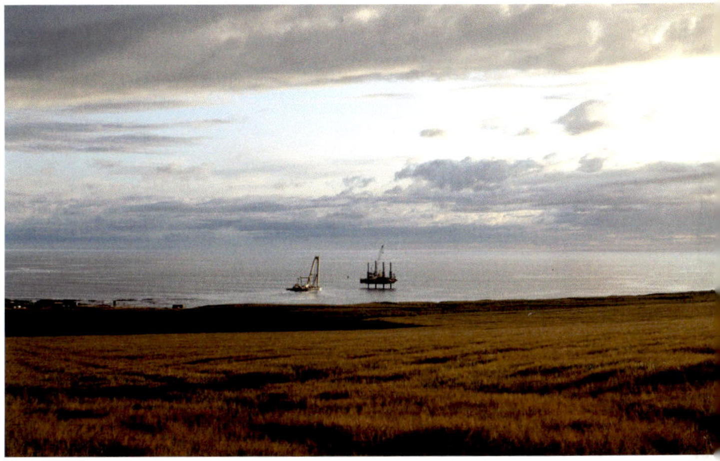

"Alien" in the sea: a wave energy generator being installed at the European Marine Energy Centre in Orkney, Scotland

Copyright: Alistair Peebles

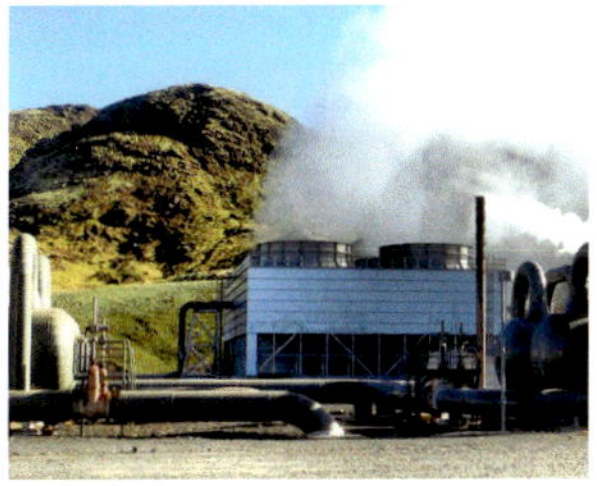

"Alien" energy: the cooling towers of a geothermal energy power plant in Iceland

Copyright: James Maguire

ESSAY

The Human Approach to Sustainability

Trine Plambech, Johanne Mose Entwistle, and Liselott Stenfeldt

The Alexandra Institute

How much energy did you consume yesterday? Was it a lot or was it only a little? How many kilowatt-hours do you use when washing your clothes? The studies that we have carried out in our sustainability group at the Alexandra Institute (Denmark) show us that only a few people would be able to come up with a clear answer to these questions, because energy consumption is the hidden consequence of all the things we do in our everyday lives.

If we want to change energy consumption, we have to understand what people do in their daily lives. This leads to the question: How can we engage people to contribute to the sustainable development of our cities using a human-centered approach? At the Alexandra Institute, we are passionate about these types of questions because, even though our civilization is so dependent on energy, it is not foremost in our minds. Yet, if there is a power outage, we immediately become aware of how great an impact energy has on our daily lives. In the pursuit of ways to save energy, several initiatives targeting the individual energy consumer have been launched, such as campaigns requesting that people use less energy or installing home devices that measure how many kilowatt-hours a family consumes. Some of these initiatives are effective; however, the outcomes may not be as much as expected and the long-term results may diminish over time. Why?

The following is an example from a study we conducted, "Energy Consumption in the Home."[1] We visited several Danish families, one of which was that of Peter, Anna, and their three girls. They consider themselves to be "green." When asked what being green means, they provided examples, such as "we never drive to the nearby store, but always take the bike," or "we never use the dishwasher for the big stuff," and "even though we do have a tumble dryer, we only ever use it for the absolutely necessary stuff." Being a family with three small children, one could assume that there might be emergency drying of items, such as children's snowsuits. However, when we asked them what the essentials actually were, Anna said (with a slight tone of regret in her voice): "I always use it for towels. I just can't stand stiff and scratchy towels. I want them soft and I want them fluffy. That's actually why we bought the tumble dryer in the first place."[2] This was not the response that we had expected, but it demonstrated something essential: Energy consumers don't really consume energy at all. They consume services. To put it quite simply: Anna wants comfort as a service, and she uses the tumble dryer to get this comfort. The fact that the tumble dryer uses electricity is hidden and circumstantial for the purpose of delivering the service.

Understanding this desire for services and comfort can illuminate why knowledge of energy data and economic incentives may not have the expected result. Such initiatives assume that if people are made aware of the amount of energy they consume, they will be motivated to reduce their usage. If individuals, such as Anna and Peter, do not see themselves as consumers of energy, they won't

[1] "Energy Consumption in the Home: An anthropological user study of needs, motivations and barriers in relation to energy consumption in the home," (The Alexandra Institute, 2009). **http://www.alexandra.dk/uk/about_us/publications/documents/energy_consumption_in_the_home_user_guide.pdf**

[2] Ibid.

respond to being approached as energy consumers. Instead, people need to be understood as multidimensional—shaped by different and often conflicting values and knowledge, as well as many other factors, which may be largely outside of their control or awareness as individuals.

The challenge thus lies in addressing divergent and competing influences—societal norms, legislation, infrastructure, technology, and design—which shape daily behavior. These influences cross and overlap disciplinary boundaries. Therefore, it is essential to establish the right mix of cross-disciplinary teams that will engage the public as co-creators, not just as passive receivers, of solutions from outsider experts.

The Land Art Generator Initiative (LAGI) submissions are exemplary proposals in which people with different competencies join forces. This cross-fertilization of ideas can result in new solutions and new ways of engaging people. The LAGI projects make visible aspects of our energy system that are not only hidden to most of us, but are often considered boring, unattractive, and technologically distant from our understanding. Insights from our studies indicate that play, social interaction, aesthetics, and design are important factors for shaping individual, daily practices. These are all aspects that LAGI incorporates and, by doing so, helps to bridge the gap between regular people and their energy services. LAGI proposes energy generators that appeal to the senses and are integrated into our urban recreational spaces. With the presence of a Land Art Generator in the public arena, people are playfully and aesthetically reminded of the energy we depend on in our lives. Civic pride is also ignited in this public display of clean, renewable energy. Together, these qualities can engage the public and invite them to take part in the sustainability agenda.

If we commit to taking a human approach—of engaging people with multidimensional needs and desires, rather than reductively as consumers—we could succeed in developing effective technological solutions. These solutions would be based on knowledge about people and complex behaviors that would create more value while reducing energy consumption. This is the human approach to sustainability.

ESSAY

Danish Wind Power: A Journey Beyond Imagination

Peter Karnøe

"You may discuss as much as you like but nuclear power is what you will get."

—Chairman of the Danish electrical utilities, 1976

"[We have] set the goal that by 2020 50% of Danish electricity production is to be generated by wind power."

—Energy Act passed by the Danish Parliament, 2012

"Wind power is now mainstream technology."

—Spokesperson from Global Wind Energy Council, 2014

As these three contrasting statements from 1976, 2012, and 2014 indicate, the status of wind power has changed in Denmark as in many other countries. According to these statements, wind power has made a dramatic shift from being unwanted by the energy establishment to becoming dominant, even categorized as mainstream. This is indeed a journey beyond imagination, and yet, not so atypical for innovations that shift the order of things in societies through what Schumpeter called "creative destruction."

This essay deals with the Danish part of this journey and uses actor-network theory as a framework for analyzing what happened between the above statements. Actor-network theory attends particularly to how technological projects, organizations, institutions like the "market" and the "economy," habits, and ideas come into existence through the actions of micro-actors, to become macro-actors that constitute societal orders. For example, in 1980, when 98% of electricity in Denmark was generated by centralized fossil fuel, the power plants represented a particular macro-actor, the result of a long struggle to put together the socio-technical bits and pieces that enabled centralized electricity generation to grow from 12% in 1923 to 98% in 1980. Wind power generated 0% electricity in 1976. In the context of the total dominance of centralized electricity generation, it did not seem to be a technological project that had much chance of success, but some had started to make narrative declarations, such as the policy statement, "there can, and there should be wind power in Denmark," or local actors posed the question, "how is it possible and desirable for me to develop wind power?"

Between 1976 and 2012, something apparently happened within the network of wind power and the networks of centralized electricity power plants. If we turn to actor-network theory (ANT), one critical lesson is that being dominant and "powerful" in a society, and being represented by relatively homogeneous voices (like the 1976 statement from the utility chairman) is the outcome of a struggle, and it is not known at the outset which technology will become or remain dominant.

In order to understand some of the changes, we can follow the making of a network, which was built gradually to form a context. Within this context, the narrative programs of action for wind power could be realized by being translated into a seamless web or a network of durable materials and relations. Rather than seeing the world as composed of pre-given objects, ANT tries to understand the ways in which they come into being. ANT seeks to understand the making of facts and functionalities by employing a constructivist and realist (empirical, "positivist") approach at the same time. Realities such as scientific truth and the functional qualities of technologies like the combustion engine or the computer chip are both constructed and made real by particular socio-material actions, such as measuring, calculation, drawing graphs, designing machines, and shaping materials (Callon, 1986a and 1986b, Callon et al. 2002, Law, 1992, Latour, 1991, 1999). Accordingly, the scientific truth (or falsity) of the functional qualities of technology are ongoing accomplishments that are achieved by making alliances that last.

In this essay, I will follow the activities of establishing wind power in the interconnected arenas of policy and regulation, design and production, research and science, and users and markets (Garud and Karnøe, 2003). Within and across these arenas, the particular patterning of micro-actor-networks emerges and generates various agency effects, and effects on regulatory, technological, and organizational competencies that in turn are used as resources to generate the very networks into which they come into existence. In other words, the resources are ordered actor-networks of heterogeneous materials whose resistance (mis-fitting) has been overcome and whose identities have been shaped temporarily to accommodate certain actions—the customized landscape that fitted itself onto the technology/statement that did not pre-exist.

1893: INITIATING THE JOURNEY AND POSSIBLE WORLDS

When did Danish wind power begin? We could start our journey with Poul La Cour from Askov Folk High School in 1893 when he began to experiment with electrical wind power as a new source of electrical energy for small towns and farmers (Karnøe, 1991).

We could also begin our story with Johannes Juul—a Danish engineer who, following the fuel shortages of WWII designed a series of two- and three-blade turbines that greatly advanced the technology—or in 1962 when the special utility commission decided that the calculations on efficiency and cost economics proved that Juul's 200 kW wind turbine design was unattractive compared to fossil fuel-based centralized power plants (Hvidtfelt Nielsen, 2001:11–20). Juul's small and weaker micro-network was beaten by the stronger actor-network of the utilities. Juul's journey ended here, but he had learned about wind power at Askov High School in 1916, and between 1947 and 1962 he invented the so-called Danish wind turbine design.

1974–1979: NEW BEGINNINGS OF THE JOURNEY II

The dominance of wind power in 2014 did not stem from one clear political ambition to solve national energy problems with a program for small-scale wind power as a natural response to the first energy crisis in 1973–1974. Several new actions occurred in relation to the macro-actor of the existing centralized electricity generation. The energy crisis became an opportunity to problematize the associations of the macro-actor that hitherto had delivered a stable supply of inexpensive electricity based upon cheap imported oil. The energy crisis opened some black boxes, such as "cheap oil dependence" and the "import problem." Nuclear power, as the preferred future energy source, became a liability and not an asset. It was not a sealed black box, but rather became a leaking black box—an opportunity for attacking the macro-actor claim that nuclear power should replace coal in the centralized power system. This was hotly debated in the Parliament, and the issue of nuclear power safety had a transformative effect on politicians and the coalition for nuclear power.

Wind power and renewable energy were not mentioned in the official energy plan of 1976. This plan, EP'76, depicted a future centralized electricity system that, by 1995, would be based on four to six nuclear power plants, construction of which was to begin in 1985.

EP'76 was immediately criticized in several newspapers and by the grassroots organization against nuclear power (OOA). Inspired by the new discourse on alternatives, the parliamentary Energy Policy Committee asked the minister to develop a plan without nuclear power. The minister refused to do this. (van Est, 1999:74)

Instead of giving up, this motivated a group of researchers to develop an Alternative Energy Plan (AE'76) during the summer of 1976 (Meyer, 2000:80). Equipped with new techno-economic calculations and charts, AE-76 challenged the official "truth" about energy demand growth, and offered alternative futures with lower energy demand supplied by a greater mix of renewable energy technologies. It contributed to problematizing the networked relations of the macro-actor by fostering intense discussion in the media and a parliamentary debate in the fall of 1976. Further, hundreds of already existing debates and meetings arranged by the grassroots movements of OVE/OOA received fresh impetus. These increasingly linked micro-actor-networks were not only social in nature—indeed, their statements for more wind power and renewable energy were materially anchored in reports filled with techno-economic models, calculations, and practical demonstrations that framed issues and stipulated new realities to be performed.

At the same time, the promoters of the nuclear power action program ridiculed the proponents of alternative energy for believing that it could generate sufficient cheap electricity for the nation's needs.

Eventually, the macro-actor network could not resist any more and things began to change in official policies. For example, in 1976 the "wind power committee" chaired by the president of the Academy of Technical Sciences (ATV) recommended a large-scale wind power (600 kW wind turbines) program. The Danish Utilities Research Unit would implement the program. Very few believed in small wind turbines, but in 1978, inspired by the United States, members of the ATV succeeded in establishing a 5-year research and development program for small-scale wind turbines (15 kW and up to 100 kW at that time).[1]

MICRO-ACTORS IN DESIGN AND PRODUCTION

While many micro-actors were involved in changing the discourse and influencing energy policy, other micro-actors were working on concrete alternative technical designs for wind turbines.

A carpenter, Christian Riisager, and a blacksmith, Karl Jørgensen, were two micro-actors in wind turbine design who did not follow the ideology of any social movement, but began experimenting with wind power to resolve their own power supply in the wake of the energy crisis. Jointly, with a network of other technically trained people, they built on Juul's design ideas. Even if they had never met, their individual micro-networks became critical in the development of the technology by inspiring many other self-builders and new startups.[2] Riisager had a particularly important role. In 1976, without asking for permission from the utilities, he hacked a wind

[1] This essay is only about small-scale wind turbine development in Denmark; the large-scale program received most of the funding, but technically did not contribute strongly to Danish wind power (See Karnøe, 1991).

[2] See Karnøe and Garud (2012) for more information about Herborg wind power and many other early entrepreneurs.

turbine into the electricity grid by means of the electricity outlet to his washing machine. The electric meter ran backwards, and as nothing happened to the neighbor's electricity, Riisager concluded that he could supply electricity to the grid without damaging it.

USER-MICRO-ACTORS OF WIND POWER ESTABLISH GUIDELINES FOR WIND POWER GRID CONNECTION

Following Riisager's successful hacking into the grid, and after he sought permission to do so, numerous newspaper articles extended his micro-actions by reporting on the event. Buses full of visitors came to see his wind turbine. Journalist T. Møller was the first to buy a Riisager wind turbine in early 1976. He had bought a country house in Jutland as a base for his left-wing newspaper, *Information*, that consumed 40,000 kWh annually. Given the uncertainty about future oil prices, he wanted to limit his future energy bill. He applied for permission to grid-connect, but had to wait because no guidelines existed. During 1976, wind power micro-actor advocacy increased in strength. With this political attention, things were spinning out of control for the Danish Electrical Utility Association (DEF), and they asked the national Price and Tariff committee to develop definite rules for connecting wind turbines to the electricity grid (van Est, 1999:78). Further, wind turbine owners received a modest payment for each kilowatt-hour they sold to the grid that exceeded their own consumption. They received an income from the savings they made by not buying electricity at the full price. In the beginning, therefore, wind power was only economically attractive for households with electrical heating.

THE ECONOMIC SUBSIDY SCHEME: A NETWORK EFFECT

Even if wind power was mentioned in the Energy Research Plan, there was no a priori functional plan for helping the market to develop, other than the guideline model. However, it was not long before an investment subsidy scheme emerged from a contingent situation, where timing among participants, problems, and solutions shaped the outcome (March et al., 1974). Rising unemployment and the Three Mile Island nuclear power plant accident (March 1979) were made contingent events. During negotiations to secure the majority of votes for an economic stimulus package, the Radical Liberal party (hostile to nuclear and favorable to wind power and renewable energy) perfectly timed their offer to vote for the bill in return for support for an investment subsidy for renewable energy (not only wind power). This program was part of a housing energy renovation package and was justified by the Minister of Housing, because the 30% subsidy was meant to "create production opportunities for the Danish industry in such a way that series production could be achieved" (van Est, 1999:79).

The subsidy scheme was contested by the "free market" advocate Ministry of Trade, by media and industry associations, and delayed until the law was finally passed in June of 1979.

The combined effect of physical hacking, new discourse, and political pressure had made it possible for wind power micro-actors to penetrate the macro-actor, and create a possible future for themselves in the Danish energy system. The various micro-actors had expanded the wind power network by three new elements: 1) the physical grid access, 2) parts of the network, and 3) the subsidy scheme and the research and development program for wind power. The material existence of wind power was still limited by the end of 1979—total installations numbered around 150 small wind turbines.

There was no initial grand plan to guide all of these actions. Instead, multiple micro-actor networks emerged in a contingent process, became linked, and gradually made the impossible seem possible—a landscape that included wind power was coming into existence. The new network effects created by the end of 1979 allowed wind power to exist as a supplement to the centralized power plant electrical system. It involved establishing new network resources, such as physical wind turbines, DEF-Guidelines, various calculative devices, subsidy schemes, and inclusion in energy policy research and development—a part of the discourse for renewables and against nuclear power. New links were forged between social movements: Riisager, Tvind, and researchers, and between issues like energy renovation, series production, and job creation.

1980s: MAKING THE NETWORK WORK, EXPANDING RELATIONS TO EXPORT

Market-wise, the 1979 subsidy-approval scheme was an important regulatory innovation that set the economics of wind power and stimulated another type of company to enter the wind power network. Entrepreneurs from within existing businesses saw an opportunity to exploit their mechanical-electrical competencies in the new technology. For example, Vestas (global leader 2012) diversified into wind power by licensing and copying design elements from Karl Jørgensen. By 1980, about 10 firms were active and registered in the sales of wind turbines in Denmark (for details about design and production see Karnøe, 1991, 1999, Karnøe and Garud, 2012). The Risoe testing and research center was funded by the small wind-power program, and the wind turbine certification procedure (users could only get a subsidy if they bought a certified wind turbine) became important in helping entrepreneurs improve the efficiency of their wind turbines while providing data for the ministry that regulated the industry.

By 1978 the users of wind turbines had already organized themselves into an Association of Danish Wind Turbine Owners. The members became an important part of the network, as a result of the competence they had acquired through their investment in this emerging and still inefficient technology. They negotiated terms with utilities, ministries, and municipalities, and they trained consultants, advising local co-operatives that owned wind turbines in the countryside. Further, they became an important part of the wind power network because they, as owners of the physical installations, were materially related to the future revenue streams over the life-time of the wind turbine.

The regulatory framework for the economics of wind power was improved (and the network was stabilized) when the state subsidy model from 1979 was supplemented with a price per kWh in 1984. With the enactment of an agreement with the electrical utilities, covering the years 1984–1994, users had a predictable investment horizon.

PHYSICAL INTERMITTENCY ISSUES

While the "red-green" political coalition supported wind turbines, the turbines only generate electrical power when there is wind. The utilities made an issue of this: "The production from a wind power generation system will be characterized by limited predictability (contrary to the stable electricity load from stored fossil energy), and frequent and often strong variations." Adding larger amounts of wind power "can result in such significant expenses, that the value of the wind power production is reduced significantly." (Karnøe, 2013). The assertion that wind power did not fit into the existing power system would become a persistent issue.

POLITICAL COALITIONS AND NEW BUNDLES OF JUSTIFICATION IN THE NETWORK

Such conclusions and claims from the electrical utility sector were not enough to break the red-green coalition of political parties that supported the regulatory framework for wind power. The population largely supported wind power policies because they generated energy, jobs, and new exports. Politically, the large-scale wind power program was run by the utilities until 1986. While the total subsidies for small-scale wind turbines were questioned throughout the 1980s, it was not enough to split the political coalition. The Danish Parliament voted nuclear power out of the future energy system in 1985, and the Chernobyl accident from 1986, together with rising exports, served to consolidate the wind power network.

From the mid-1980s, the wind power network was strengthened by a surprising set of links to new issues: export income and jobs for Denmark. Exports of small (55–75 kW) Danish wind turbines increased from 30 to 2,000 per annum between 1982 and 1985, making the wind turbine famous in Denmark. The booming export market had a positive effect on the domestic market as it shaped expectations and market policies at a point in time when the existence of wind power in Denmark was fragile and contested for being subsidized and irrelevant to base-load energy-production. The 1984–1994 agreement with the utilities was only part of the regulatory framework. There was no long-term agreement about the subsidies, and the agreement had to be negotiated and approved at the annual December negotiation of the state budget. Every year it was nerve-racking for users and producers of wind power. However, the red-green coalition of parties could now defend the wind power subsidies with reference to manufacturing exports and new jobs, as well as increasing domestic energy production and reductions in pollution and acid rain).

EXPANDING THE NETWORK: THE PHYSICAL BASE AND ANCHORING THE NETWORK IN THE PEOPLE AND IN FUTURE REVENUE STREAMS

The many cooperatives turned wind power into “the people’s project,” and by end of the 1980s, some 3,000 privately owned wind turbines had been installed—already representing a stunning 2% share of Danish electricity generation that year. However, the cooperatives that owned most of the 3,000 wind turbines comprised some 100,000 people who for various reasons invested money in wind power, mostly assisted by the Wind Turbine Owners Association. Even if wind power anti-programs contested the subsidy schemes heavily, there was a growing popularity of wind power among owners and in the population in general. This provided both pressure and legitimacy to the red-green coalition. Further, given these investments and expected revenues from all these new installations, wind power had a stronger ticket to the future. Wind power expansion could be stopped, but it would be expensive to buy all the stakeholders out.

During the 1980s, wind power was contested and supported at the same time, but it became increasingly difficult to find politically acceptable arguments for closing it down. The emerging network had been able to stand against various attacks, and to expand its alliances.

1990s: NEW ALLIES, ATTEMPTS TO STOP THE JOURNEY, ACCELERATION

Politically, the 1990s began with the new energy plan “Energy 2000”—an action plan for sustainable development. Sustainability had become a hotly-debated issue in Danish energy policy since the Brundtland report in 1987. However, wind power was not yet promoted in this context. Suddenly, a new election and a new government shifted the priorities. The coalition of parties behind the greener policies lost their majority in the 1991 election. Over the next two years, 1992–1993, the newly appointed Ministers for the Environment and Energy brought momentum to new problems and concerns stemming from the use of wind turbines, such as noise and visual landscape pollution—issues raised by people living near the wind turbines. These “hot issues” problematized the entire future of wind turbines in the Danish landscape until these concerns could be investigated. Private wind power installation almost stopped in 1993.

Again, contingencies mattered, and the anti-wind power government’s majority was surprisingly short-lived: a political scandal ended it before it could put a stop to wind power. The new election brought the red-green coalition of the 1980s back into government in 1993—now with a majority.

1994–2001: NEW ALLIES IN THE NETWORK: AUKEN AND THE STRONG COMMITMENT TO CLIMATE CHANGE AND CO_2-REDUCTIONS

After the failed attempt at reducing wind power drastically, the new Minister for Energy and Environmental Affairs, Svend Auken, forged strong new linkages between energy policy; renewable energy, most notably wind power development; and CO_2-reductions. The wind power network was not associated with the work of the UN

Climate Convention in Berlin in 1995, which was linked to binding CO_2 reduction targets for industrialized nations. Denmark signed the Kyoto Protocol in 1997. The EP'21 had maintained the broad policy goal of creating a robust and efficient energy system, with a high share of domestic sources (oil and gas) and renewable energy (p. 10), and to use energy policy to strengthen new industries like wind power. During Auken's time in office, wind power became bigger (wind turbines were by then 600–800 kW), spurred by the policy goal of 500 MW of onshore wind power by 2005. By 1998, the ministry had pushed through a new agreement with power companies, requiring them to install 750 MW in offshore wind farms and banning the introduction of new coal fired power plants.

The regulatory subsidy scheme was changed to a feed-in tariff (FIT) model with guaranteed prices for wind power owners and was re-set for 1994–2000. Significant efficiency improvements in wind turbine technology (brought about by the larger swept areas of the 600–800 kW turbines) resulted in high profits from wind turbine investments. As in the 1980s, the media and some political parties contested the subsidies and high profits. The growing numbers of wind turbines in Denmark stimulated the creation of a new NGO: the Danish Society of Wind Turbine Neighbours, established in 1999. While Auken's regime was a kind of love affair with wind power to combat climate change and to create new jobs and exports, the legitimacy of wind power was again actively being challenged, especially as it related to the economic subsidies resulting in higher energy prices for electricity consumers. Other issues, such as bird deaths and noise pollution from wind turbines, were also controversial and were now being used in the same way as the safety problems that had been associated with nuclear power.

For some new groups, wind power had become a "monster," and anti-wind programs were formed; however, the pro-wind network alliances resisted these problematizations.

THE PHYSICAL BASE AND ANCHORING THE NETWORK IN FUTURE REVENUE STREAMS

From 1994 to 2001, wind power underwent very strong growth, and combined private and utility installations increased wind power's share of Danish electricity generation from 3% to 18%. This share of Danish electricity generation had become the largest in the world and was linked to large revenue streams stretching into the future. The ownership-base also shifted from being only cooperatives to include individual farmers advised by the investment advisors from the Danish Agricultural Association to make very profitable investments. The combined effect of the ownership base, the CO_2-free electricity generation and income streams was having a stabilizing effect on the network. As wind power continued to replace CO_2-polluting electricity generation, it became increasingly irreversible. A change in course would have required a great amount of compensation to owners for their expected revenue streams.

2001–2004/08: NETWORK WEAKENING, REJECTION OF CLIMATE CHANGE, NO PRIDE IN SUBSIDY BASED INDUSTRY—LITTLE NEW BEGINNING 2004

The new neoliberal government from November 2001, led by Anders Fogh Rasmussen (AFR), immediately started a process of problematizing and dismantling the strong associations between climate change, CO_2 reductions, and the role of wind power in the mix of preferred energy technologies that was sustained by Auken's so-called "energy and environmental mafia." Neoliberal thinking and negative issues around wind gained momentum in the discourse. Within one year, the planned offshore wind farms were cancelled, and many public agencies with knowledge of environmental problems were closed down. AFR and his coalition made sure that the climate change and environmental rationale for supporting renewable energy was challenged with the establishment of the Environmental Assessment Institute, directed by the controversial political scientist Bjørn Lomborg.

The attack came at the same time as the physical imbalance in the Danish electricity system, which became a hot issue once again. Danish media reported that "Denmark is flooded with excess electricity," and grid-stability was at stake. The phrase "flooded with electricity" refers to the total power generation within a particular geographical area being larger than the demand load for the same area, which may drive prices down unless it can be exported at very low prices or for free. The situation was relieved by the transmission cables to Norway; however, the energy was often exported at extremely low prices.

Wind power development almost stopped until 2004. What already existed could operate, but there was no future for wind power in the neoliberal project for the society.

"I WAS WRONG"

During 2008, AFR and the coalition became increasingly isolated with his rejection of climate change and his climate-related policies. Strong conclusions in the IPCC 2007 report, the Stern Review in 2006, Al Gore's movie, and many prominent international and Danish business leaders all supported action on climate change. In November 2008, AFR shocked the public by his statement at the Liberal Party Annual Meeting "I am sorry—I was wrong about climate change—we must change policies." This was one year before Denmark would host COP 15 in December of 2009. Danish energy policies began to change again. AFR left to become the Secretary General of NATO, and the Minister for Climate Affairs initiated a governmental Climate Commission. By 2010 it concluded that the Danish energy system could become 100% free of fossil fuel by 2050. Under the strategy, wind power was one of the central technologies, second only to improved energy efficiency (demand-side savings). Partly on this basis, the government published a broad and general program for "Energy Strategy 2050." It was not concrete, but it was once again a return of wind power, now in alliance with a new agenda for the future energy system.

NEW ALLIES AND ACTORS FOR WIND POWER

New actors in the network were also moving toward wind power. In 2006 DONG, the largest (state owned) Danish energy company, revolutionized its strategy, stating that by 2030, 80% of its business would come from renewable and only 20% from fossil fuel energy. In 2013, 50% of DONG's profits came from wind power investments, which represented only 20% of their total activities.

REFERENCES

Michael Callon, "Some Elements of a Sociology of Translation: Domestication of the Scallops and the Fishermen of St Brieuc Bay," in *Power, action and belief: a new sociology of knowledge?*, ed. J. Law (London, Routledge, 1986), 196–223.

Michael Callon, "Techno-economic Networks and Irreversibility," in *A Sociology of Monsters: Essays on Power, Technology and Domination*, ed. J. Law (London, Routledge, 1991), 132–161.

Michael Callon, Cécile Méadel, and Vololona Rabeharisoa, "The Economy of Qualities," in *Economy and Society*, 31–2 (2002), 194–217.

P. Karnøe, *Danish Wind Turbine Industry—A Surprising International Success: On Innovations, Industrial Development and Technology Policy* (in Danish; Copenhagen: Samfundslitteratur, 1991).

R. Garud and P. Karnøe, "Bricolage versus Breakthrough: Distributed and Embedded Agency in Technology Entrepreneurship," in *Research Policy* (2003), 32.

P. Karnøe and R. Garud, "Path Creation: Co-creation of Heterogeneous Resources in the Emergence of the Danish Wind Turbine Cluster," in *European Planning Studies*, vol. 20, no. 5 (May 2012).

P. Karnøe, "Large-Scale Wind Power Penetration: Breaking Up and Re-mixing Politics, Technologies, and Markets," in *La Revue de l'Énergie*, no. 611 (January–February 2013).

2012–2014: RETURN OF CLIMATE CHANGE, DENMARK 100% RENEWABLE BY 2050, AND 50% WIND POWER BY 2020

Once again, a new coalition government including parties from the red-green coalition of the 1980s, combined Auken's commitment with the new issue of the future 100% renewable energy system. After one year of intensive work, the new Minister for Climate and Energy, Martin Lidegaard, presented a unique majority vote across the political spectrum for a concrete new national energy plan. The new plan preserved the goal of a carbon-free Denmark in 2050, with 50% of electricity generation coming from wind power in 2020.

The present controversies are no longer about wind power in the energy system, but about the social contract for financing the cost of transition, and about combining subsidy schemes with market pricing, and how markets must be designed to solve the problems of electricity balancing (Karnøe, 2013, Lund, H. et al., 2011). The controversies surrounding CO_2-prices, internalizing pollution costs in fossil fuels, and the re-design of electricity markets to accommodate wind power, have shown that markets are not pre-existing price-making arrangements, but are rather constructed entities and an intrinsic part of the network for any energy technology.

The narrative programs of action for wind power have gradually created a landscape for their own existence by translating themselves into a network of more durable materials and relations. Increasingly, the wind power network is composed of and driven by massive problems—CO_2, large industrial players, big finance, developers, and electrical utilities—all of which may have contributed to making the wind power network more robust. Big actors are coming together with increasingly high-capacity wind turbines—now 2 MW, 3 MW, 5 MW, and even 8 MW wind turbines are in the market. Big actors, big wind turbines, and big finance require a stable landscape to secure the returns on their investments. These shifts in the network of wind power cannot be dealt with in this essay.

CONCLUSION

The journey of Danish wind power is indeed beyond imagination, and the reality of the outcome attests to an ongoing accomplishment of alliances which were repeatedly broken and re-built, and gradually became more and more robust. This happened in a process where the following groups and conditions all emerged somewhat independently: the many distributed micro-actors like Riisager and others experimenting with the technology; the social movements and their visions; the alternative calculations and depictions of new futures; the ATV group; the first wind turbine owners; some politicians concerned with Danish energy policy; nuclear power, and high unemployment. Despite their independent emergence, they were all gradually translated into alliances that made wind power happen in Denmark.

Further, the motives behind the successful network cannot be associated with one narrative program or vision for action. Doing so risks betraying the early actions and multiple voices that fought the small battles before victory could be achieved. The wind power network is better seen as a narrative program that itself was composed of smaller narrative programs of action that together allowed different identities to be realized in a transformative process where some of the original visions have been left behind as new have been added.

Actor-network theory attends particularly to the processes where new ideas and technological projects come into existence through the actions of micro-actors that, through alliances, eventually may become macro-actors like Danish wind power, in the sense that wind power constitutes a dominant facet of the Danish societal order. In other words, wind power is constituted by ordered actor-networks of heterogeneous materials whose resistance (mis-fitting) was overcome while new identities were shaped temporarily to accommodate certain actions. The accomplishment of the new reality of wind power was characterized by contingencies, shifts, and transformations that occurred over time, but what emerges as important for the wind power network is the alliances with a politically sanctioned regulatory framework that gives it priority and secures revenue streams for investments. In this way wind power is no different from investments in and consumption of fossil fuels and nuclear power, which are all subsidized in various ways as demonstrated in recent OECD-reports. Here we find some important clues to the struggles involved in what Schumpeter labeled "creative destruction" in the era of climate change, the struggles to change the relative strength of the macro-actor networks supporting fossil fuel-based production and consumption practices.

The socio-technical landscape or network that hosted wind power neither pre-existed nor was it static. For example, in the beginning wind power was small wind turbines, relatively small amounts of investment, relatively modest energy-environmental policy ambitions, etc. Recently, wind power has become increasingly associated with big wind turbines, big finance, and big problems like climate change.

The story of Danish wind power as a whole has been a combination of concerned groups who turned into micro-actors, who were then able to convince the government to shift from nuclear power to a support for renewables by putting forward a regulatory framework that allowed company entrepreneurs to sell their wind turbines to new users and buyers. However, as this essay has shown, the devil of this history is in the details of motives and translations that repeatedly formed and shifted the identities of machinery and people over decades.

The emergence of the Danish wind power network demonstrates that it is possible to overturn even the most powerful networks, because they are made up of nothing else than strong alliances that can be weakened and transformed. Following actor-network theory, the dominance of technologies in societal orders does not stem from any magic or natural development process. It requires at least as much hard work as it did for those who established these networks in the first place, but it may require even more, as the dominant technologies tend to black-box or protect the weak points of their networks, and because many of us contribute to their maintenance by our own consumption. However, in Denmark many micro-actors started to act differently—supporting new networks while not completely abandoning the old ones.

Bruno Latour, "Technology is Society Made Durable," in *A Sociology of Monsters—Essays on Power, Technology and Domination*, ed. J. Law (Routledge, London, 1991).

Bruno Latour, *Pandora's Hope: Essays on the Reality of Science Studies* (Harvard University Press, 1999).

John Law, "Notes on the Theory of the Actor-Network: Ordering, Strategy and Heterogeneity," in *Systems Practice*, 5 (1992), 379–93.

H. Lund, ed., "Coherent Energy and Environmental System Analysis," *Main report* (Aalborg University, Department of Planning, 2011).

N. I. Meyer, "Politik og Vedvarende Energi," in *Vedvarende Energi i Danmark 1975–2000 (Renewable Energy in Denmark 1975–2000)*, eds. E. Beuse et al. (Aarhus: Organization for Renewable Energy, 2000), 75–110. **http://www.ove.org**

K. H. Nielsen, "Tilting at Windmills," 2nd Edition (PhD diss., History of Science Department, Aarhus University, and Department of Organisation, Copenhagen Business School, 2001).

K. H. Nielsen, F. Petersen, and H. Siggaard Jensen, *Til Samfundets Tarv—Forskningscenter Risøs historie* (Forskningscenter Risø, 1998).

Rinie van Est, *Winds of Change: A Comparative Study of the Politics of Wind Energy Innovation in California and Denmark* (Amsterdam: International Books, 1999).

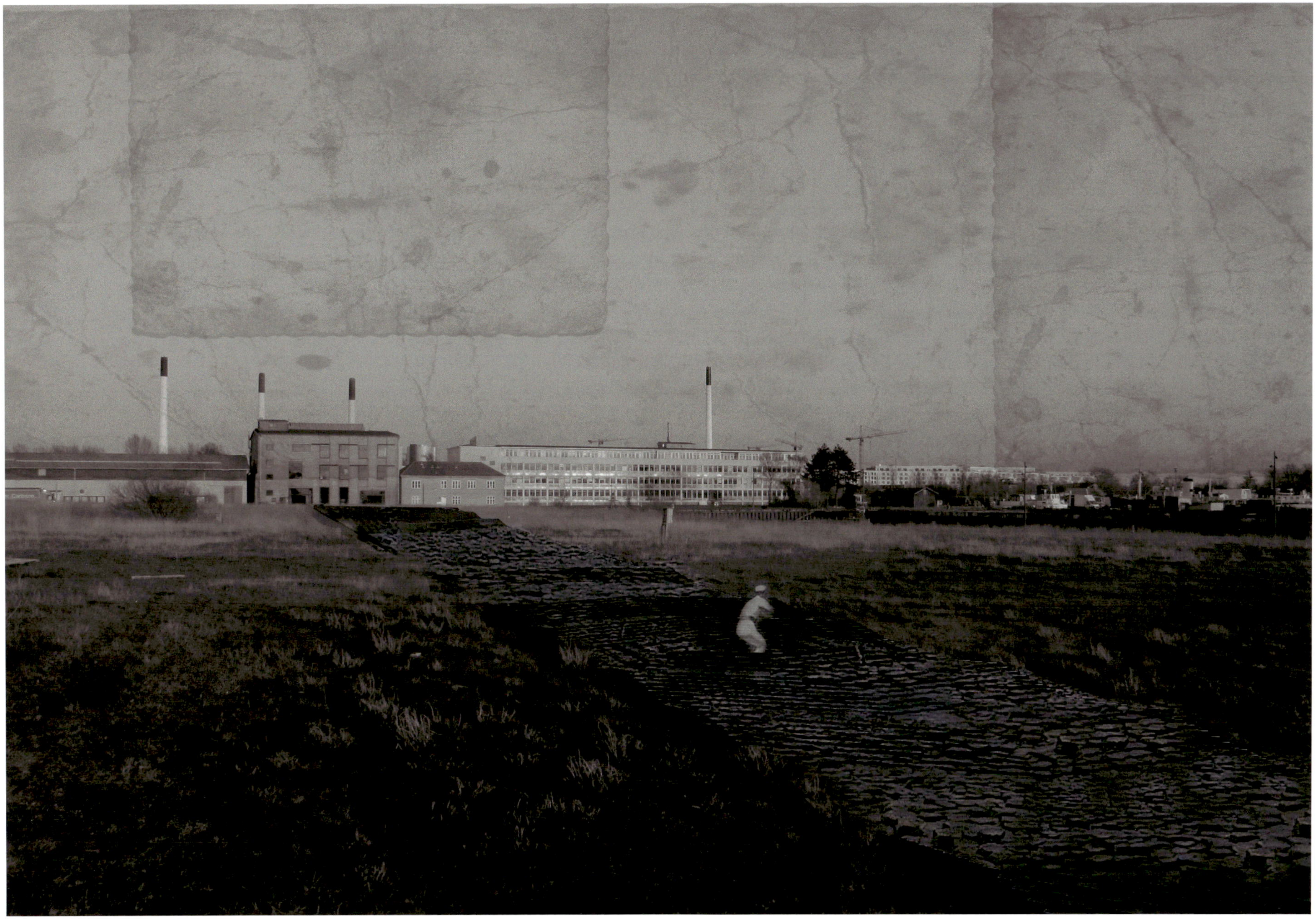

Stone Wave by Nicolas Violette, Christofer Cornevaux

See page 158

ESSAY

Working Together

Chris Fremantle

REFERENCES

David Haley, "Art, Ecology and Reality: The Potential for Transdisciplinarity in the Proceedings of Art, Emotion and Value," (presented at the 5th Mediterranean Congress of Aesthetics, 2011). **http://www.um.es/vmca/proceedings/docs/52.David-Haley.pdf**, accessed 12 March 2014.

Wallace Heim "Evaluation Report DEFRA Climate Challenge Fund CCF9 Project code AE017 Greenhouse Britain: Losing Ground, Gaining Wisdom," (2008). **http://greenhousebritain.greenmuseum.org/evaluation**, accessed 12 March 2014.

Grant Kester, *The One and the Many: Contemporary Collaborative Art in a Global Context* (Duke University Press; Durham and London, 2011).

Joachim Sauter, interview in *Data Flow: v. 2: Visualizing Information in Graphic Design* (Berlin: Gestalten, 2010).

Curriculum for Excellence: Building the Curriculum for Skills for Learning, Skills for Life and Skills for Work (Scottish Government, 2009).

Superflex, *Superflex: Tools* (Verlag der Buchhandlung Walther Konig, 1999).

Supergas website: **http://www.supergas.dk/introduction**, accessed 12 March 2014.

Jeremy Till, "The Negotiation of Hope" in *Architecture and Participation* (London and New York: Taylor & Francis, 2005).

Why is "working together" so vital right now? It is at the heart of ecoart practices and at the heart of the Land Art Generator Initiative. It is one of the features that distinguish these practices and programs. LAGI asks architects, designers, and artists (a.k.a. "creative practitioners") to work with scientists, engineers, inventors, land managers, ecologists, manufacturers, and communities.

There are several elements to working together. Teamwork is considered to be an important skill. In fact, it is part of the national curriculum in Scotland (2009). Participation has become mainstream in art, design, architecture, and new media. Interdisciplinarity is the *mot du jour* in academic research. Collaboration and creativity, participation and knowledge have become powerful words in the discourse today, but they are double-edged. Do they reinforce existing marketization, or do they open up new forms of public space?

LAGI invites teams to form and work together, ideally with communities, to develop new solutions for our societies' energy systems and to imagine new structures for generating renewable energy at the mid-scale—the scale that relates to settlements. LAGI wants us to embrace renewable energy as a beautiful part of the places we live in.

"Embracing" is a good word in this context, because we need to embrace renewable energy. It is also a good word for the particular sense of working together, because creative and techno-scientific practitioners need to embrace each other's skills and expertise, knowledge and methods. LAGI is not looking to decorate existing energy installations or plop down energy-producing sculptures. This embrace must not be uncritical. The future of energy must be renewable, and it must be socially just. The BBC reported when the renewable energy system on Eigg, an island off the west coast of Scotland, came online. What wasn't reported was the social justice built into the system. Renewable energy is limitless over time, but limited at any point in time. On Eigg, every house and business has a cut-out switch, which stops an individual from using too much energy at any particular moment. This is a form of community collaboration, which is significant and which addresses the "tragedy of the commons" (the tendency for people to act in their own short term interests even if this has long term negative consequences for the community). On Eigg people embrace each other with social as well as environmental justice.

If we want to understand why working together with artists might be important, it is worth looking to the practice of Helen Mayer Harrison and Newton Harrison. These eminent ecological artists responded to an invitation from David Haley (Director, Ecology in Practice, Manchester Metropolitan University, UK) to consider the impact of global warming on the island of Britain. The result was the project *Greenhouse Britain: Losing Ground, Gaining Wisdom* (2006–2008), which was funded by the UK government as part of its Climate Challenge program. In the independent evaluation of that project, Wallace Heim comments on interviews she conducted with the Harrison's project collaborators:

> They all reported that the experience was illuminating, informative, challenging, imaginative, liberating. Their respect for the cross-disciplinary knowledge of the Harrisons was high, including both the science, the land-use planning and the architectural aspects, including Newton Harrison's ability to ask "the right questions."

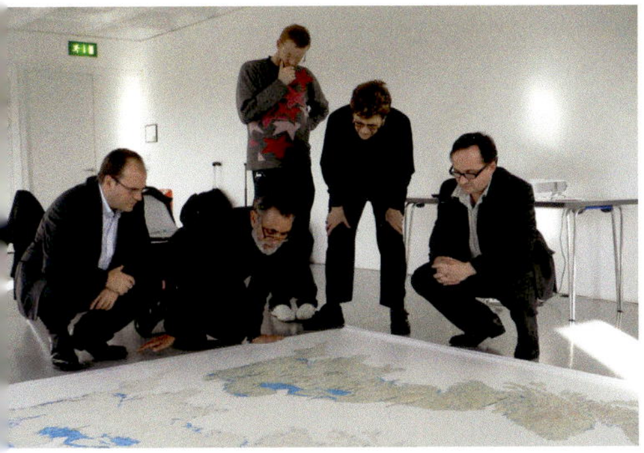

Greenhouse Britain: Losing Ground, Gaining Wisdom. 2006–2008. Discussing sea level rise at Arnolfini, Bristol. (left to right) Tom Trevor, Newton Harrison, Martin Clark, Helen Mayer Harrison, and Chris Fremantle

Photo by David Haley

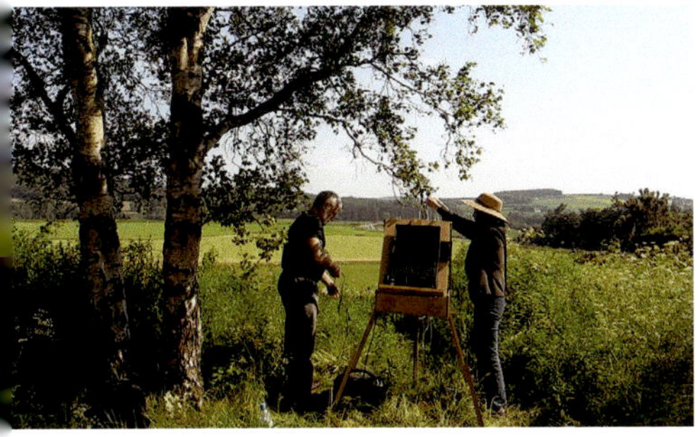

Plein Air: Tim and Reiko working with trees in Aberdeen. 2010–2014. Reiko Goto and Tim Collins, with Michael Baldock, Carola Boehm, Matt Dalgliesh, Trevor Hocking, Chris Malcolm, and others

> Further, they had been taken on a journey, relieved of the strictures of their respective disciplines and work practices, and had found it in some way transformative of their way of considering climate change and possible adaptations to it. But, from their responses, the exercise was not just one of being relieved of limitations, but one which was highly informed, creative, and reflective, not merely of their own methods of work, but of more conventional responses to climate change. They reported feeling supported, mentored, and reported an appreciation of what this kind of process of "art" can achieve in providing the context, the time and space for imagining possible futures, for rehearsing what may happen. (2008, p. 9)

The words that Heim chooses to characterize the experience of collaborating with the Harrisons are also used by others when speaking about the quality of collaborative relationships between artists and scientists. LAGI is seeking to provide a context, time, and space for that quality of informed, creative, and reflective practice to imagine possible futures and rehearse what can happen as we embrace renewable energy.

There are dangers, however, in focusing on an idealized form of collaborative practice. What Heim's description does not suggest is that the result of *Greenhouse Britain* is a problem solved. Rather it is making sense of our new circumstances and exploring what some futures might look like.

In his essay *The Negotiation of Hope* (2005), Jeremy Till addresses John Forester's argument that the role of designers in particular should be understood as "sense-making" rather than "problem-solving." Till states: "Central to Forester's argument is that such a move from the problem to sense-making necessarily brings with it an acknowledgement of the contested social situation in which the design process is first initiated...."

When we step outside our specialized spaces, whether the galleries, concert halls, and theaters of artists, or the labs of scientists and engineers, we are negotiating our practices. Increasingly, we are negotiating with communities as well as other professions. Creative practitioners working with ecological systems, human habitation and development, energy and resource generation, and so on, quite specifically embrace other ways of working, and in particular other methods. They can enter into deep relationships. There is a sharp edge here, because this involves dealing with other living things, not just inert materials. Therefore, this embrace has to be respectful, has to have an ethical dimension, and has to be caring.

To understand what this might mean, Tim Collins and Reiko Goto's project *Plein Air* (2010–2014) required that they work with engineers to develop a range of sensing technology. This technology enabled the public to perceive trees breathing and, in collaboration with musicians and audio artists, to transform the data streams of that breathing into acoustic experiences. Collin's and Goto's concern was to encourage empathy, using technology to heighten awareness.

In being collaborative, we are often being interdisciplinary—working between, with, across, into, and beyond disciplines and between different forms of knowledge and practice. Sometimes conversations explore the similarities between artists and scientists,

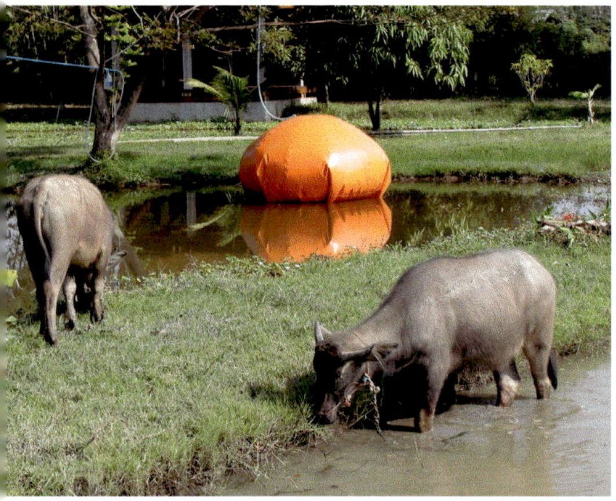

Installation photo of *Supergas/ User/ The Land*, Chiang Mai, Thailand. 2002.

Photo courtesy of Superflex

designers and engineers, but a discipline is a specialization. With specialization comes skill and expertise. Ecoart always requires multiple and varied skills and expertise. There are many dimensions to this. Creative practitioners tend to have thematic interests, such as water, biodiversity, urban greenspace, brownfields, phyto-remediation, farming, orchards, and permaculture. Ecoartists will name their collaborators and will report, and sometimes document, the dialogues.

I'm increasingly concerned about the terms "collaboration" and "interdisciplinarity" because these words might be obscuring the basic act of "working together." However, not all "working together" is the same. David Haley (2011), using the analysis of Basarab Nicolescu, suggests some ways of thinking about the differences. A group of people with different specializations can all work on the same question. This might be called "*multi*-disciplinary." If those people exchange methods, so that the specializations become hybrid, then that might be called "*inter*-disciplinary." Then there are circumstances where different specializations come together to focus on a problem, setting aside any hierarchies of specializations, and this might be called "*trans*-disciplinary" (the prefixes post- and extra- have also been used). Haley argues that the repositioning of specializations, clarified by this terminology, is vital to address 21st century questions. I would argue that ecoart is inherently interdisciplinary—it is not just the knowledge domains that are embraced. If you look at a lot of ecoart, it actually sits in grey areas between art and design—not clever product design, but design in the sense of clear communication of information, clear construction of process resulting in impacts. Joachim Sauter opens up the issue when he states:

> In short: the result of design work has to be understood immediately and should be directly legible by as many as possible. This means it has to be told in a language that everyone understands. Artwork however is produced using an individual and personal language and it is mainly not meant to be understood immediately or by everyone. The process of understanding an artwork by deciphering is very important. It forces one into a much deeper dialogue with what is presented. In design work it is the opposite—if there is a fire, you don't want to decipher an exit sign. It goes without saying that the borders are blurry and that you find both approaches in both fields. (2010, pp. 250–251)

Perhaps the Danish collective Superflex might exemplify this issue. In addition to their work 2000 Watt Society Contract, which relates to the collaboration on Eigg mentioned above, Superflex's Supergas project sits in this blurry, in-between space. The Supergas website Introduction page states:

> In 1996–1997 Superflex... collaborated with biogas engineer Jan Mallan to construct a simple, portable biogas unit that can produce sufficient gas for the cooking and lighting needs of an African family. The system has been adapted to meet the efficiency and style demands of a modern African consumer. It is intended to match the needs and economic resources that we believe exist in small-scale economies. The orange biogas plant produces biogas from organic materials, such as human and animal stools.

First, note that the engineer is credited in the first sentence. Second, the Supergas project appears to conform to Sauter's description of design. In Tanzania, Cambodia, Thailand, Zanzibar, and Guadalajara, the project's function is clear. When seen in an art exhibition, however, for example at the Louisiana Museum, Denmark (1997) or at Marres, Centre for Contemporary Culture, in Maastricht (2011), it becomes a kind of personal language that requires deciphering. In those contexts, it becomes an "issue-based" work of art. There is a third position from which it also needs to be deciphered. As Mallans states in an interview: "That's also different from industry. In industry you don't ask whether there is any money. Of course there is. But here you know there's no money." (1999)

Creative practitioners working on environmental and ecological projects, including those contributing to LAGI, might be attempting to operate, like Superflex, in both of Sauter's modes. Their works often operate at more than one level—to understand immediately what the project is doing and make it directly legible, but also to enter into a deeper dialogue through a more personal relationship with the work.

In exploring collaborations between artists and communities, Grant Kester is interested in the politics of collaboration:

> In the most successful collaborative projects we encounter instead a pragmatic openness to site and situation, a willingness to engage with specific cultures and communities in a creative and improvisational manner..., a concern with non-hierarchical and participatory processes, and a critical and self-reflexive relationship to practice itself. Another important component is the desire to cultivate and enhance forms of solidarity.... (2011, p. 125)

Kester's defining characteristics are leitmotifs. In particular "solidarity" is a political word (perhaps more so if you are connected to Poland and grew up in the 80s), but it signals the importance of respect and justice in the process, echoing openness to site and situation, reinforcing and engaging with specific cultures and communities, and embedding an alternative politics.

Kester's phrase "a critical and self-reflexive relationship to practice itself" opens up space for the practice to inhabit the blurry space between clarity and directness on the one hand, and depth and personal language on the other.

The reason we might need to rethink our understanding of creative practice, as suggested at the start of this essay, is because the most provocative examples of ecoart, and LAGI in particular, are characterized by a shared process rather than an autonomous one. The artists are not adding decoration to something that engineers have designed, and the designers are not simply designing the logo for the product. There's a deep understanding that to make sense of our energy challenges and to intervene effectively takes multiple intelligences, multiple practices, multiple creativities working together.

ESSAY

Visual Art and the Ecological Challenge: About Some of the First Creators of Ecological Art and Their Influence

Else Marie Bukdahl

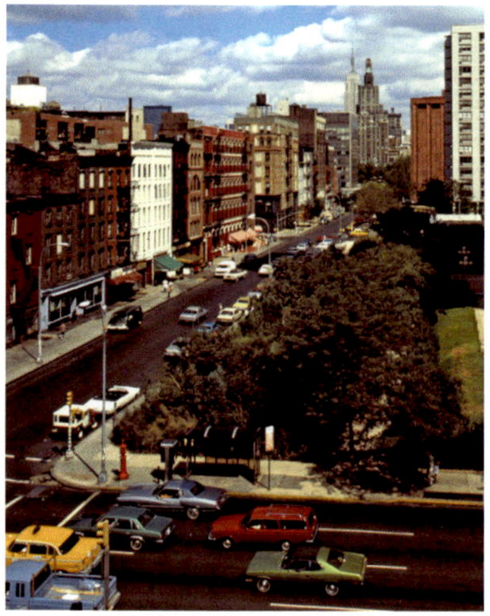

Alan Sonfist. *New York City. Time Landscape*. 1965–Present.

Photo by Alan Sonfist

Visual art, much more so than poetry, has focused on interpreting and visualizing the changing aspects of nature and the complicated, organic growth processes that underlie it. It is within garden art and the attempts to create an earthly paradise—first and foremost Chinese, Arab, and English Romantic gardens—that some of the earliest endeavors to protect nature and show greater care in the use of its resources become apparent.[1]

It was, however, only in the 1960s that Environmental Art and Ecological Art started to emerge. American artists established both movements. Environmental Art, or Land Art, founded by artists like Robert Smithson and Walter de Maria, is an art form using the site itself as the starting point and basis for creative activity, whether on land, in the water, or in the air.[2]

At the same time, a number of other prominent artists, like Michael Singer, Alan Sonfist, and Helen Meyer and Newton Harrison, were occupied with forming the landscape from the standpoint of nature itself. They can be regarded as some of the first creators or forerunners of Ecological Art, later called Sustainable Art or Green Art. Individually, they are quite different, but what they have in common, as Michael Auping remarked in 1983, is that their artworks "directly or indirectly, reflect the concerns of increasing numbers of biologists, ecologists and legislators attempting to develop a legal ethic that addresses man's relationship to the land; an ethic that views natural objects, such as trees, mountains, rivers and lakes as having a legal right to existence."[3] These artists created works that were in nature and/or restored nature, but changed it as little as possible. They also wanted to stimulate a greater awareness of nature and the earth, creating a focus on respect for nature and the desire to preserve it. European artists, like Joseph Beuys and Friedensreich Hundertwasser, worked towards the same goal and created impressive ecological art in their countries.

Alan Sonfist is an artist/designer working with natural landscapes, attempting to reveal hidden layers of meaning in the earth. His most famous environmental project is a reconstructed forest entitled *Time Landscape*, which was conceived in 1965 and is still ongoing. It was developed and installed in successive stages on the corner of West Houston Street and LaGuardia Place in Greenwich Village. He described it as follows:

> The reconstructed forest was a way of going back into my childhood forest in New York as it would have been, initiated in Greenwich Village. I transplanted living tree species such as beech, oak and maple and over 200 different plant species native to New York, selected from a pre-colonial contact period in New York. These are still there on site. Besides experiencing the indigenous trees of New York City, *Time Landscape* allowed me to experience and interact with foxes, deer, snakes, eagles and this was part of my experience.[4]

More recently, in 2005, Sonfist created an impressive ecological landscape project called the *Lost Falcon of Westphalia* on Prince Richard's estate outside Cologne in Germany. It is only possible to see the silhouette of the hovering Falcon properly from a bird's eye view. The walls of earth are reminiscent of the mounds of nearby Celtic fortifications.

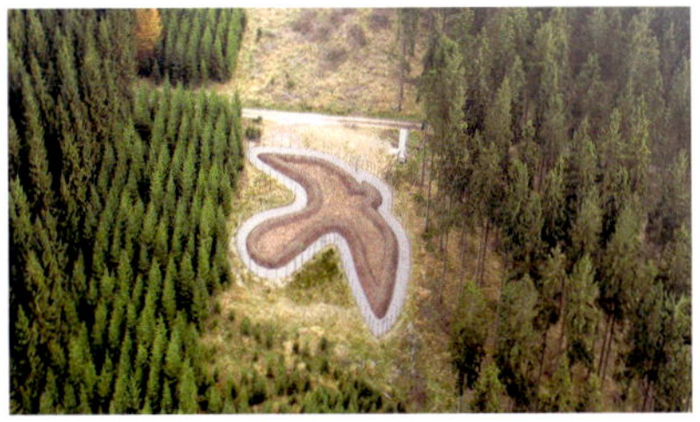

Alan Sonfist. *Lost Falcon of Westphalia.* 2005.

Photo by Alan Sonfist

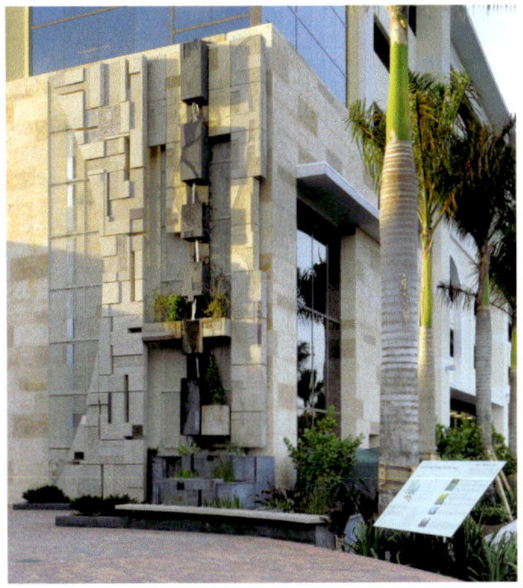

Michael Singer. *The Sculptural Biofiltration Wall.*

Photo by David Stansbury, courtesy of Michael Singer Studio

Since the end of the 1960s, Sonfist's Green Art projects have inspired younger artists to understand that "we are only one of many internal structures that exist in nature,"[5] which encourages them to work with ecological art.

The collaborative team of Helen Mayer and Newton Harrison (often referred to simply as "the Harrisons"), are among the leading pioneers of the Ecological Art (Ecoart) movement. The Harrisons are historians, diplomats, ecologists, investigators, emissaries, and art activists. An early work they created together was a long-term work called *The Lagoon Cycle* (1974–1984). In this project, they observed and experimented with *Scylla serrata* lagoon crabs, which they transferred from a lagoon in Sri Lanka.[6] Jack Burnham describes their main aim: "Eventually Harrison and his wife want to reintroduce the utilitarian into art at an extremely refined level. And in the process they hope to provide an antidote for the prevailing cynicism of the Art World by making art the non-verbal teaching system it once was."[7] The Harrisons have been particularly significant for the Bio Art of our times, an art form that is developing in many surprising ways.

In 1982, Joseph Beuys launched the "7000 Oaks" project at Documenta 7 in Kassel, Germany, for the growing of 7,000 trees, which he called a "social sculpture"[8] and which he described as follows: "I believe that planting these oaks is necessary not only in biospheric terms, that is to say, in the context of matter and ecology, but in that it will raise ecological consciousness—raise it increasingly, in the course of the years to come, because we shall never stop planting,"[9] A part of the project has been realized in New York on West 22nd Street between 10th and 11th Avenues through the Dia Center for the Arts, and it continues to stimulate younger artists today.

Friedensreich Hundertwasser launched ideas of green architecture, ecology, and "urban gardening" from the start of the 1980s. He visualized them not only in his paintings and installations, but also in buildings that can be seen today, for example *The Hundertwasser House* in Vienna (1983–1986). "It features undulating floors ('an uneven floor is a melody to the feet'), a roof covered with earth and grass, and large trees growing from inside the rooms, with limbs extending from windows."[10] His aim with this building was to make the world a better and greener place. His earliest work with sustainability and ecology inspired many contemporary artists and architects working with green technology, for example William McDonough + Partners and Terunobu Fujimori.[11]

Michael Singer is among the seminal figures of the Environmental Art Movement working in the 1970s in remote, delicate natural areas, like the Everglades, threatened Long Island saltwater marshes, and the bogs and forests of New England. His intention was, he emphasizes, for the projects to appear "as if they were nature's own work." He regards art as a balanced visual dialogue with the conditions found at the locality, as can be seen in *First Gate Ritual Series* 4 /79 (1979).

By the end of the 1980s Michael Singer was known for his innovative approach to garden design, architecture, infrastructure, and planning. He frequently posed the question: "How do we balance and integrate human needs with those of nature, how do we live with nature rather than against it?" His response has been to create ecological and sculptural gardens, like the poetic Concourse C. Denver International Airport (1994) in Colorado, and to transform

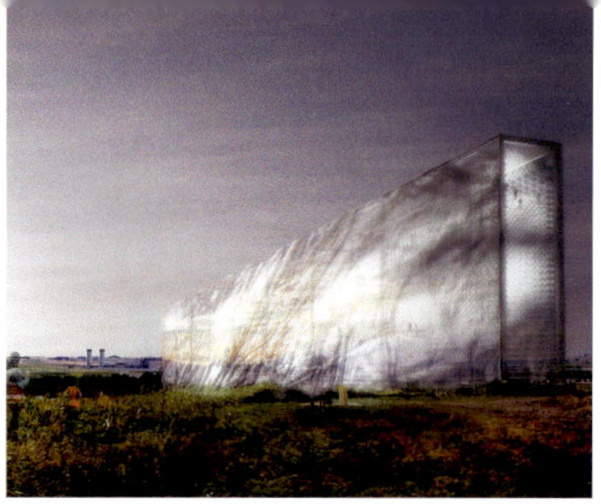

James Murray and Shota Vashakmadze, *Scene-Sensor // Crossing Social and Ecological Flows*. 2012.

infrastructure projects, like the Solid Waste Transfer and Recycling Center (1989–1993), in Phoenix, Arizona, into enthralling buildings by means of the aesthetic application of green technology.[12]

The Sculptural Biofiltration Wall for the Seminole Coconut Creek Casino in Florida is one of his most recent green projects. Built in collaboration with his colleagues in 2012,[13] it illustrates impressively his ability to create dynamic interactions between art, the urban environment, and sustainability. The large wall is highly visible at the southeastern corner of the large new parking structure. It is covered by an intense and richly expressive sculptural depiction comprising an astonishing diversity of structures, which form an extremely varied play of light and shadow, creating a unique life and lyrical atmosphere. These artistic structures are enhanced by a variety of climbing vines, creeping ficus, and aquatic plants that will, in time, partially cover the sculptural elements. This will create a living wall with a pattern of green hues that will enhance the surfaces with a bright and expressive range of earth tones. At the base of the wall is a pond with inviting sculptural bench caps in the foreground and a variety of water plants and colorful flowers. The sculptural wall is "conceived as a living system designed to regenerate the surrounding environment by improving water quality, enhancing habitat viability, and informing and inspiring the public about ecological systems. The project filters retention pond water and also harvests rainwater." This ecoart project regenerates the environment because it "fosters biological systems including plants, fish, and beneficial bacteria that naturally cleanse water." In addition, it generates energy, because "the 23 kW parking roof deck solar photovoltaic canopy arrays provide enough power to supply 2 to 3 average Florida homes."[14] The sculptural wall provides the otherwise anonymous surroundings with a new identity and transforms them into an impressive and inviting place with a fully functioning ecological system that cleanses and restores the environment. It also demonstrates, on a small scale, the essence of Michael Singer's closely combined artistic and ecological strategies, which have inspired artists and architects for over 30 years—not least through his many years of teaching at a variety of schools of arts and architecture.

Over the past three decades, more and more artists have worked towards finding artistically satisfying solutions regarding the relationship between art and ecology. They have applied and transformed new instrumentalist technologies in their artistic production in many imaginative and original ways. These practitioners include the media artist Maria Michails, who works with new combinations of art, technology, and ecology,[15] and Nohra Corredor, who in 1996 began to combine ecological artworks with technology, because she wanted to develop a model for online Ecological Arts institutional displays.[16] New relations between art and technology are also found in the Bio Art of Shiho Fukuhara and George Tremmel.[17] In the projects created by these artists and their colleagues, a respect for nature and mankind is combined with inventive artistic goals.

1 E. M. Bukdahl, *The Re-enchantment of Nature and Urban Space. Michael Singer Projects in Art, Design and Environmental Regeneration* (Aalborg, Denmark: Utzon Center, 2011), 37–41.

2 Elizabeth C. Baker, "Artworks on the Land," in *Art in the Land. A Critical Anthology of Environmental Art*, ed. Alan Sonfist (New York, 1983), 73–84. Baker also presents her opinion on the effect artworks by Robert Smithson, Walter de Maria, and Michael Heizer may have on the ecology of these sites and discusses what relation this type of artwork may have on everyday life.

3 Bukdahl, *Re-Enchantment of Nature*, 37–41.

4 John K. Grande, "Interview with Alan Sonfist" (January 12, 2008). **www.umintermediai501.blogspot.dk/2008/01/interview-with-alan-sonfist-john-k.html**.

5 Cited in *Art in the Land. A Critical Anthology of Environmental Art*, ed. Alan Sonfist (New York, 1983), 103.

6 Ibid.

7 Jack Burnham, "Contemporary Ritual: A Search for Meaning in Post-Historical Terms," in *Arts 47*, no. 5 (March, 1973), 41.

8 Norbert Scholz, "Joseph Beuys 7000 Oaks in Kassel," in *Anthos*, no. 3 (Switzerland, 1986), 32.

9 Johannes Stüttgen, *Beschreibung eines Kunstwerkes* (Düsseldorf: Free International University, 1982), 1.

10 Andrea Rygg Karberg and Camilla Jalving, "Hundertwasser. The Phenomenon," in the catalogue for the exhibition *Hundertwasser* (Arken Museum of Modern Art, 2014), 36–38.

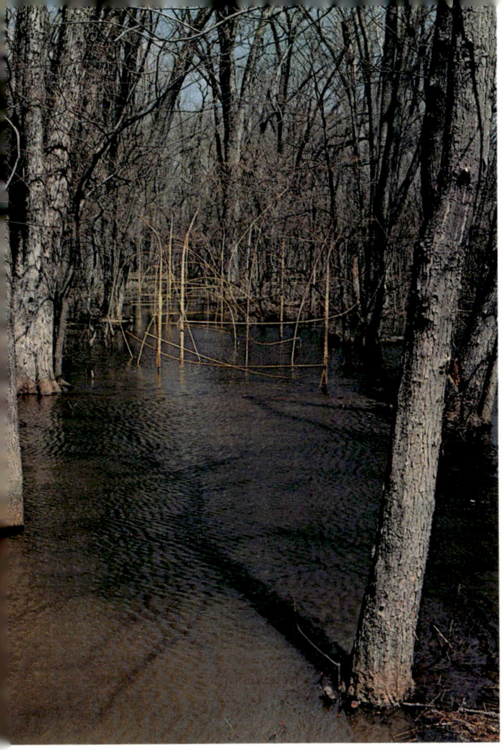

Michael Singer. *First Gate Ritual Series 4/79.* 1979.

Photo courtesy of Michael Singer Studio

A more comprehensive overview of artists' desire and ability to experiment with technological and artistic tools can be gained by studying the many interesting ecological art projects that have come out of the *Land Art Generator Initiative* project. The artists involved in this innovative competition are bringing respect and awareness of the environment through art. This includes the work of James Murray and Shota Vashakmadze, who created *Scene-Sensor // Crossing Social and Ecological Flows*, the first-place design for LAGI 2012, in which they developed interconnections between nature, urban space, sculpture, ecology, and engineering. James and Shota describe their project as follows:

> Between the north and east mounds, a channel of ecological flows defines a corridor of optimum wind flow. To sense and harness the energies inherent within this conduit, the project proposes a wind-mapping screen.[18]

LAGI and a range of other art projects, including Michael Singer's biofiltration wall, reflect a new tendency within ecological art: to engage engineering technologies in order to participate in and provide innovative artistic contributions to the creation of a transition to a more sustainable future. Artists creating ecological art projects have always been aware that, because verbal language never corresponds entirely with artistic expression, visual art is capable of grasping perspectives or revealing traces and significances that philosophers and scientists cannot communicate with their tools alone. Ecological art is very important to public space because it can help cleanse the air, water, and earth; it can generate renewable energy; *and* it can inspire us to resist the destruction of nature.

The different forms of ecological art offer a local and global perspective while transcending the existing boundaries of the contemporary art world in an inspiring way. Ecoart will continue to enable us to become more active in efforts to protect nature, create sustainable growth, and solve ecological issues, which will ultimately benefit nature and our society. We will be drawn into a fascinating artistic experiential space that will widen the perspectives of our daily lives.

11 **www.wien-vienna.com/hundertwasser.php**

12 Bukdahl, *Re-Enchantment of Nature,* 47–50, fig. 26 and 53–59, figs. 28–29.

13 *Michael Singer Studio Team:* Jason Bregman, Jonathan Fogelson, Alan Chapman, and other colleagues.

14 Michael Singer's description of the integrated ecological systems in *The Sculptural Biofiltration Wall*: **www.ecoartsofla.org/wp-content/uploads/2012/05/MSI_Seminole_BioWall_Summary_v2.pdf**

15 **www.eyebeam.org/people/maria-michails**

16 Ibid.

17 Monika Bakke, "Art for 'Plant Dignity.' Bottom-up Ethics in the Biotech Age," in *Towards the Third Culture. The Co-Existence of Art, Science and Technology* (Gdansk: Laznia Centre for Contemporary Art, 2011), 97–103.

18 Caroline Klein, Elizabeth Monoian, and Robert Ferry, eds., *Regenerative Infrastructures: Freshkills Park NYC, Land Art Generator Initiative* (Munich: Prestel Verlag, 2013), 56–61.

ESSAY

Generating Futures: LAGI as an *Imaginatorium*

Lea Schick and Anne Sophie Witzke

> "It is the broadest, the greenest and the most long-term energy agreement that has ever been reached in Denmark. In our everyday political work, the parties are different shades of red and blue. But together we have laid down the foundation for a green future."
>
> —Martin Lidegaard, (former) Minister for Climate, Energy, and Building, at the approval of the 2012 energy agreement.

The 2012 Energy Agreement committed Denmark to doubling the production of wind energy by 2020. With 5,000 wind turbines (one per 1,000 inhabitants) and a world record of 30% of the country's energy production coming from wind, Denmark is densely populated by wind turbines. As "sculptures" ornamenting the Danish landscape, these windmills have become both national landmarks and objects of public controversy. The Land Art Generator Initiative (LAGI) takes place within this context and contributes to the strengthening of the government's roadmap to a green and sustainable future. Compared to the Danish government, however, LAGI has a novel and quite different take on what power plants could look like, how they are formed, and by whom. In this essay, we will argue that LAGI—working as an *imaginatorium*—expands the space in which energy futures can be imagined and made. With a multiplicity of ideas for aesthetic power plants, LAGI lays a foundation for imagining multiple ways of forming our future energy landscape, in which wind turbines are accompanied by other forms of energy generation.

LAGI can be conceptualized as both a competition that collects ideas for a variety of artworks, where each can be valued for their individual aesthetic qualities, and as an art project in its own right. In this essay, we focus on LAGI as an artwork that performs a distinct artistic strategy. LAGI takes the shape of a conceptual apparatus—a generator, as the title of the project suggests—that enables the ongoing production of ideas for site-specific artworks. Even though a winner is selected, LAGI is less about emphasizing the one winning sculpture than about demonstrating a catalogue of ideas for alternative power plants. This multiplicity indicates a creative potential waiting to be explored.

Like many contemporary art projects engaged in building a more environmentally responsible society, LAGI operates within an expanded field of art, blurring boundaries between art, design, architecture, and engineering. LAGI thus departs from the idea of art as solely an object of aesthetic contemplation—art is seen as a tool for action and environmental change as well. LAGI builds on the concept of "social sculpture," formulated by the conceptual artist, politician, and environmentalist Joseph Beuys in the 1960s. His central idea was that art and human creativity hold the capacity to transform society, and that all of us can contribute creatively to this process: "Everyone is an artist," as Beuys famously stated. Similar to social sculpture, LAGI expands the gallery and museum

[1] A. Barry, "On Interactivity: Consumers, Citizens and Culture," in *The Politics of Display: Museums, Science, Culture,* ed. S. Macdonald (Routledge, 1998), 98–117.

space, moving into public space where the artworks engage with "real life," contribute to solving problems, and have an effect on the world. LAGI thus transcends traditional art discourse and makes aesthetic tools with transformative capacities.

By fusing art and renewable energy and by inviting people to submit ideas on this important and otherwise very expert-led subject, LAGI intervenes in governing versions of sustainability and renewable energy, where efficiency, predictability, and identifiability normally dominate the way energy production is imagined and actualized. Contrary to the Danish government's energy development, based primarily on wind turbines, LAGI demonstrates how a variety of beautiful, spectacular, and thought-provoking power-plant sculptures could inhabit the country, thereby illustrating how energy production might otherwise be. When the many ideas generated through LAGI are presented alongside one another—in this very book, in the exhibitions, and in the website portfolio—an imaginative space unfolds *in-between* the creative scenarios, a space in which the spectator is invited to imagine alternative landscapes of energy generation.

Inspired by the American physicist Frank Oppenheimer, who in 1969 opened The Exploratorium, the world's first interactive science museum in San Francisco,[1] we will coin the engaging and creative space that LAGI creates as an *Imaginatorium*—a space that sparks the imagination of the viewer. Oppenheimer's Exploratorium was a democratic project that, like LAGI, mixed art with science, not only to show that science is a creative discipline, but also to empower the public and make them engage with and learn about science in new ways. Like the Exploratorium, we believe that LAGI can engage the public in thinking and rethinking the infrastructural foundation of society by creating imaginary spaces for exploring the concrete, physical form of renewable energy power plants. Through its generative capacity, LAGI thus emerges as an imaginatorium that allows us to envision the world as a more fascinating, manifold, and environmentally sound place to live.

WINNING ENTRIES

First Place Winner
The Solar Hourglass

Second Place Winner
Quiver

Third Place Winner
eMotions: Energy Motions and Art Emotions

“The overall objective of a sustainable society is, in the end, a collective objective. Taking over ownership of energy production is not an end in itself. It is the taking ownership of the collective solution that is interesting. The initiatives of individuals should be a part of that.”

—Natalie Mossin

Head of the Danish Architects’ Association and Partner at Smith Innovation

The Solar Hourglass

TEAM:
Santiago Muros Cortés

TEAM LOCATION:
Buenos Aires, Argentina

ENERGY TECHNOLOGIES:
concentrated solar power (thermal beam-down tower with heliostats)

ANNUAL CAPACITY:
7,500 MWh

Rather than using sand to measure time, *The Solar Hourglass* uses the power of the sun to electrify hundreds of homes while providing a breathtaking setting for inspiration and relaxation. The installation consists of an upper and a lower bulb. Dozens of people can gather on the bottom bulb during the day, sheltered by the shade of the top bulb.

The project works as a solar central receiver, consisting of an arrangement of small flat mirrors that concentrate their reflection of solar energy on a tank holding a heating medium. Sun-tracking mirrors (heliostats) on top of the upper bulb reflect solar heat onto a cone-shaped set of smaller mirrors that concentrate these reflections and direct them down the neck of the installation.

The concentrated beam of solar heat then reaches a receiver containing heat transfer fluid (HTF) of molten nitrate salt, which is heated to temperatures over 600°C. The HTF is conveyed through a heat exchanger where water is turned into steam to run a 6.2 MW turbine generator. A small percentage of the steam produced is released back into the neck of the hourglass, thus making the solar beam visible to the public.

The higher heat capacity of the molten salt allows for the system to store heat and produce energy during cloudy weather or even at night.

All the energy production and transformation components are hidden inside the bottom bulb and inaccessible to the general public, making it safe for public engagement.

At night, the beam turns off, and the thin layer of OLED (organic light-emitting diode) that covers the installation, lights up on the surface of both bulbs, transforming the hourglass into a pair of elegantly curved planes that shine on the edge of Refshaleøen.

The Solar Hourglass reminds us that energy is just as precious and fleeting as time, and thus we should take care of it, appreciate it, and not waste it. The installation aims to send an optimistic message to those who visit: we still have time to make things right with the environment, and if we act now, it is not yet too late.

"An hourglass to remind us, that as long as we take care of our environment... the energy will never run out."

—Santiago Muros Cortés

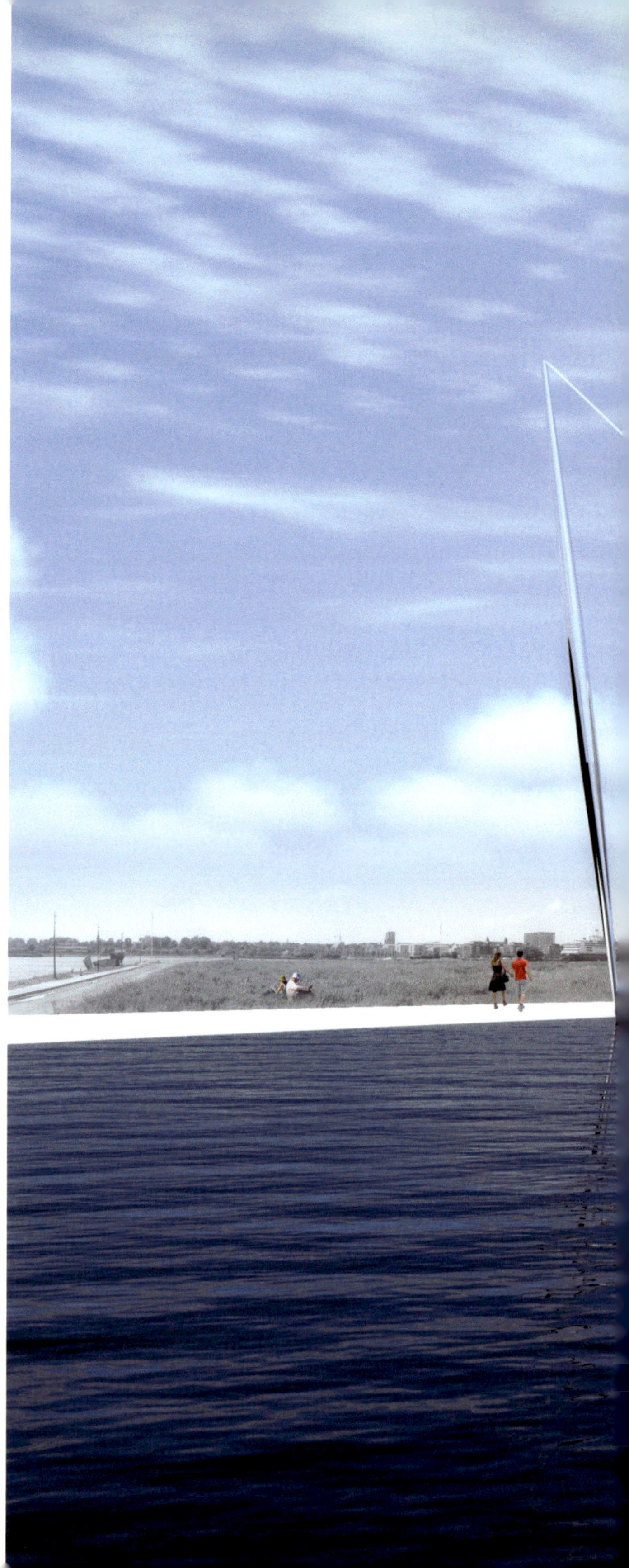

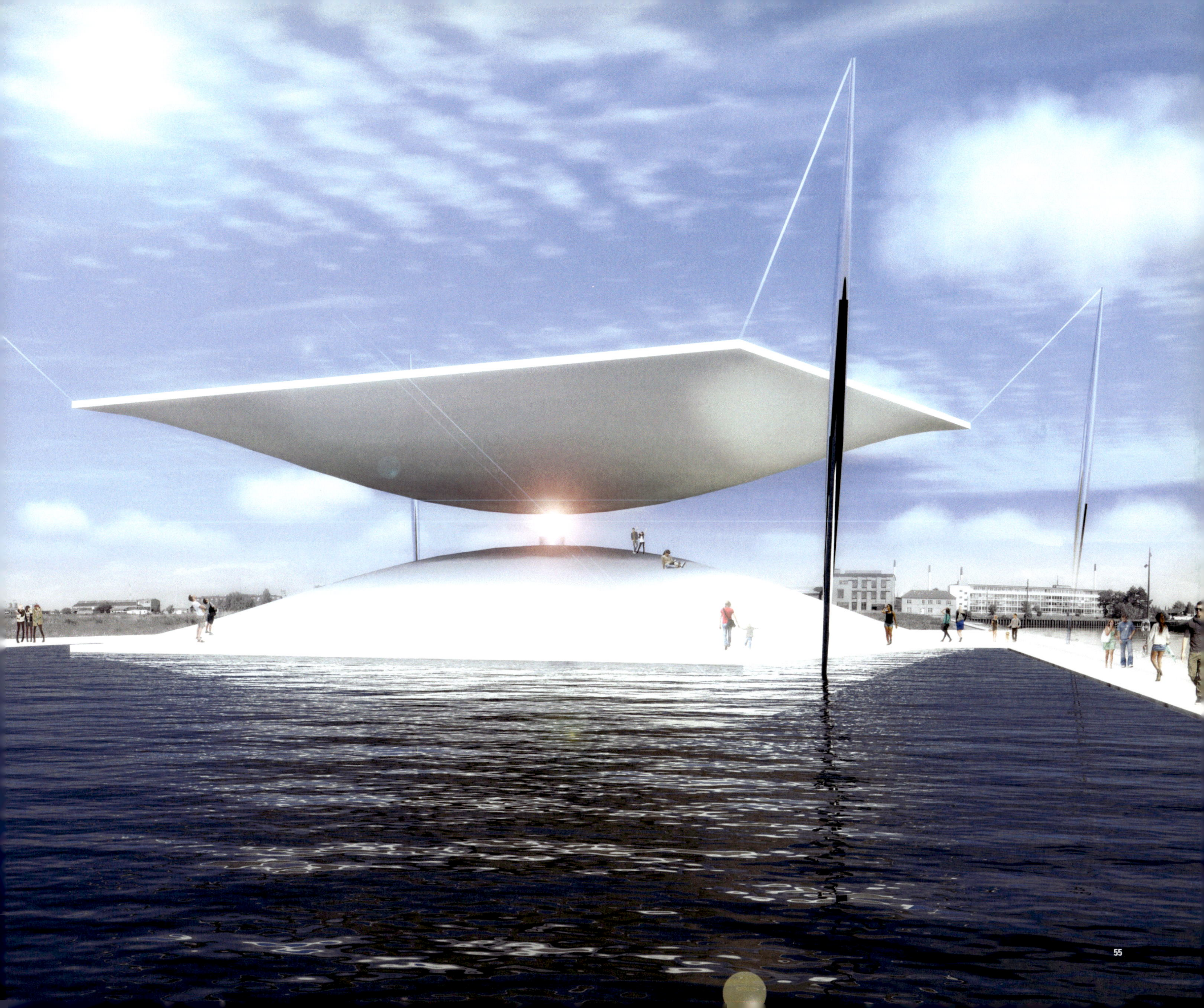

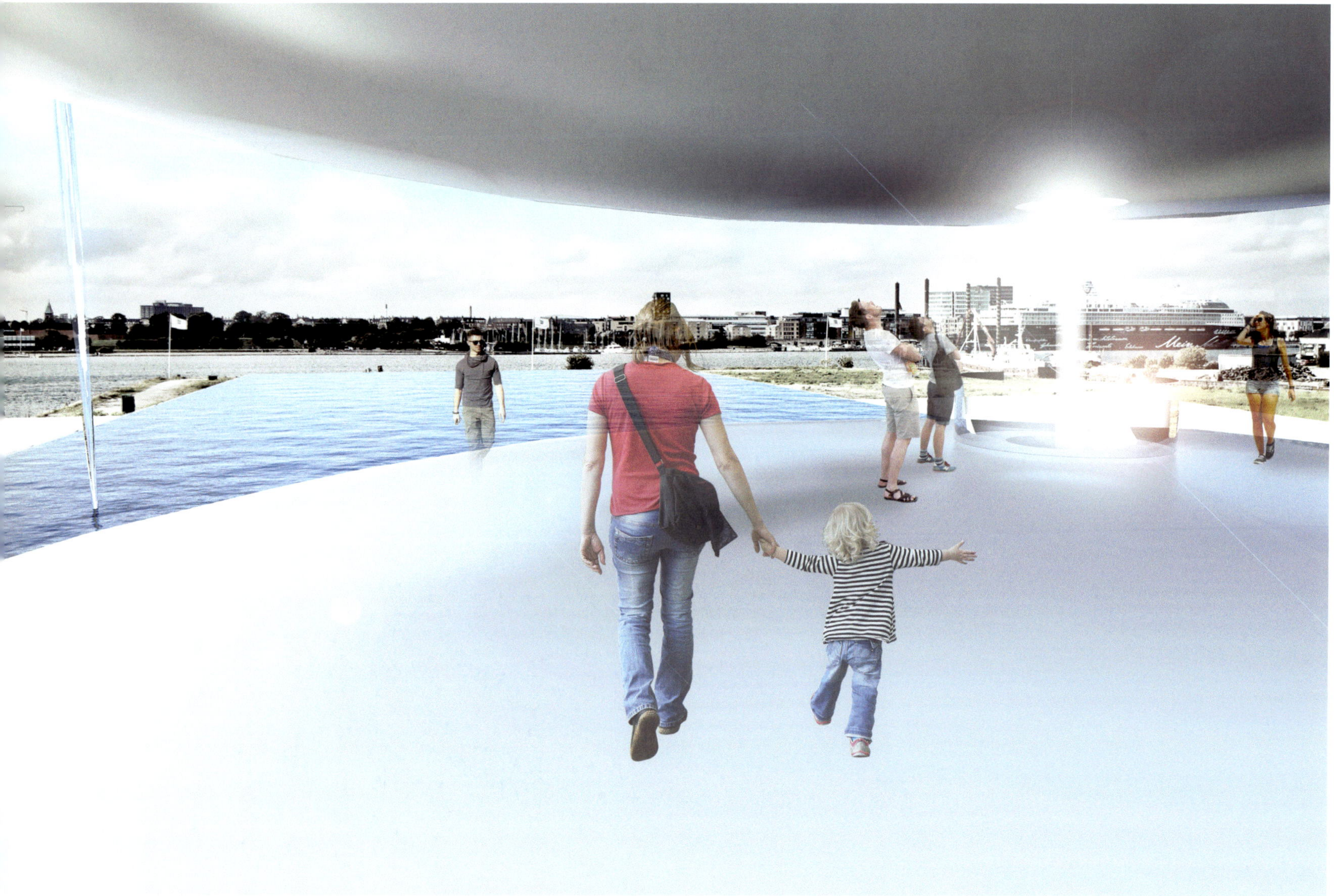

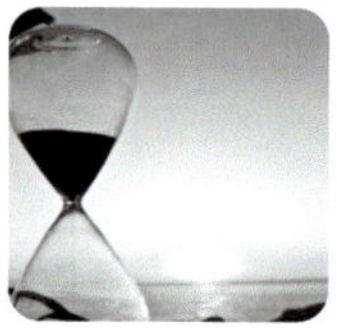

Think of an hourglass

...how it measures time in such an elegant and simple way.

Now imagine if we could capture energy in a similarly beautiful way...

... and as we do so... remind ourselves that there's still time to improve our repertoire of sustainable infrastructure.

The concept behind the installation is to create an environmentally friendly and sustainable power-producing hourglass that resembles not only the shape of these beautiful artifacts but also the way in which they work, using a solar beam instead of sand as a trickling material.

In this way, the project not only converts heat into electricity, but it also makes a visual spectacle out of it, providing an outstanding platform from which to enjoy incredible views of Copenhagen's harbor and the installation itself.

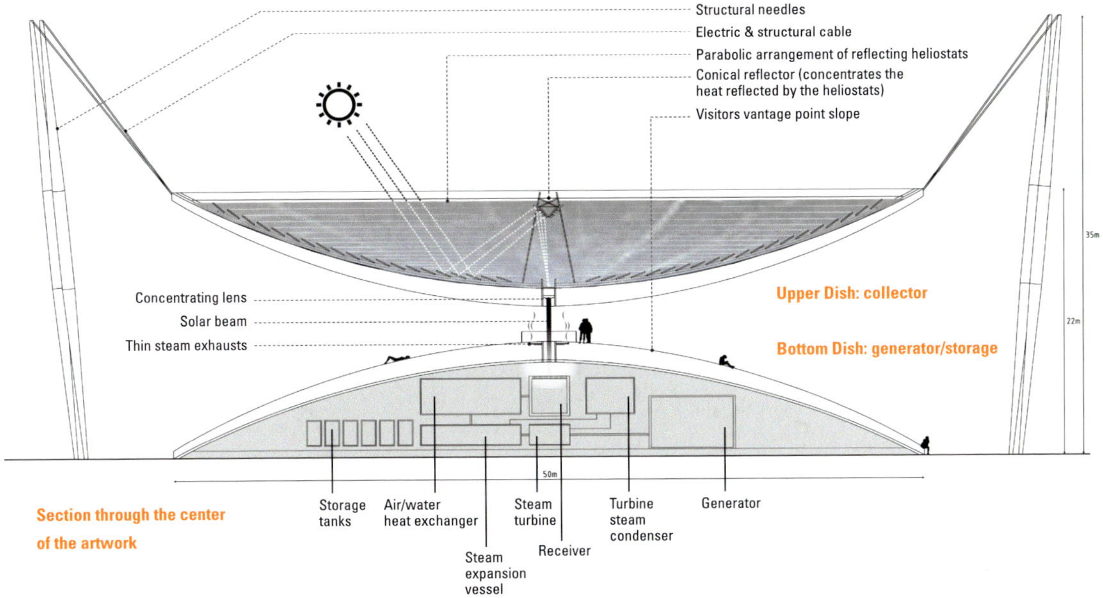

Section through the center of the artwork

Production Potential

With a 32% capacity factor and with a diamater of 50 m, the 1960 sq² parabolic set of heliostats would concentrate enough heat to produce 6.2 MW, which would be enough to provide electricity to 860 homes.

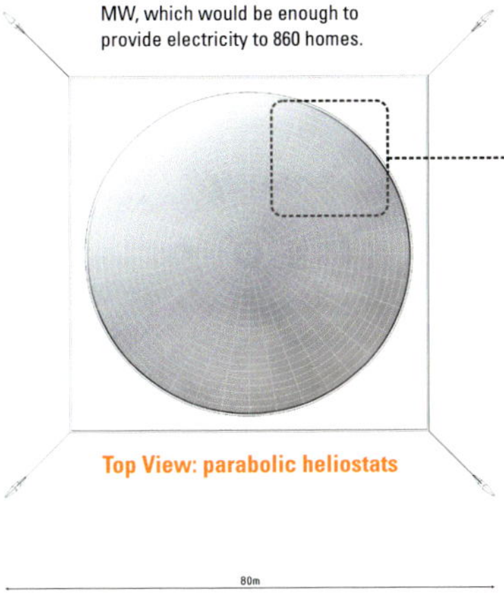

Top View: parabolic heliostats

Heliostat field

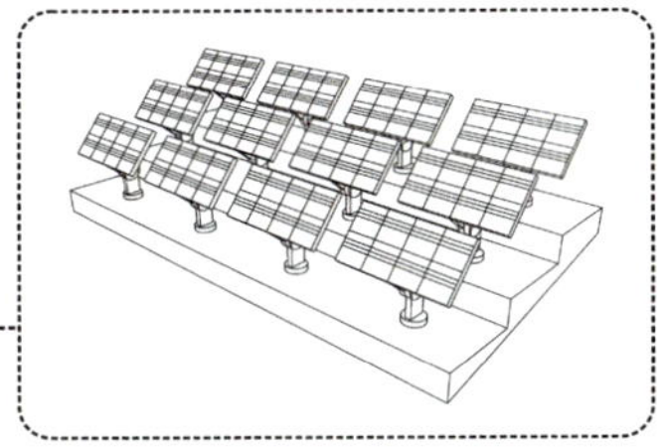

Heliostat articulation

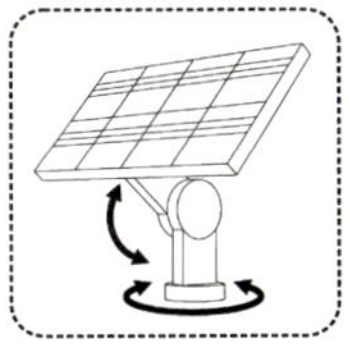

Each heliostat has individual articulations that allow it to follow the sun's movement with an outstanding degree of accuracy, thus achieving high efficiencies. Computer software controls the heliostats' movement based on a calendar to ensure that this level of precision is consistent all year long.

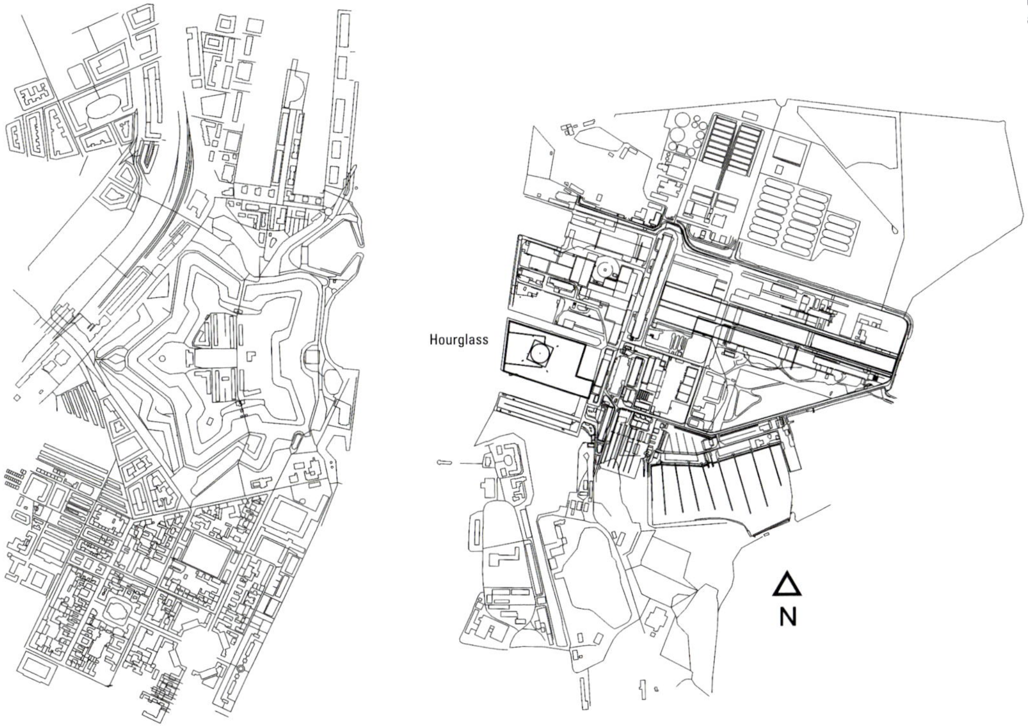

Site plan in context

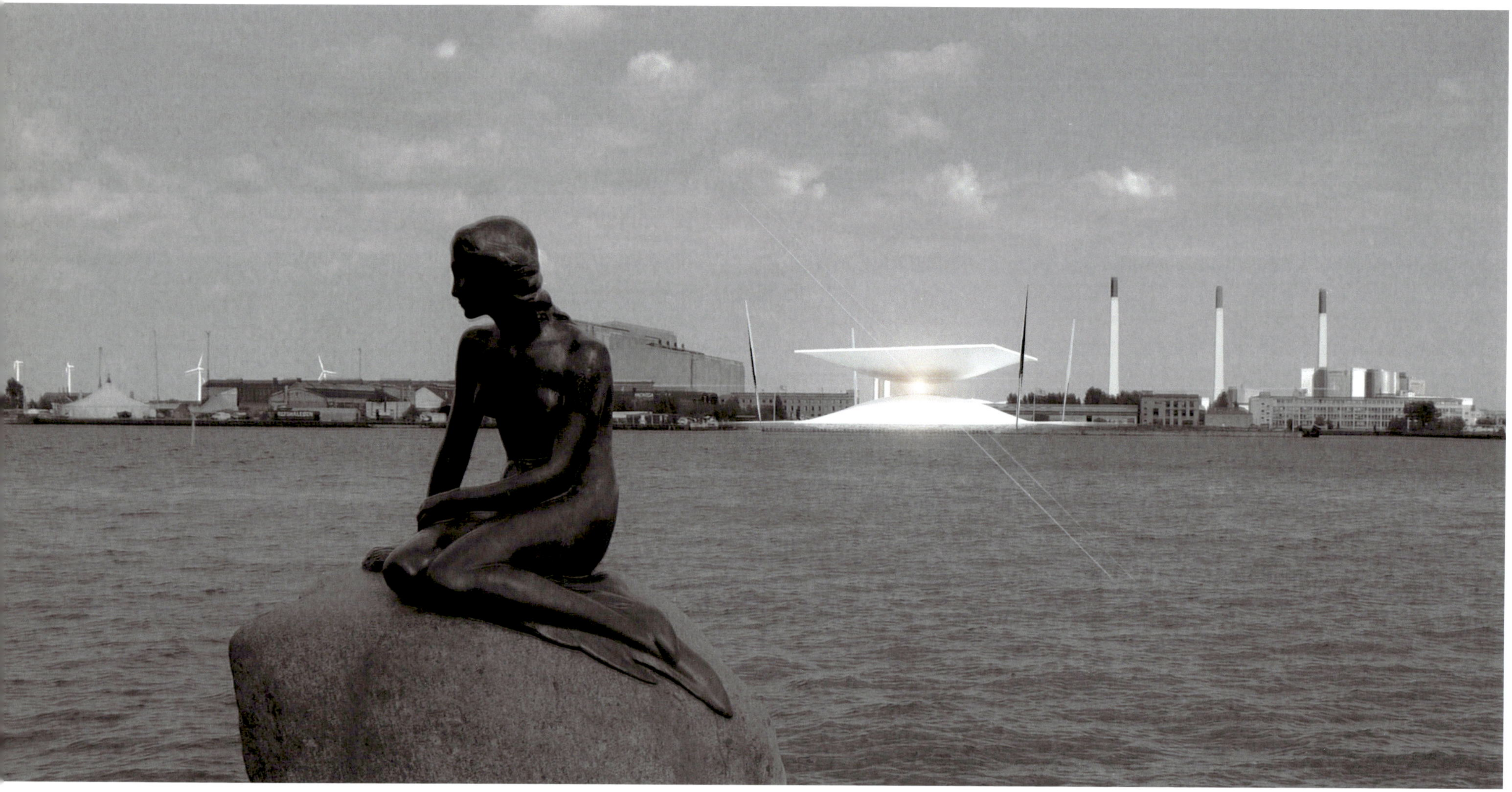

The Solar Hourglass represents one more step in our search for sustainable, feasible, and beautiful ways of producing clean, renewable, and environmentally friendly energy. Along with the earth, the hourglass will keep rotating, giving us with every turn, another beam of hope for a brighter and more sustainable future.

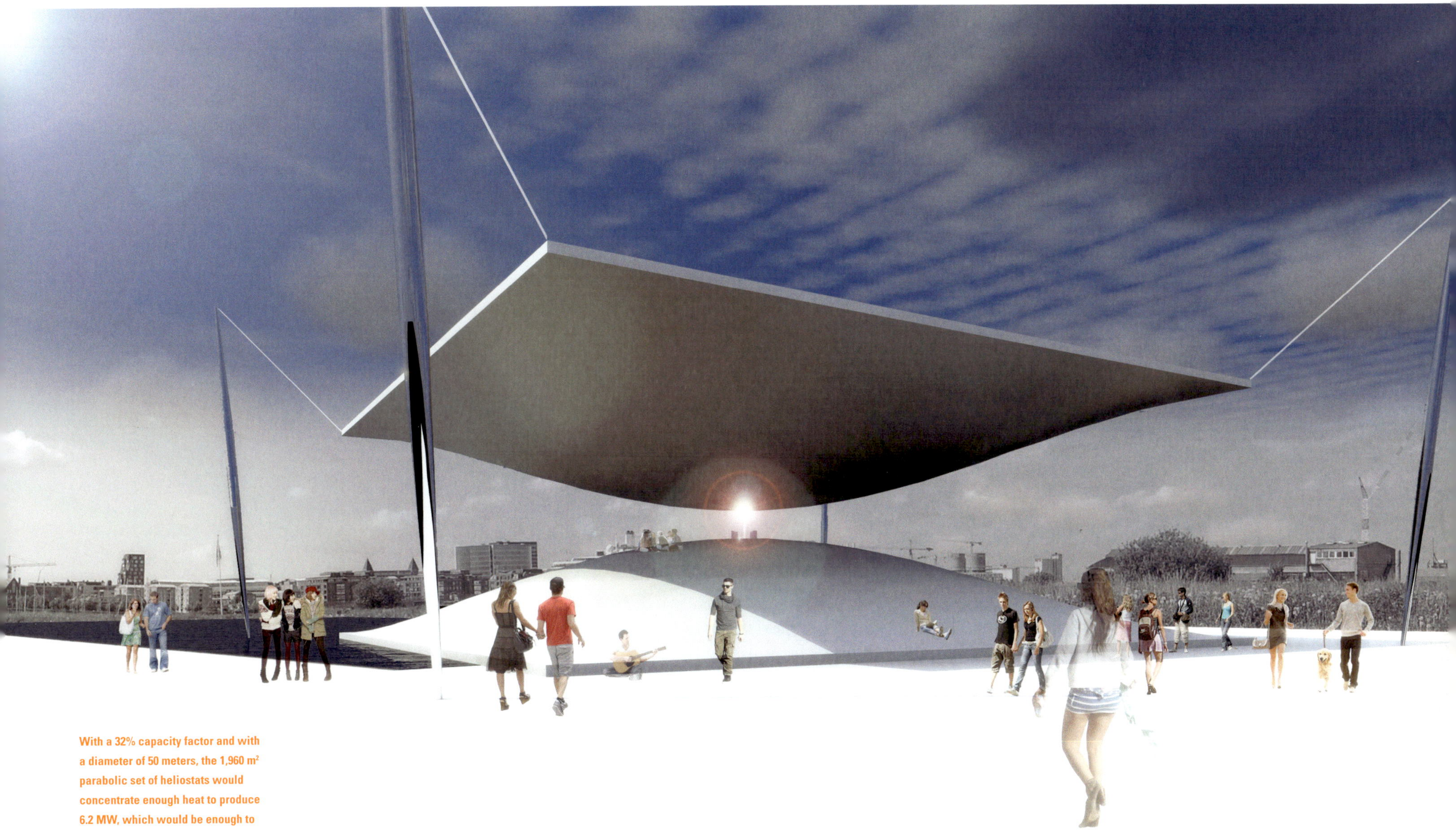

With a 32% capacity factor and with a diameter of 50 meters, the 1,960 m^2 parabolic set of heliostats would concentrate enough heat to produce 6.2 MW, which would be enough to provide electricity to 860 homes.

Quiver

TEAM:
Mateusz Góra, Agata Gryszkiewicz

TEAM LOCATION:
Warsaw, Poland

ENERGY TECHNOLOGIES:
biofuel, aeroelastic flutter (Windbelt™)

ANNUAL CAPACITY:
550 MWh (223 MWh bio, 327 MWh Windbelt™)

A wall of greenery with a gap in the middle suggests a way "into the wild." We are entering a garden without knowing what we will experience. The one sure destination is a monolithic, ephemeral tower, shining on the horizon. Along the way through a field, we discover that the greenery offers multiple spatial experiences with a variety of niches, corridors, and chambers. Closer to the tower, we notice that it quivers in the wind, as does the surface of the land. Leaving the garden, we enter the tower's forest-like structure. In front of us, a framed panoramic view of Copenhagen, and above us the sun and the sky can be viewed through an undulating filter.

Quiver has two main elements: the garden and the tower. The garden is a field of Miscanthus grass-energy crops that grows to a maximum of four meters tall and can be harvested twice a year. Changing planting schemes offers diverse functional patterns and spaces for citizens and visitors. The grass grows well in blighted soils, assisting in their remediation over time by accumulating pollutants into its root systems.

At the far end of the park is the tower—a landmark that welcomes boats arriving to the city. The footbridge brings visitors to the top of the installation and to an expansive view of Copenhagen. The panorama is in continuous dynamic flux as the Windbelts™ flutter in the wind.

During the night, the tower has another active role, calling to mind a lighthouse. LED lighting reflects a visual representation of the current wind conditions in Copenhagen.

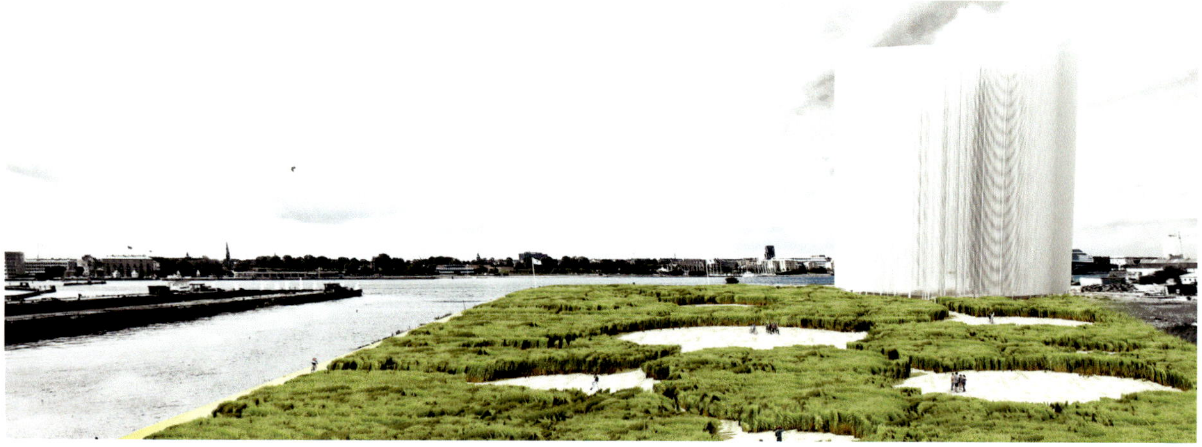

Perspective view revealing a pattern of the cut-out garden

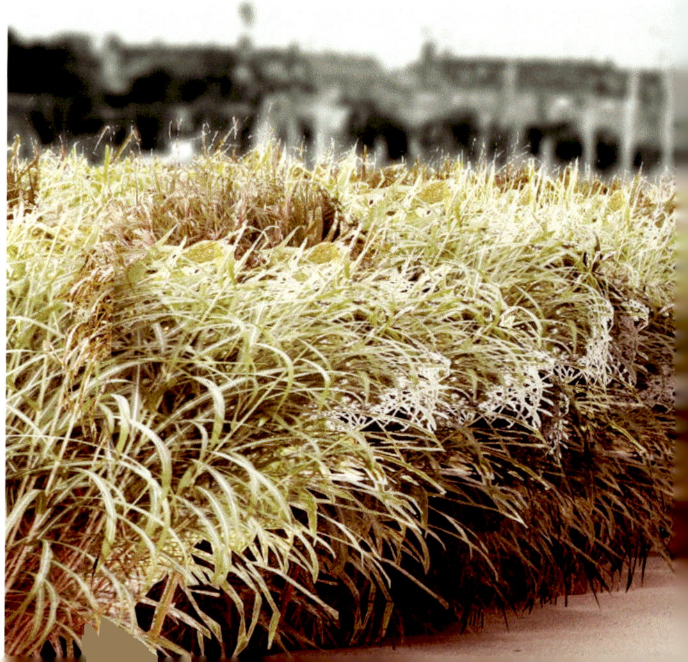

Halfway through the journey—exiting a clearing with the tower on the horizon

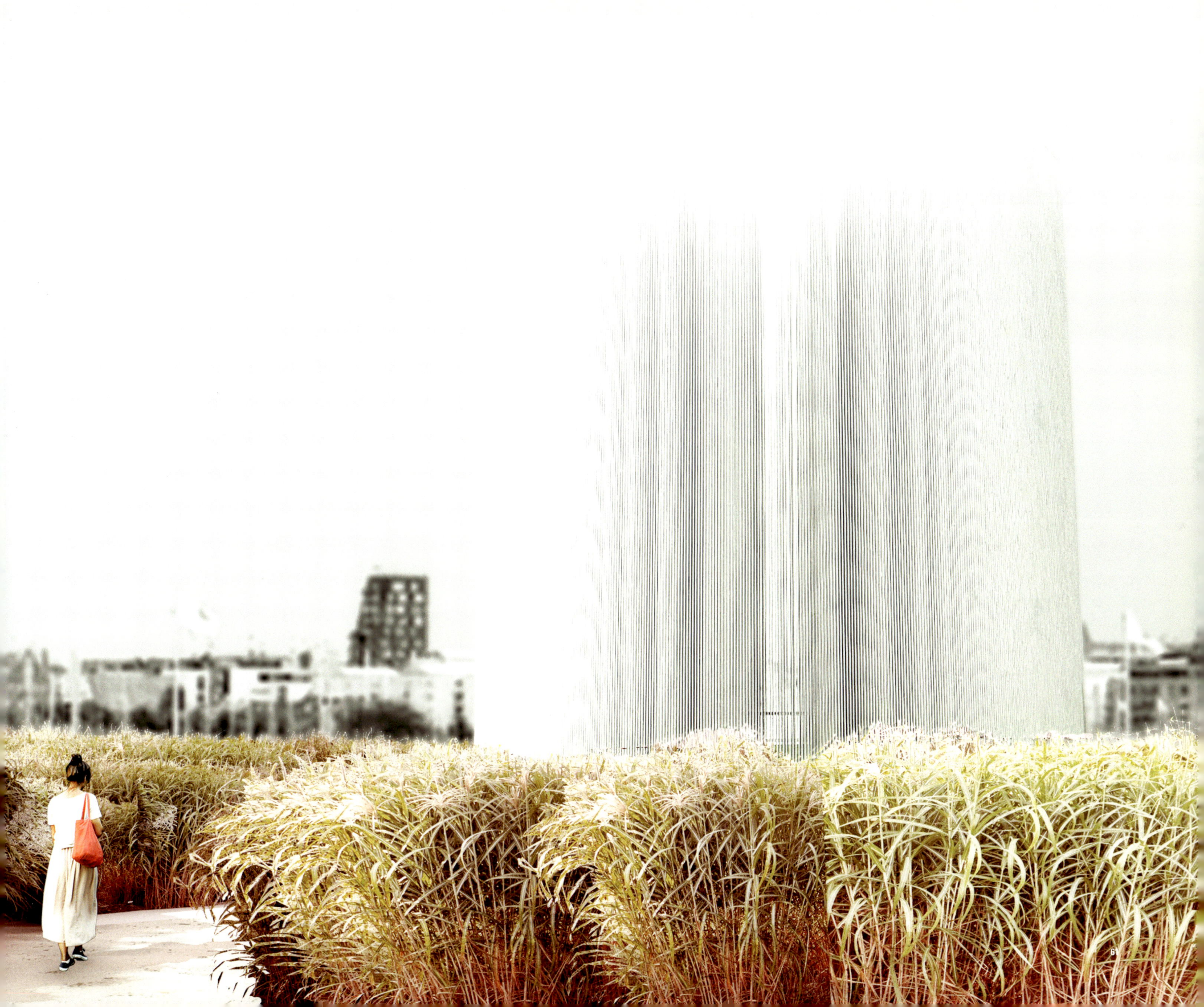

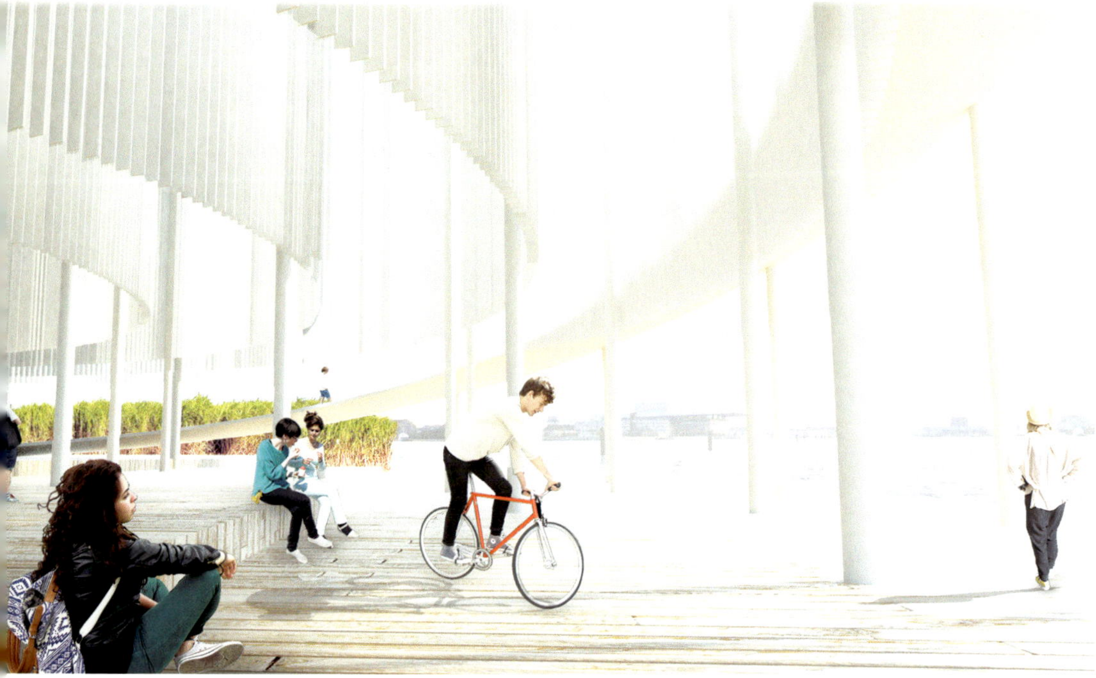

At the end of the journey through the garden, a public space emerges within the installation. A stepped wooden square has a direct connection to the water.

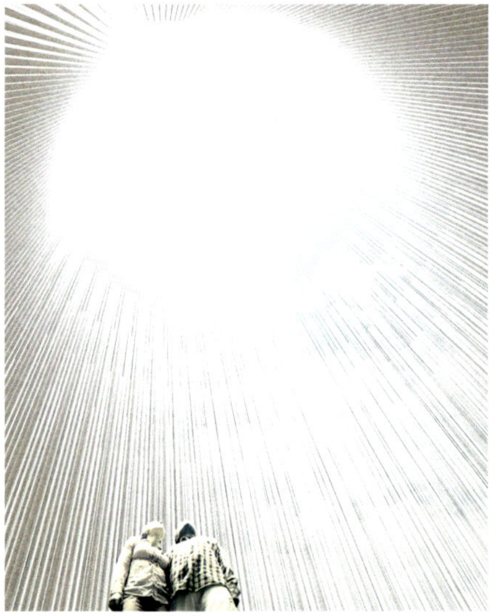

Experience under the tower

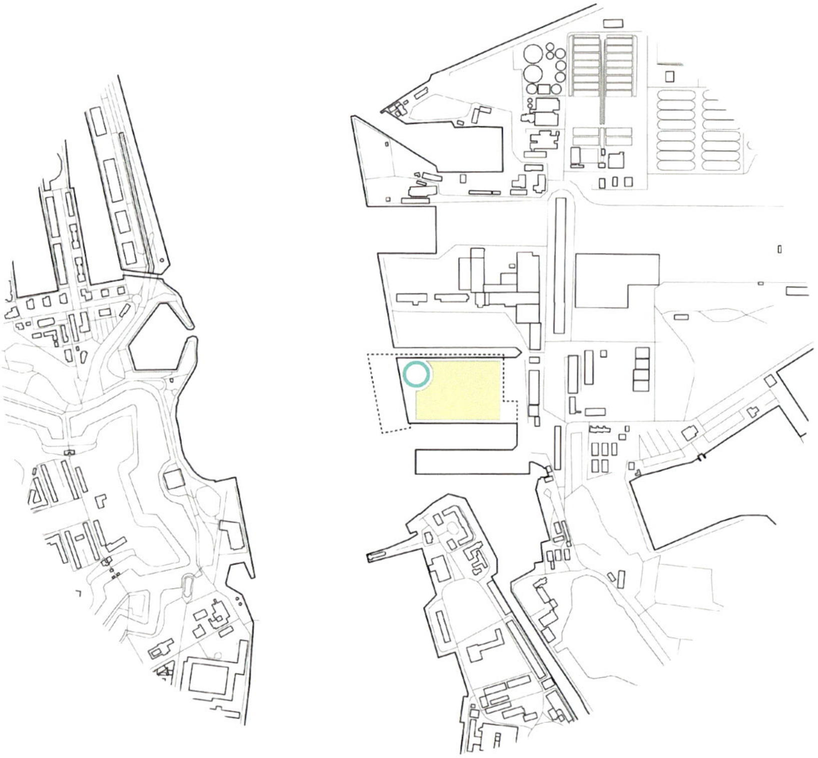

View from across the harbor

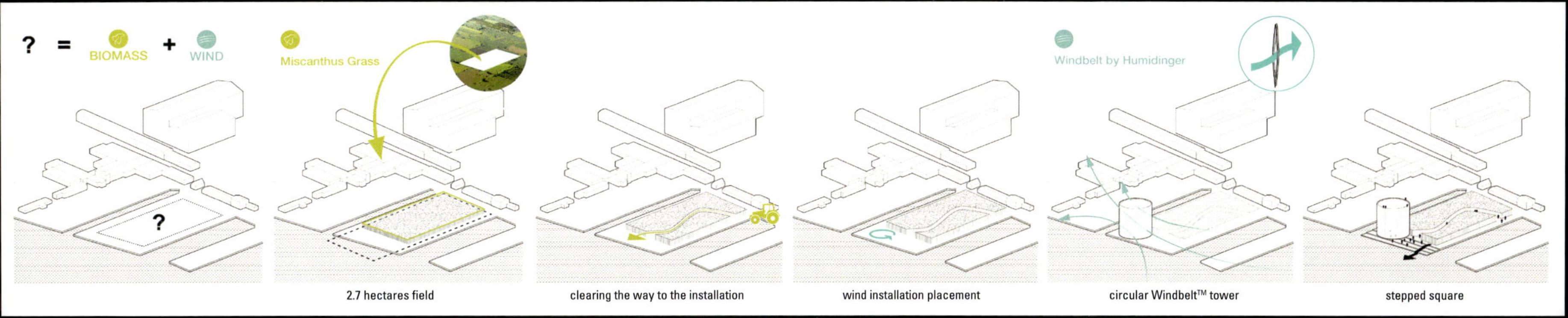

Design evolution and energy solutions

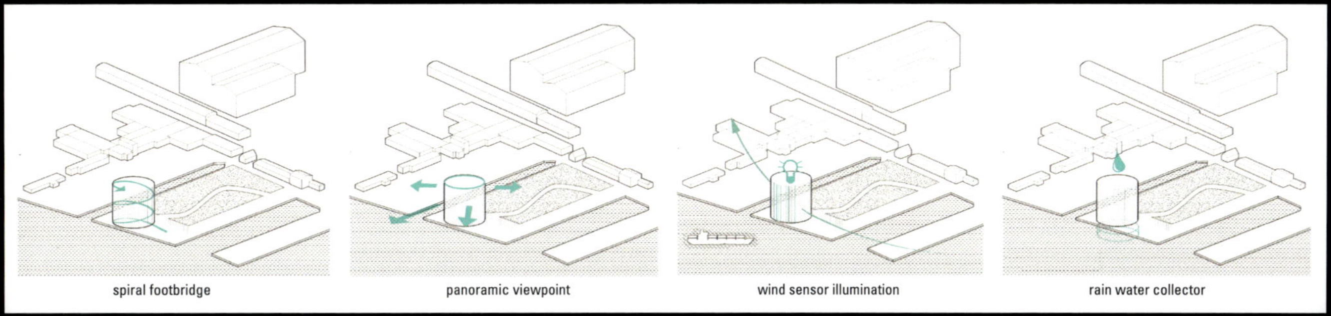

Wind tower functions in addition to energy harvesting

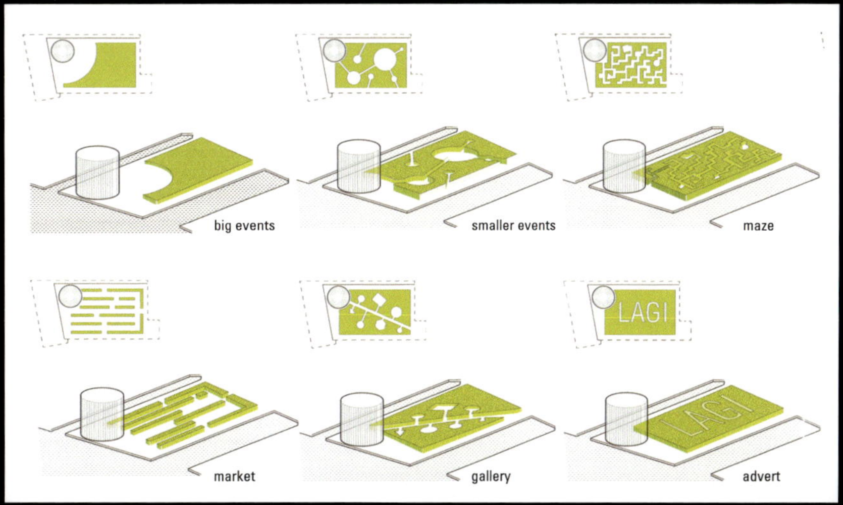

Biomass cut-out garden pattern possibilities

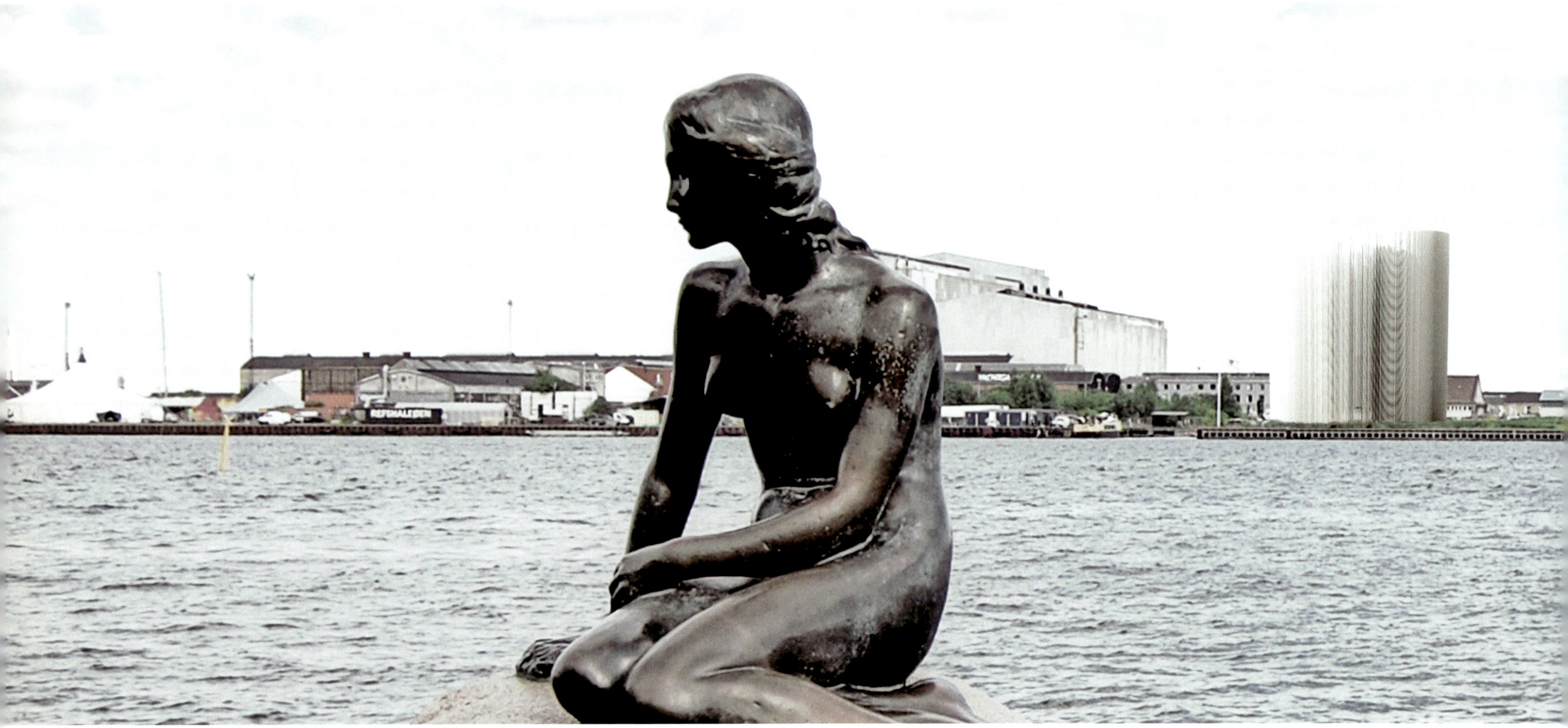

View from across the harbor

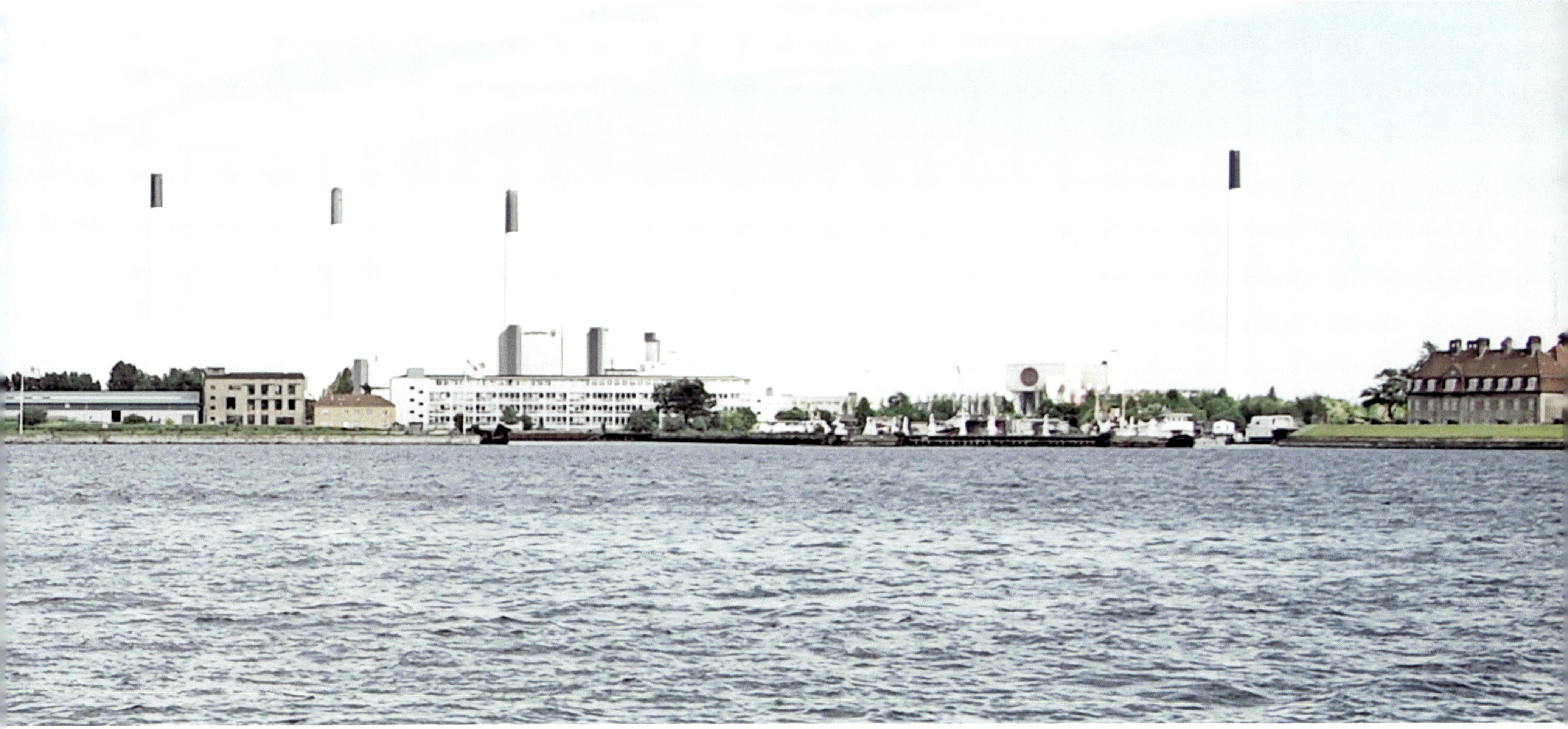

eMotions:
Energy Motions and Art Emotions

TEAM:
Antonio Maccà, Flavio Masi

TEAM LOCATION:
Padova, Italy

ENERGY TECHNOLOGIES:
photovoltaic panels, micro-scale vertical axis wind turbines (VAWT) and horizontal axis wind turbines (HAWT), stacked ceramic multilayer actuators, piezoelectric wind energy systems

ANNUAL CAPACITY:
2,000 MWh

eMotions showcases artistic interpretations of disparate Danish ecosystems, each varying in materials, energy technologies, dimensions, and textures. The installation places on display Denmark's biodiversity while engaging different communities within the city.

The artwork is also meant to evoke the image of a large generator, its viewing loop representing the generator's "belt" and the infinite production of clean energy through its micro-scale generators. The generators are listed below, each representing a type of ecosystem.

River is the public connecting loop with a piezoelectric floor made up of stacked, ceramic, multilayer actuators (SCMA). The actuators absorb energy from the movement of the people and convert it into electricity.

Beach is the harbor bath with a wooden amphitheater, realizing the transition from land to water as a terraced landscape where people can relax and reach the sea. The circles on the facade of the generator symbolize the beach waves and integrate horizontal axis wind turbines (HAWT), varying in dimensions depending on the internal diameters.

Marine houses many activities, such as swimming, scuba diving, and kayaking. The intersecting pattern of sinusoids symbolize the overlapping sea waves and integrate vertical axis wind turbines (VAWT).

Sand Dune represents the aeolian sediment transport of the migrating sand dunes. Its facade is composed of thin-film semitransparent photovoltaic panels and horizontal-axis wind turbines.

Lake draws its inspiration from the superimposition of waves onto the still surface of a lake. The structure supports semitransparent photovoltaic panels, which enclose a water playground.

Agricultural is made up of semitransparent photovoltaic panels that evoke the rotational, open-field production of agricultural crops. This "urban agriculture" garden showcases the biodiversity of plant life in Denmark, bringing knowledge of the country's varied foods and plants to the public.

Arctic calls to mind a snowfall in a typical frozen arctic landscape and hosts an outdoor ice skating rink during the winter. The floating snowflakes are oscillating piezoelectric energy systems, converting wind-induced vibrations into electrical power.

Grassland is composed of piezoelectric energy systems moved by the wind. The green area offers recreational open grassland for sports activities.

Forest is symbolized by a pattern of overlapping tree silhouettes that support vertical axis wind turbines.

Urban hosts a renewable energy museum showcasing a history of sustainable energy research and incorporates round microcrystalline photovoltaic panels of different diameters.

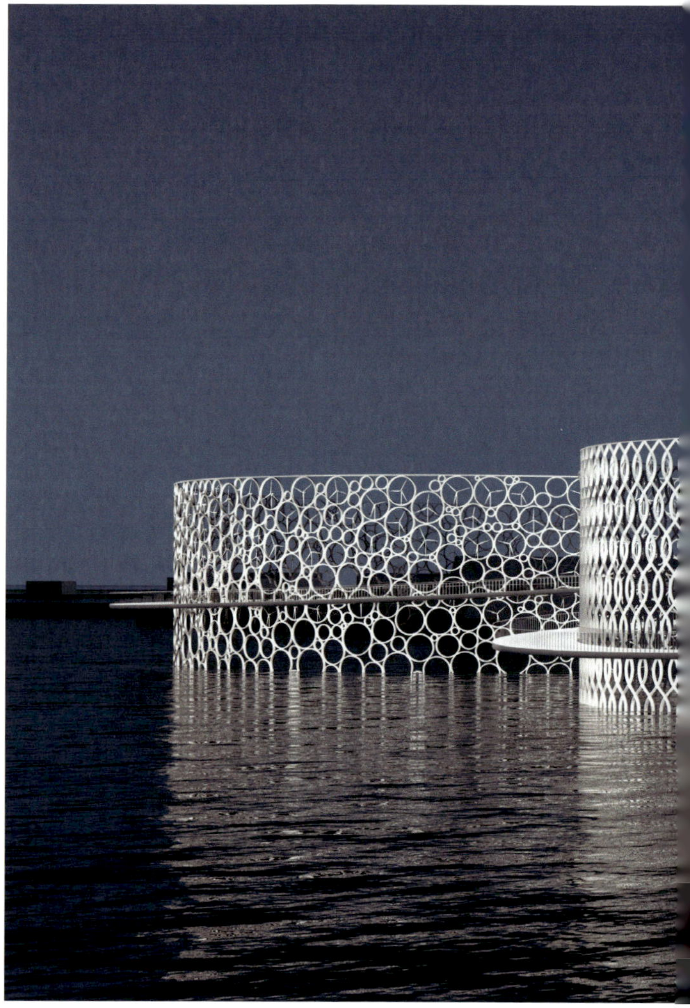

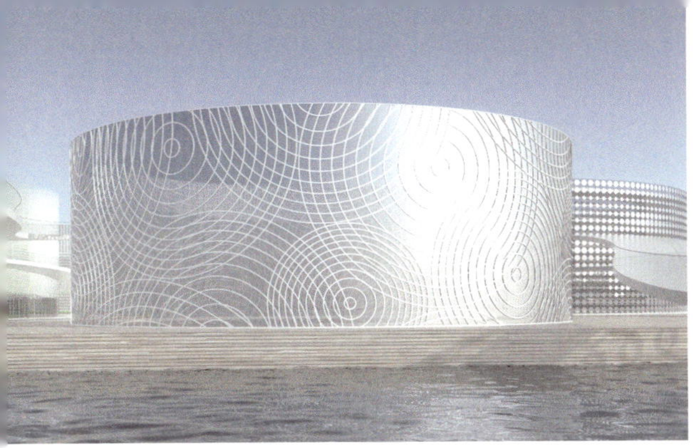

The "Lake Ecosystem" draws its inspiration from the phenomenon of the superimposition of the waves on the still surface of a lake. The structure representing the interferences of the overlapping concentric wave propagations supports semitransparent photovoltaic panels, which enclose a water playground.

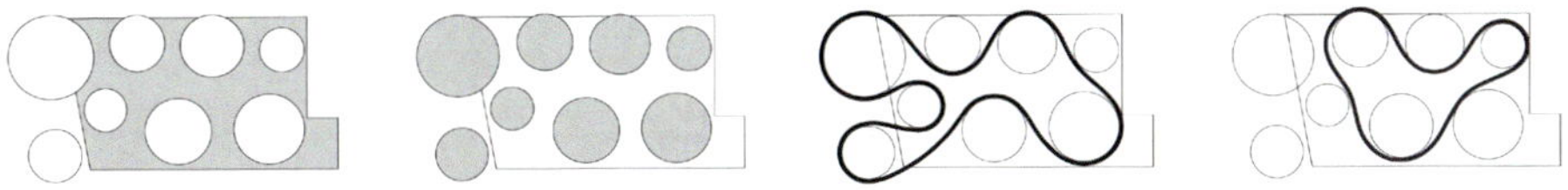

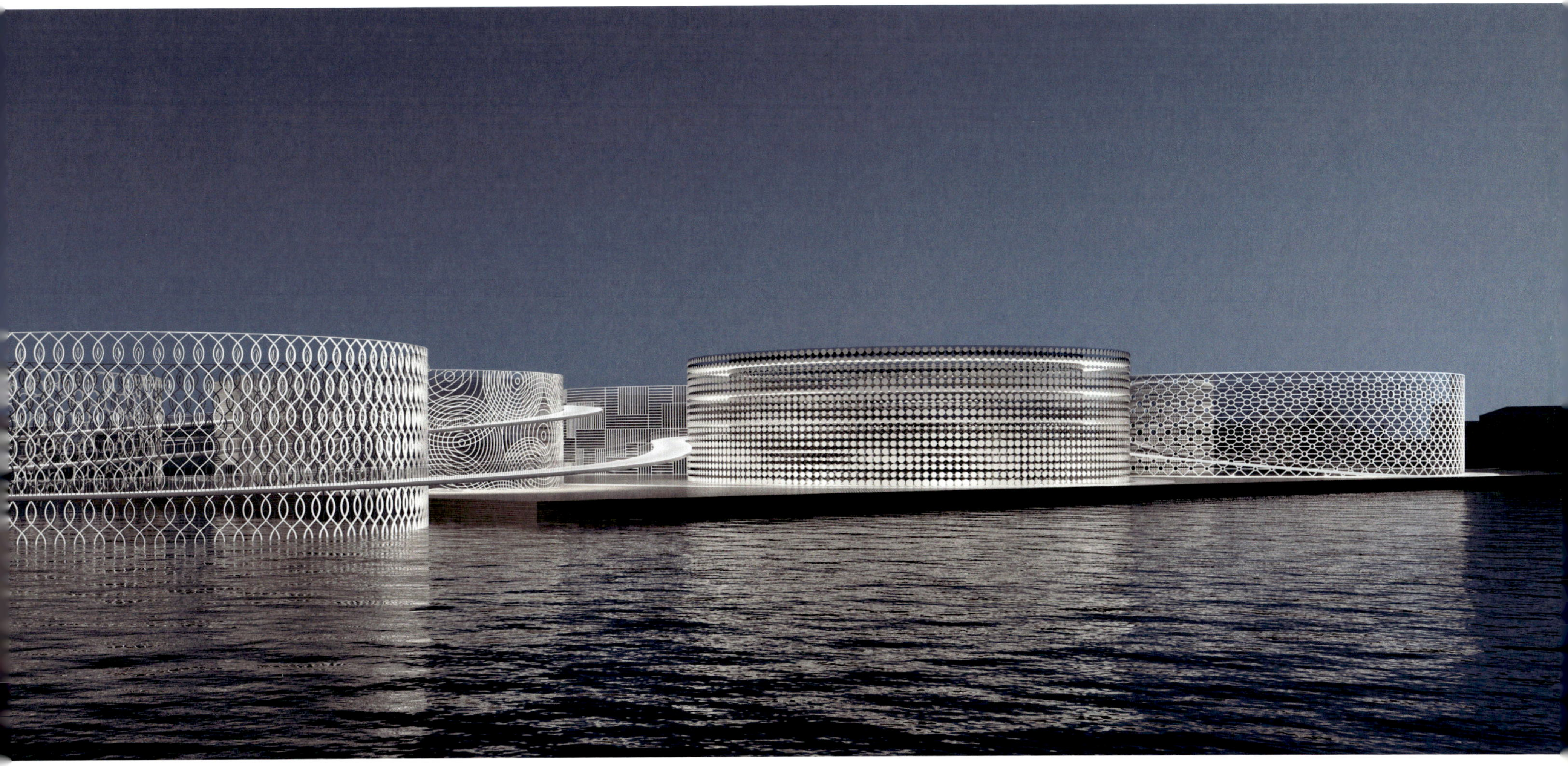

View from the harbor

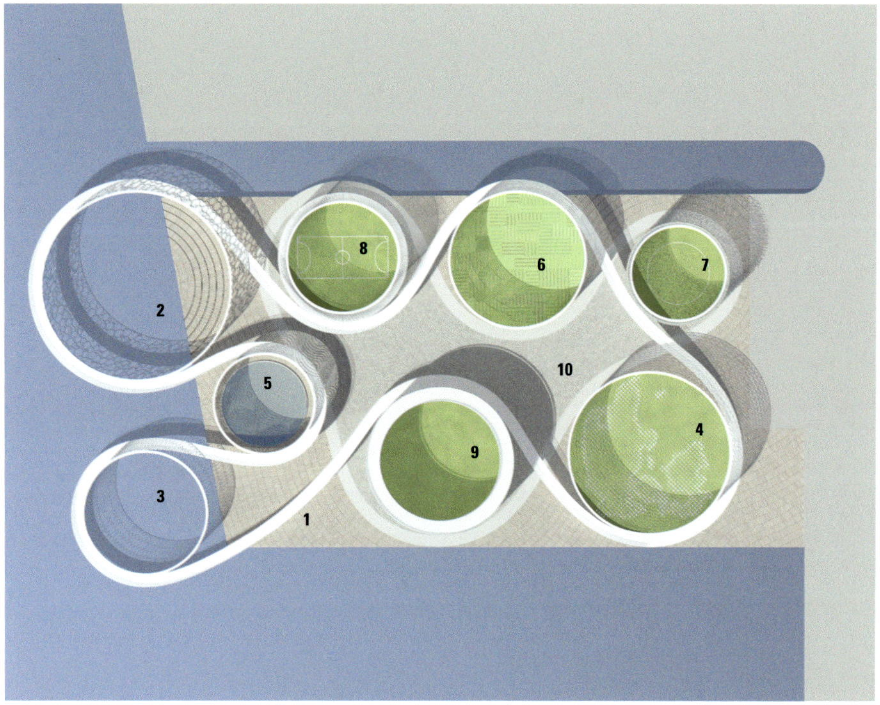

The Ecosystems: 1 River, 2 Beach, 3 Marine, 4 Sand Dune, 5 Lake, 6 Agricultural, 7 Arctic, 8 Forest, 9 Urban, 10 Energy Square

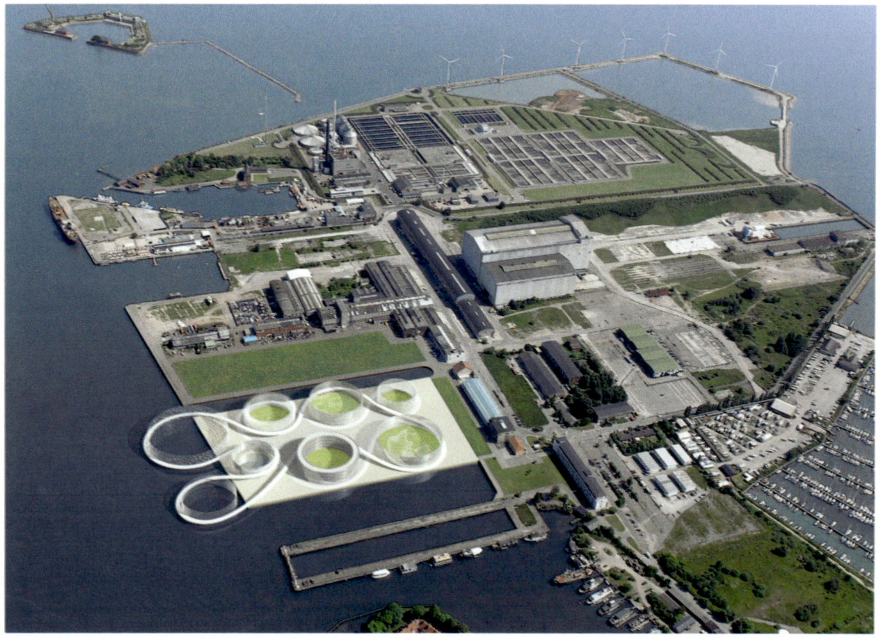

Aerial view of *eMotions* at Refshaleøen

The "Sand Dune Ecosystem" represents the aeolian sediment transport of the migrating sand dunes. Its facade is composed of thin-film semitransparent photovoltaic panels and horizontal-axis wind turbines.

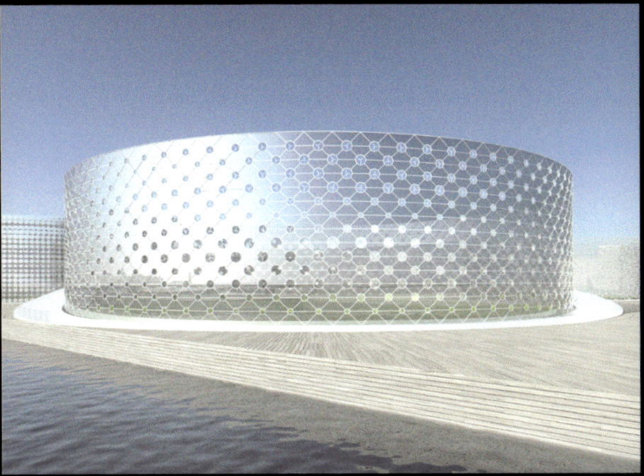

The "Beach Ecosystem" is the Harbor Bath. A wooden amphitheater creates a transition from land to water as a terraced landscape.

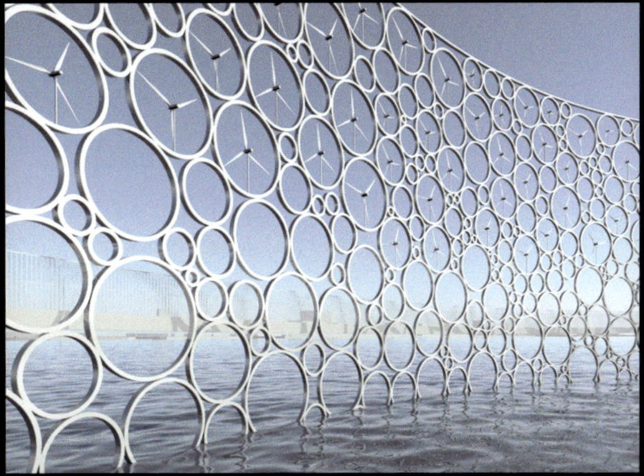

The "Forest Ecosystem" is symbolized by a pattern of overlapping tree silhouettes that support vertical axis wind turbines.

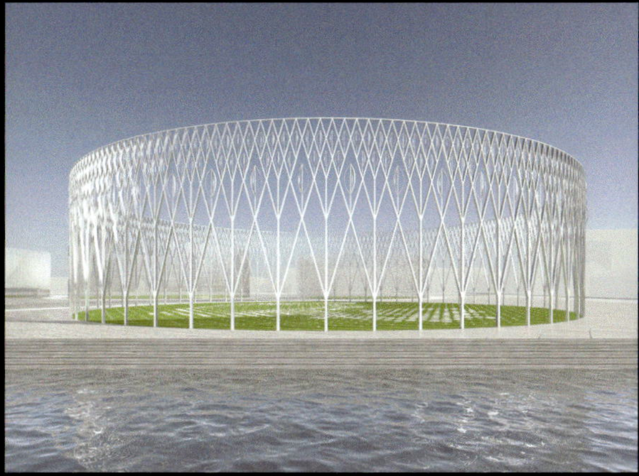

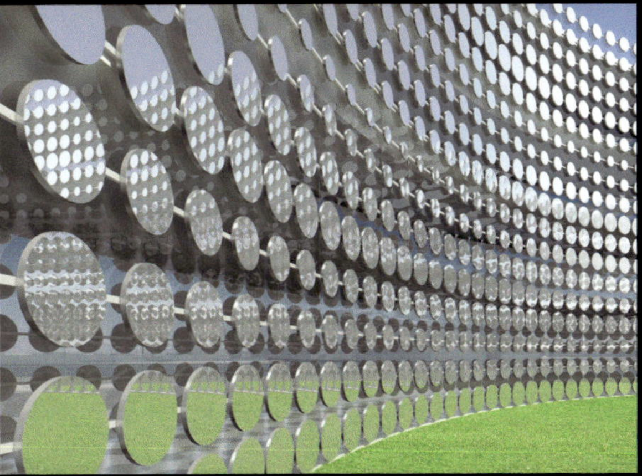

The "Urban Ecosystem" hosts a "Renewable Energy Museum," dedicated to Hans Christian Oersted and Niels Bohr, representing the motion of electrons in a solenoid with round photovoltaic panels.

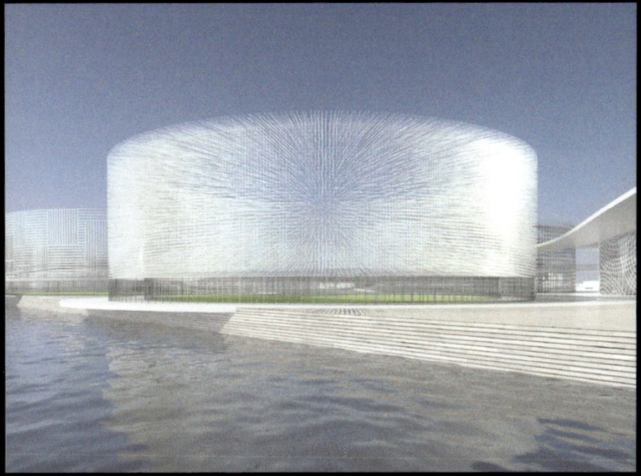

The "Arctic Ecosystem" calls to mind a snowfall in a typical frozen arctic landscape. The floating snowflakes are an oscillating piezoelectric energy system.

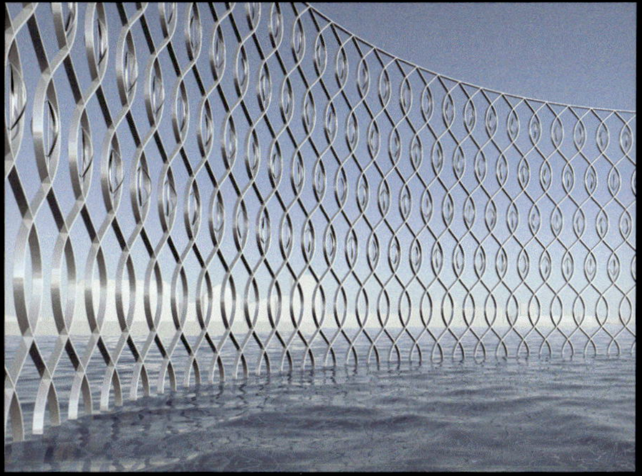

The "Marine Ecosystem" houses many activities like swimming, scuba diving, and kayaking. The intersecting sinusoids symbolize the overlapping sea waves and integrate vertical axis wind turbines.

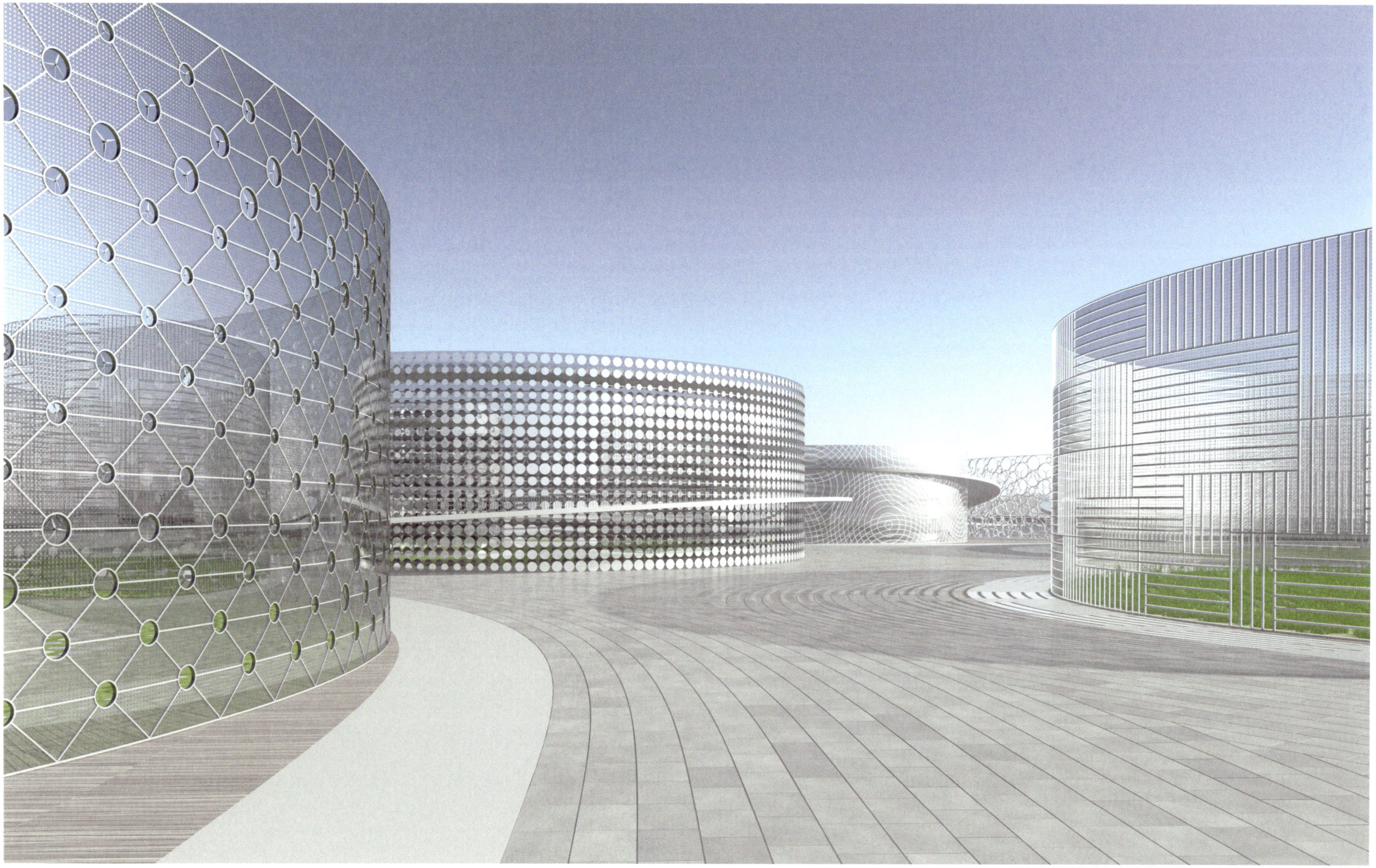

View walking through the energy loop

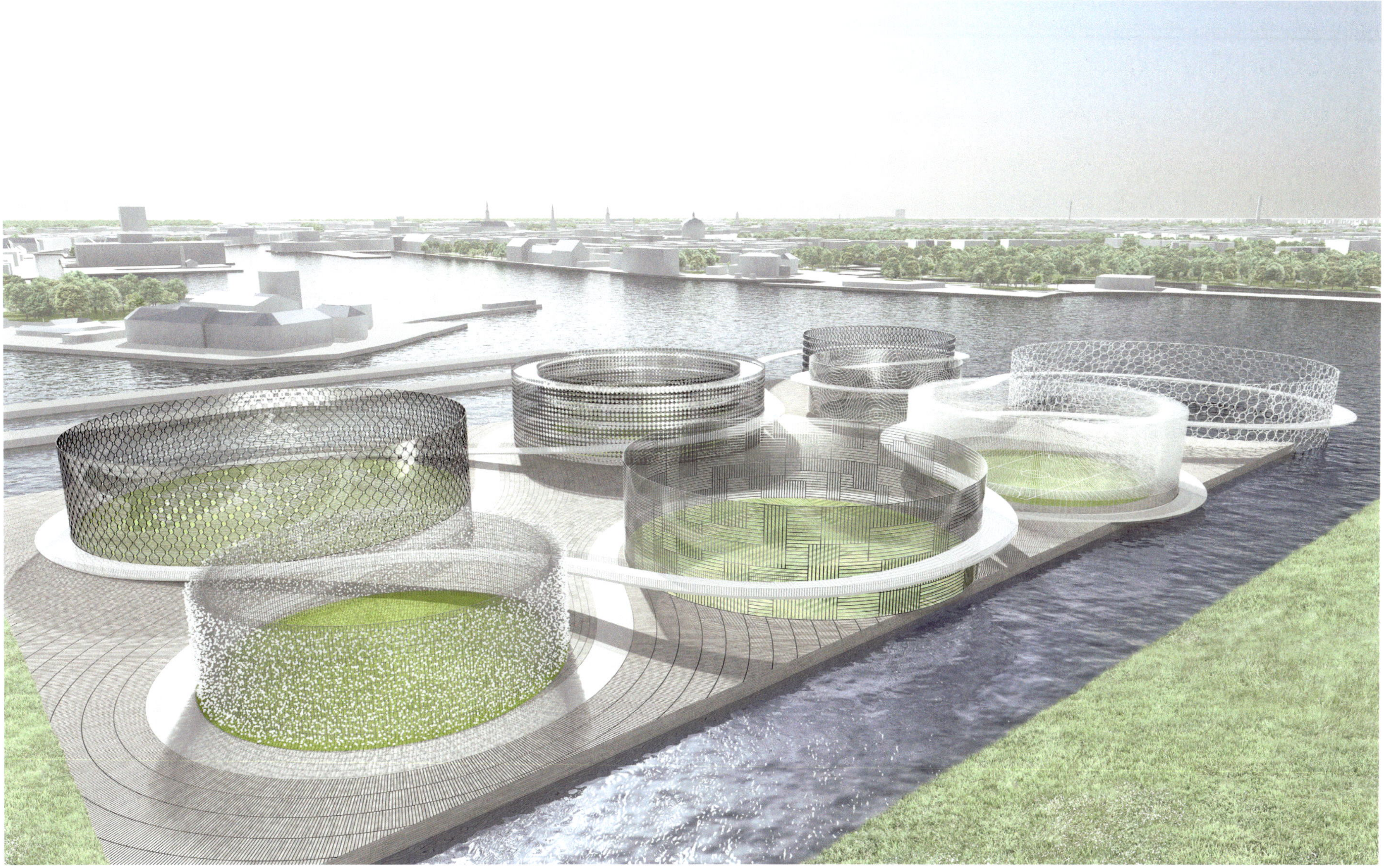

View with the Copenhagen
skyline beyond

SHORTLISTED ENTRIES

Beyond the Wave

The Place of Tomorrow // Aesthetic Representation of Copenhagen's Future Plan

LightSponge

Aetherius

Wind Decoder

The Sound of Denmark

Zephyrus Park

GRID Slide

Echo of Wind: Visualizing the Beauty of Wind Patterns

Onshore Power

Windshape

The Ephemeral Machine

SunBath

The Cloud

Balance I Imbalance

Wind Harp: Wind Energy as Music

Energy Duck

Oscillating Platforms

Super Cloud

Algaescape

Sail

Stone Wave

"The artists are not adding decoration to something that engineers have designed, and the designers are not simply designing the logo for the product. There's a deep understanding that to make sense of our energy challenges and to intervene effectively takes multiple intelligences, multiple practices, multiple creativities working together."

—Chris Fremantle

Beyond the Wave

TEAM:
Jaesik Lim, Ahyoung Lee, Sunpil Choi, Dohyoung Kim, Hoeyoung Jung, Jaeyeol Kim, Hansaem Kim (Heerim Architects & Planners)

TEAM LOCATION:
Seoul, South Korea

ENERGY TECHNOLOGIES:
organic photovoltaic (OPV), kinetic harvesting (piezoelectric)

ANNUAL CAPACITY:
4,229 MWh

Inspired by Len Lye's diverse and dynamic kinetic art, the wavy expression of ribbons and flexible poles that comprise *Beyond the Wave* creates tangible and intangible movements through a healing environment. The allocation of the poles and ribbons are based on Copenhagen's wind rose analysis and soil survey map. Therefore, the frequency, density, and spacing between the poles are determined by the wind strength and intensity.

The strength of the wind influences the varying movements of the flexible poles. The ribbon that interconnects the poles symbolically becomes a "wave," representing the encounter between the water and the wind. The system utilizes the power of the sun while also harnessing the forces within the support structures to produce additional energy. The site is composed of an array of poles to allow spatial settings for various human activities and movements.

The ribbons consist of transparent, organic solar material that responds to the movement of the wind. The OPV panel attached to 1.5-meter wide ribbon generates energy, which is partially used for OLED lighting. The display panel in the lower part of the pole indicates the amount of energy generated and reduction of CO_2, showcasing energy saving effects in real-time.

Organic photovoltaic panel is capable of producing 100 W of power per square meter of surface area.

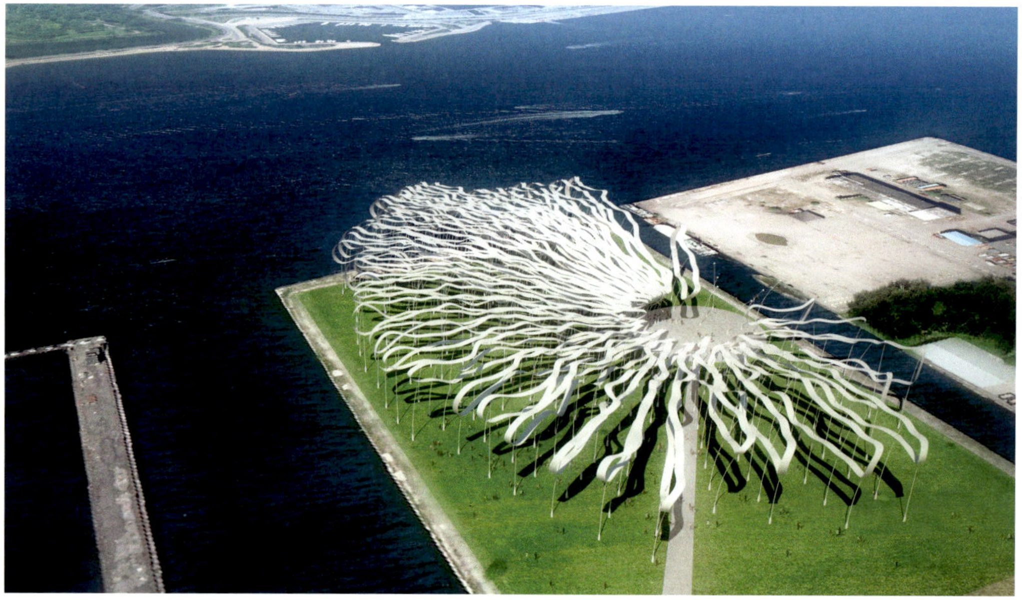

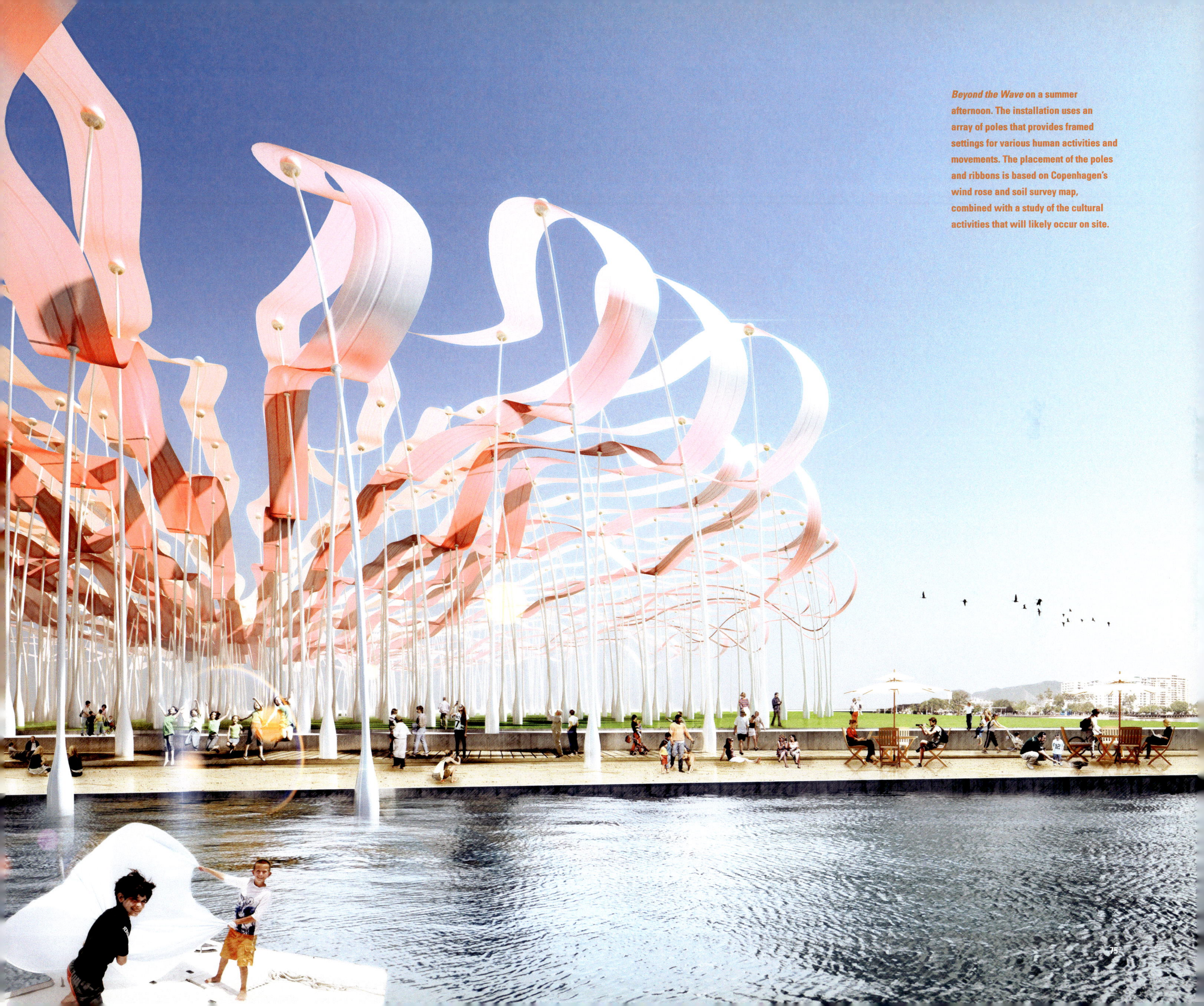

Beyond the Wave on a summer afternoon. The installation uses an array of poles that provides framed settings for various human activities and movements. The placement of the poles and ribbons is based on Copenhagen's wind rose and soil survey map, combined with a study of the cultural activities that will likely occur on site.

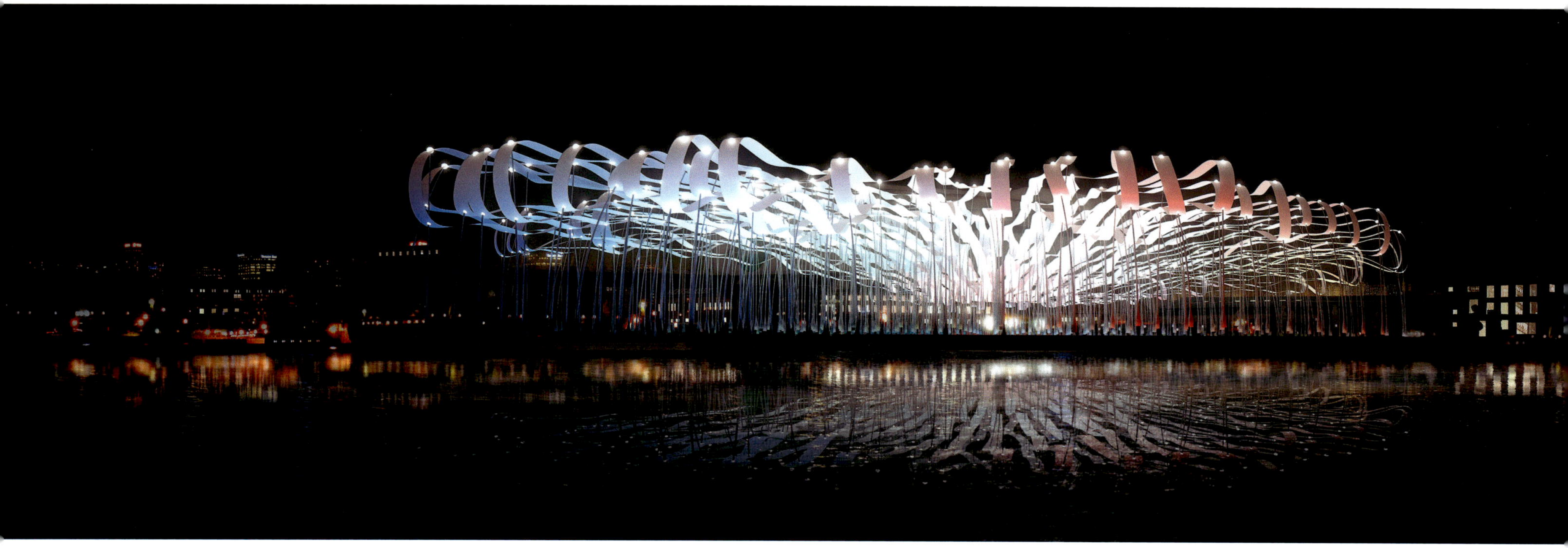

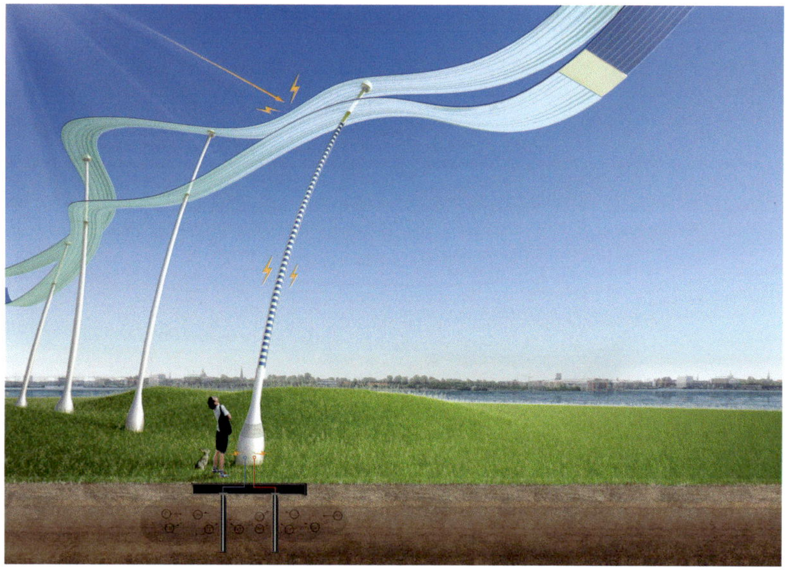

Electrokinetic Remediation: Installation of electrodes in the soil that induce a conductive physical/chemical reaction with the addition of electrical current. The contaminants are then extracted and removed through this technique. Existing soil pollutants are concentrated and removed in the cathodic direction through the flow of electro-osmotic fluids. The electrophoresis (electro-osmosis) is the phenomenon of the charged particles present in the soil moving in an electric field towards a particular direction of electrodes. Cation (positively charged) elements, such as heavy metals, will move in the cathodic direction to be removed, while organic and inorganic anions (negatively charged) move in the anodic direction.

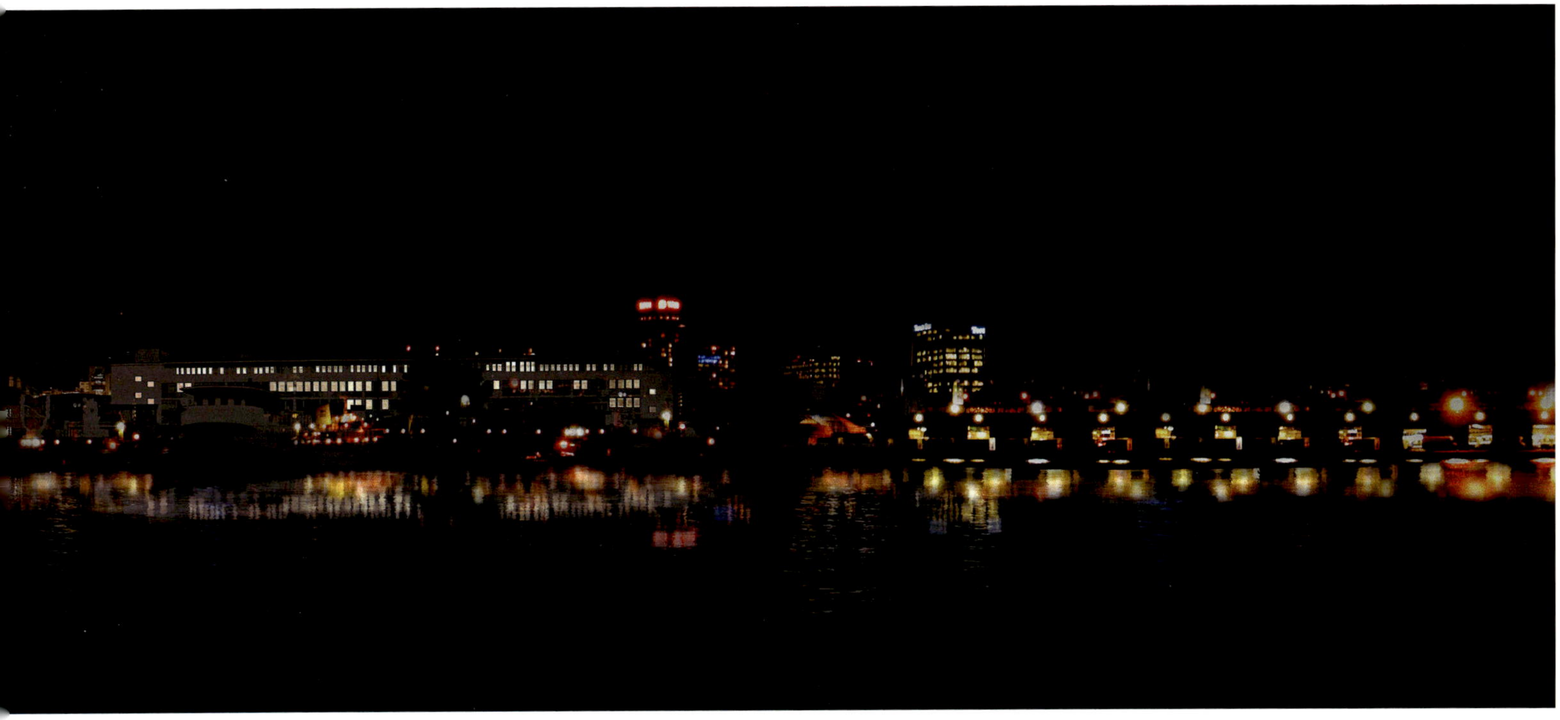

The organic light-emitting diodes panel is characterized by its lightweight, flexible, and translucent properties, which creates colorful lighting at night.

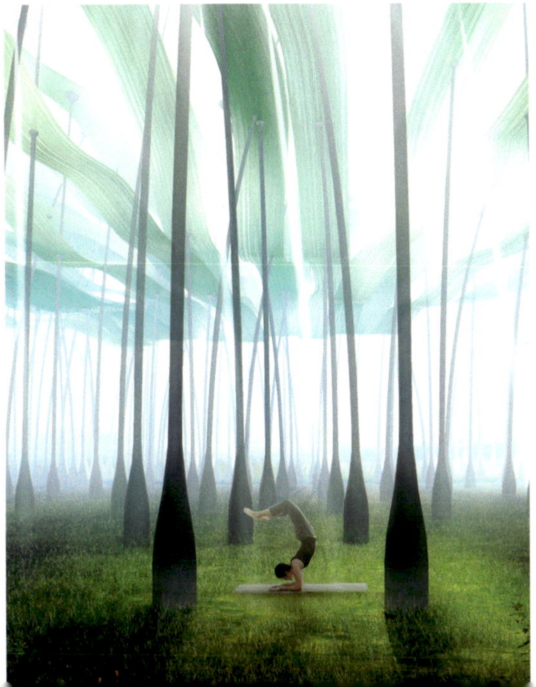

In the morning, the soft wispering of the wave forms creates a meditative setting.

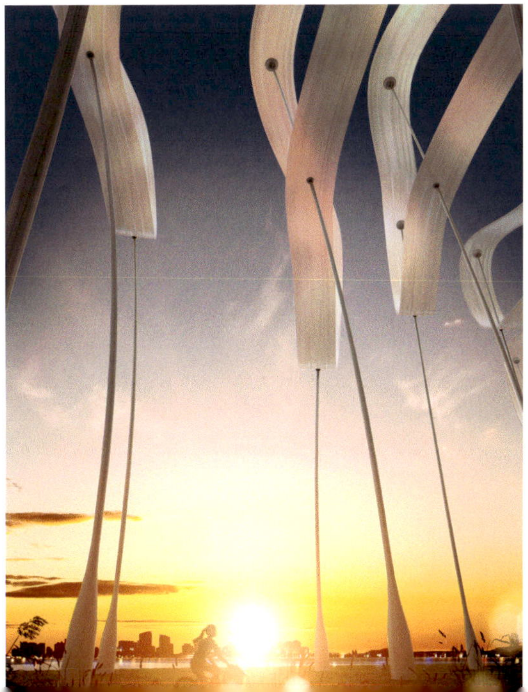

Transparent pulp with added resin: This new technology will be applied to transparent OLED substrates and the substrates of organic thin-film photovoltaic cells. The transparent sheet has a thickness of 100 micrometers, and a light transmittance of 87.8%.

The Place of Tomorrow //

Aesthetic Representation of Copenhagen's Future Plan

TEAM:
Amir Shouri, Fereshteh Tabe

TEAM LOCATION:
New York City, USA

ENERGY TECHNOLOGIES:
piezoelectric fabric, piezoelectric discs

ANNUAL CAPACITY:
1,200 MWh

The Place of Tomorrow is a new type of public garden—a poetic, natural space that demonstrates the future of clean energy.

Oscillating textures of fabric shades are framed in homage to Copenhagen's future energy-plan. As they wave in the wind, the shades make visible the wind's forces and aesthetically display the real-time production of electrical energy.

As a reflection of the 8,000 workers who once worked in the Sønder Hoved shipyard, the garden is designed under 8,000 linear fabric shades of different lengths. The energy fabric dances freely in the air, while below park visitors enjoy swimming in the clear blue water. Cyclists move between lounging couples, enjoying the rippling effects of the sun through the fabric above.

The longer fabric shades produce more electricity and are representative of the future (when most of Denmark's electricity will be produced by wind), and the shorter shades represent the early 21st century (when only 22% of electricity was produced by wind). Overall, 8,000 pieces of electricity-generating fabric benefit from the movement of electrons in nanowires to produce enough energy to power the LED lights of the project and to offset 900 households on a windy day.

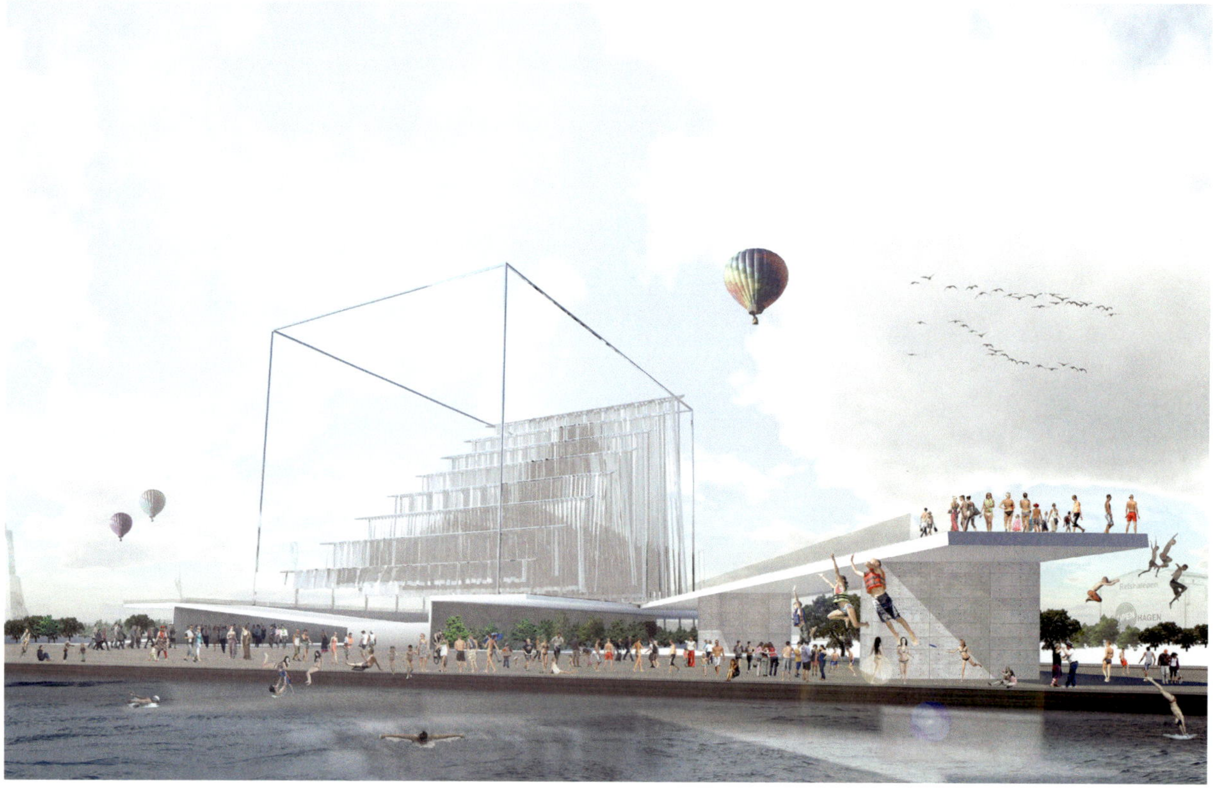

The garden under the fabric shades is surrounded by two platforms that are connected by a ramp. The lower platform expands the site and connects the pedestrian and bicycle routes. The higher platform is a place for people to recreate and to dive into the harbor water in the summertime.

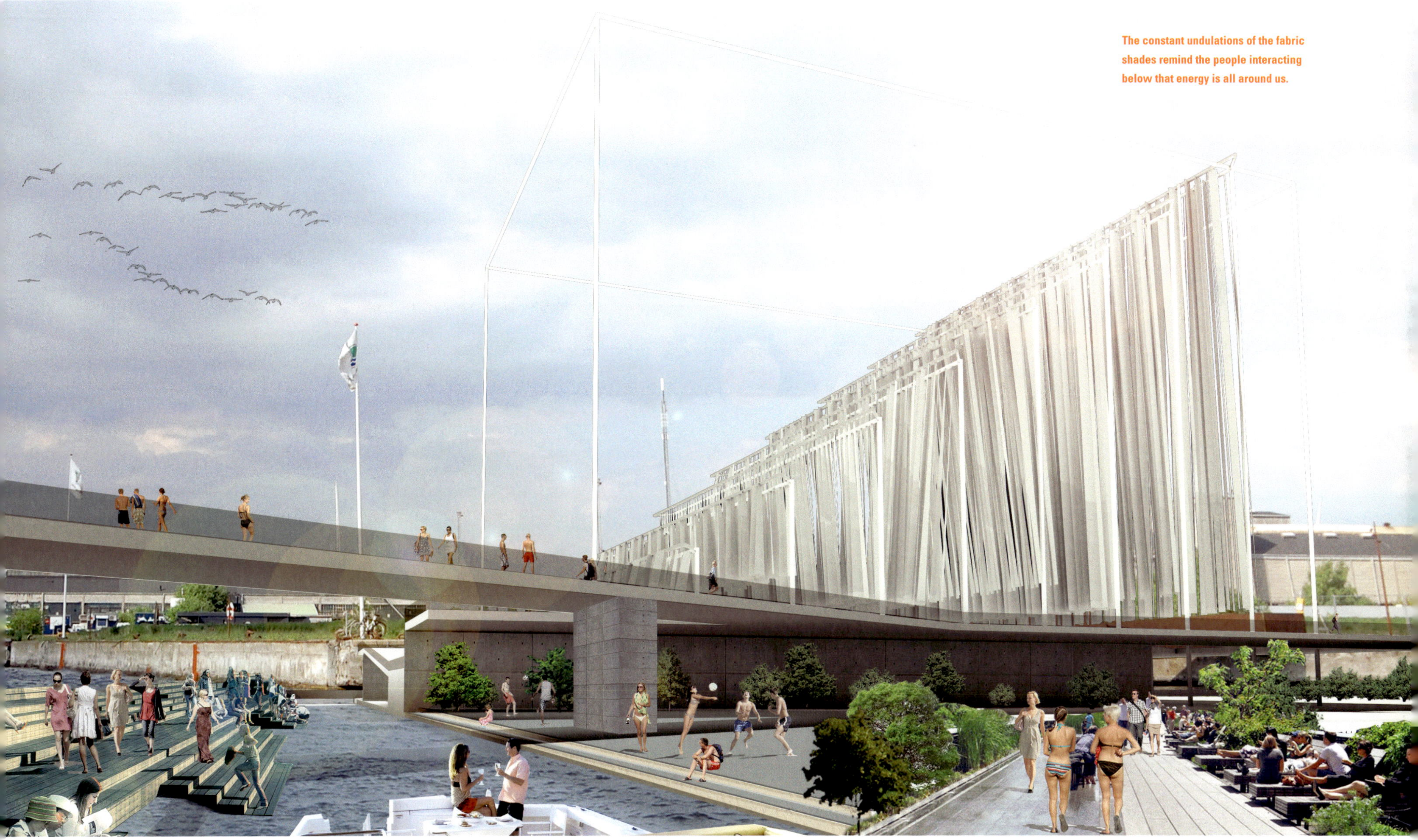

The constant undulations of the fabric shades remind the people interacting below that energy is all around us.

The Place of Tomorrow // Aesthetic Representation of Copenhagen's Future Plan

1. Installation area
2. Blue deck
3. Public garden
4. Interaction deck
5. Proposed water taxi terminal
6. Existing water taxi terminal
7. Existing concrete path
8. Proposed area for future developments
9. Open garden

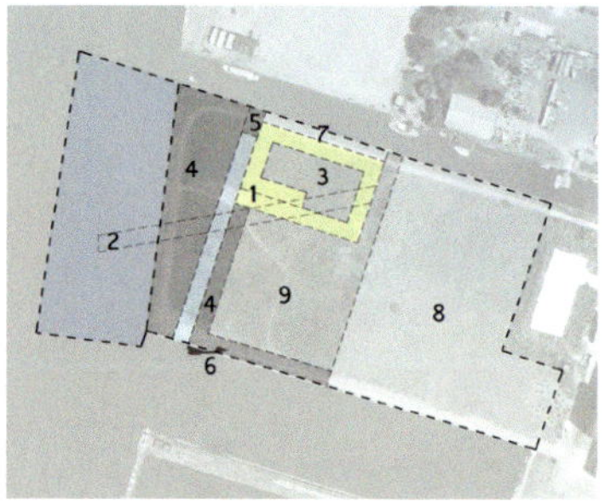

Site Plan

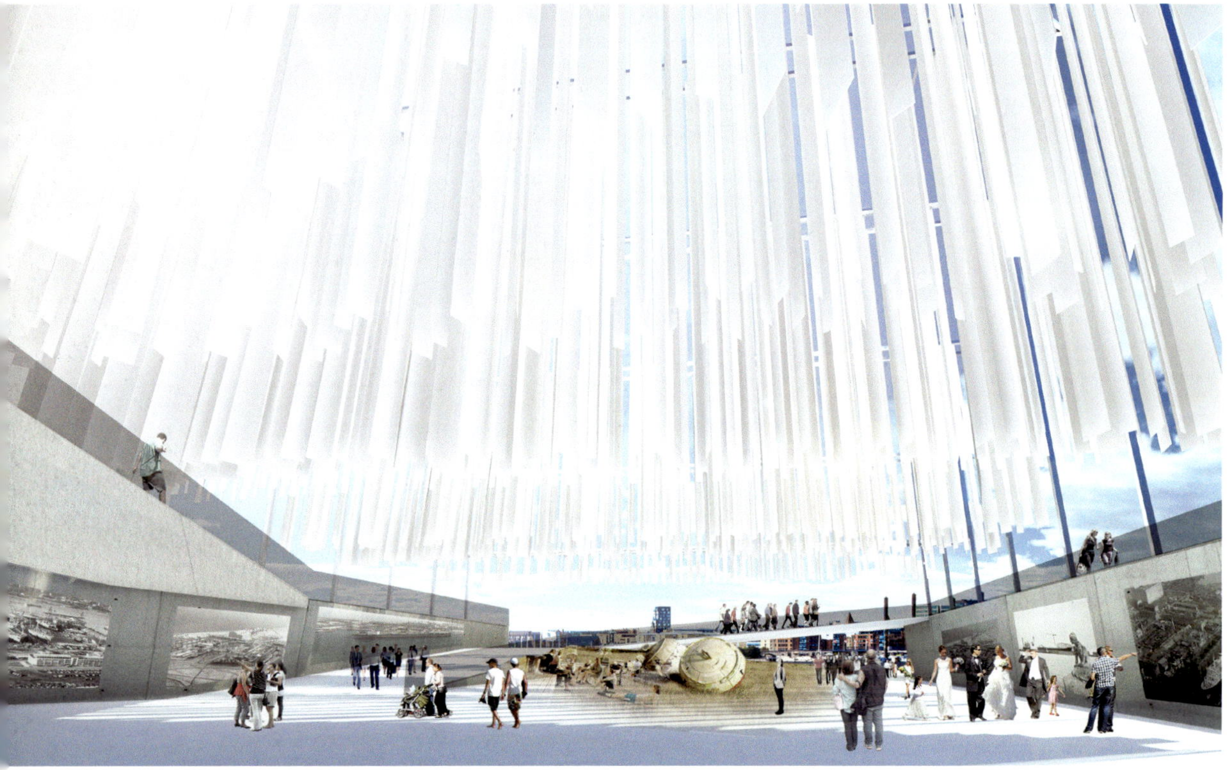

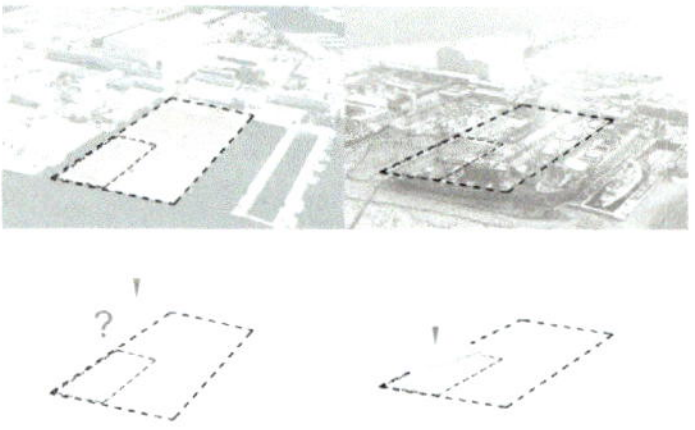

2014
Basin area used as an installation space

1996
Basin area was once used to work on ships

Historical icon vs. contemporary artwork

In the public garden, directly below the fabric shades, an exhibition provides visitors with a connection to the industrial history of the site.

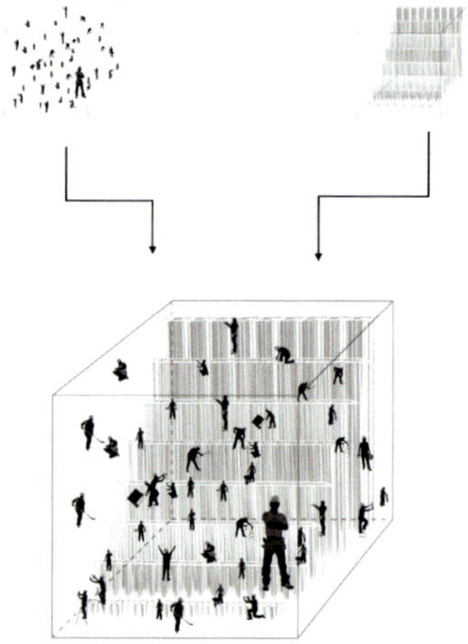

8,000 workers in the shipyard of Sønder Hoved

8,000 pieces of electricity-generating fabrics

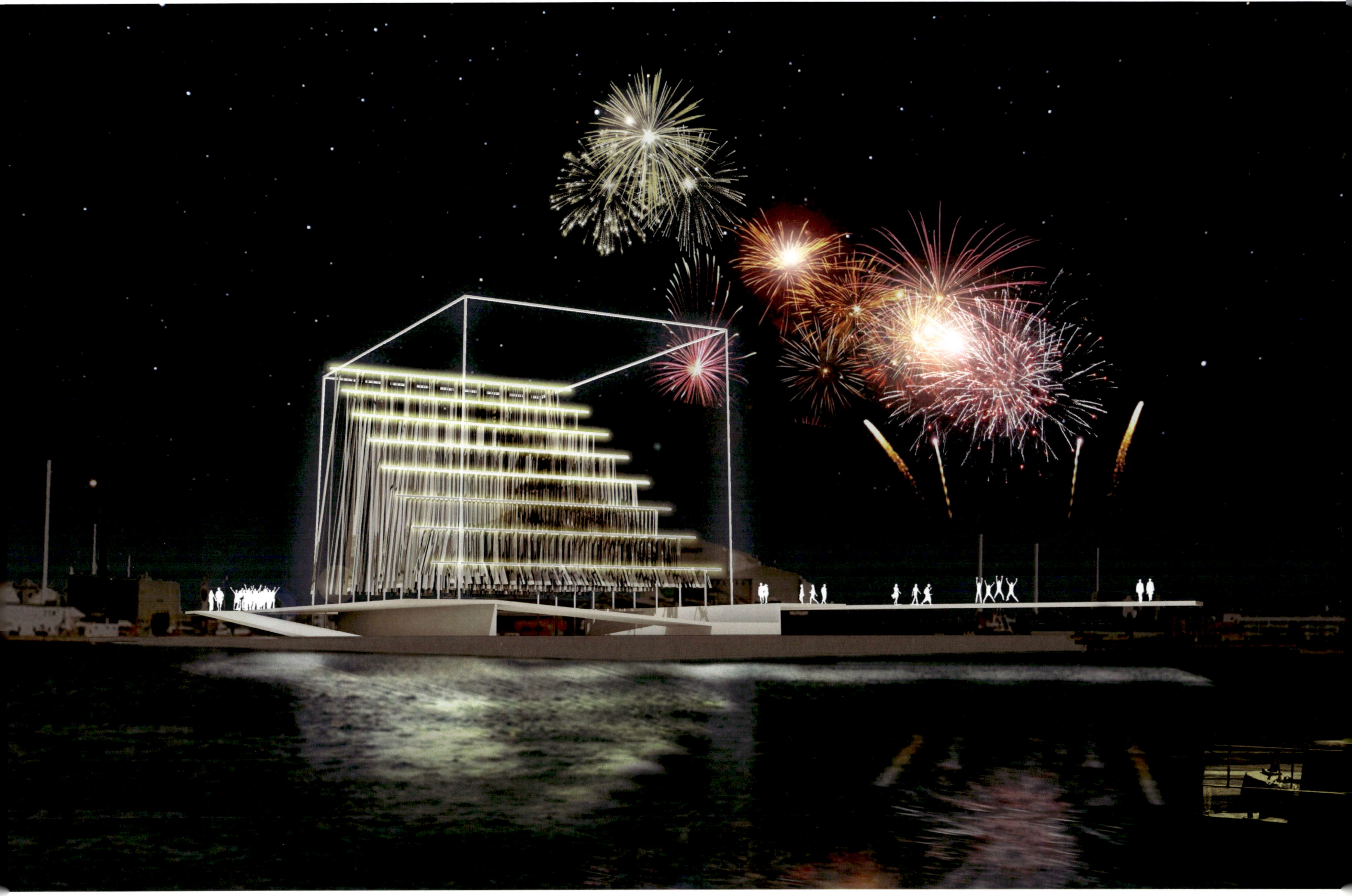

Celebration of a clean energy future

LightSponge

A mere glimpse into a new world can ignite the imagination with great possibilities and mature ideas that have lain dormant.

Assuming a passive role, the common sea sponge might serve as the underwater background to such tales as the Little Mermaid, cleaning the water with its uncelebrated daily activity. *LightSponge* takes up the untold story of the sponge, lifting it out of obscurity into the light of day.

The technology of printed photovoltaic cells is based on an organic photoreactive polymer that can be applied to flexible substrates. Beyond the capacity for generating energy from sunlight, this application on plastic sheets provides interesting mechanical properties: thin, flexible, lightweight, and translucent.

A study of this material prompted the development of a structural organization that derives directly from its unique properties. The three-dimensional cellular structure of a sponge provides ideal conditions for sunlight absorption due to its high surface area, particularly in view of the panels' transparency and dual-sidedness.

LightSponge is an ideal structure for the local water taxi terminal, which serves as the gateway to Refshaleøen. The terminal would passively demonstrate solar technology to a wide and diverse audience in a casual and unimposing manner.

During the daytime the terminal acts as a transient shelter for passengers while inconspicuously harvesting sunlight. At night, it comes to life with a dazzling LED light performance employing a portion of the power it has absorbed throughout the day.

TEAM:
Asaf Mayer, Diego Prilusky, Michal Lerman, Adi Glaser

TEAM LOCATION:
Tel Aviv, Israel

ENERGY TECHNOLOGIES:
organic thin-film photovoltaic

ANNUAL CAPACITY:
60 MWh

It takes only a mere glimpse into a new world to spark the imagining of great possibilities and to mature the ideas that have always been in the background.

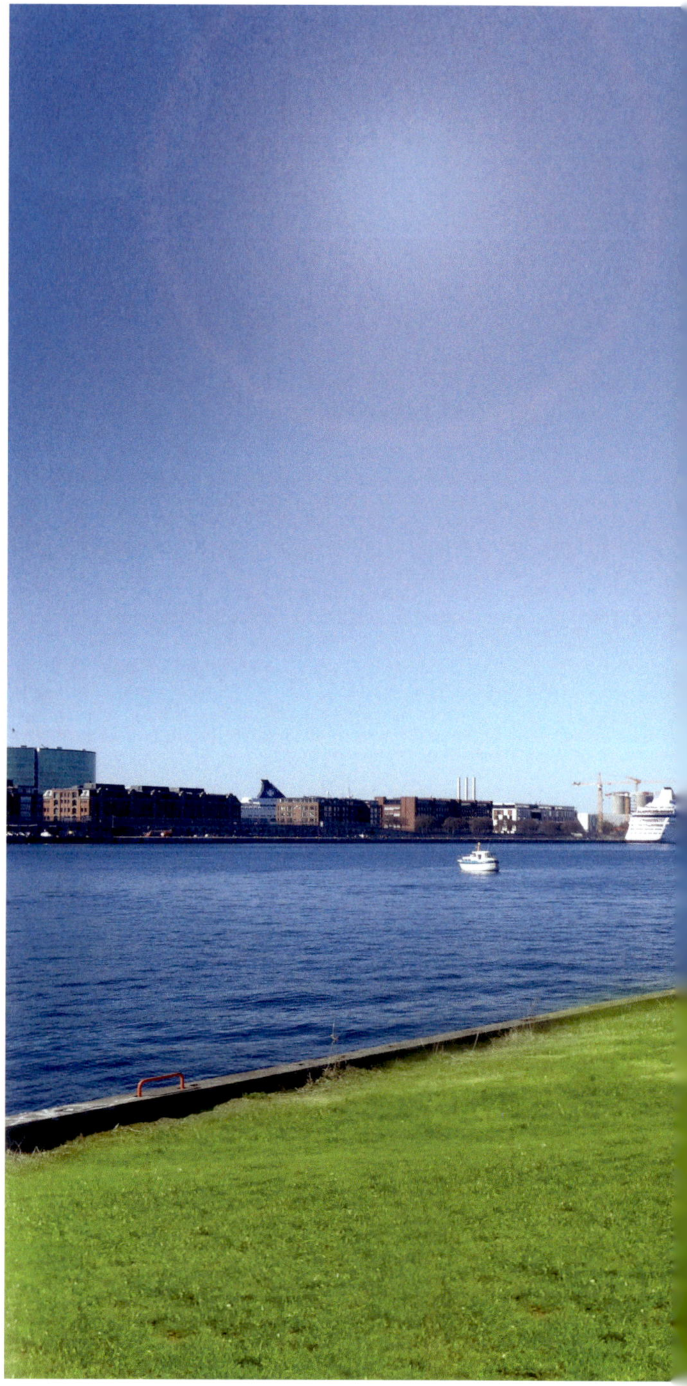

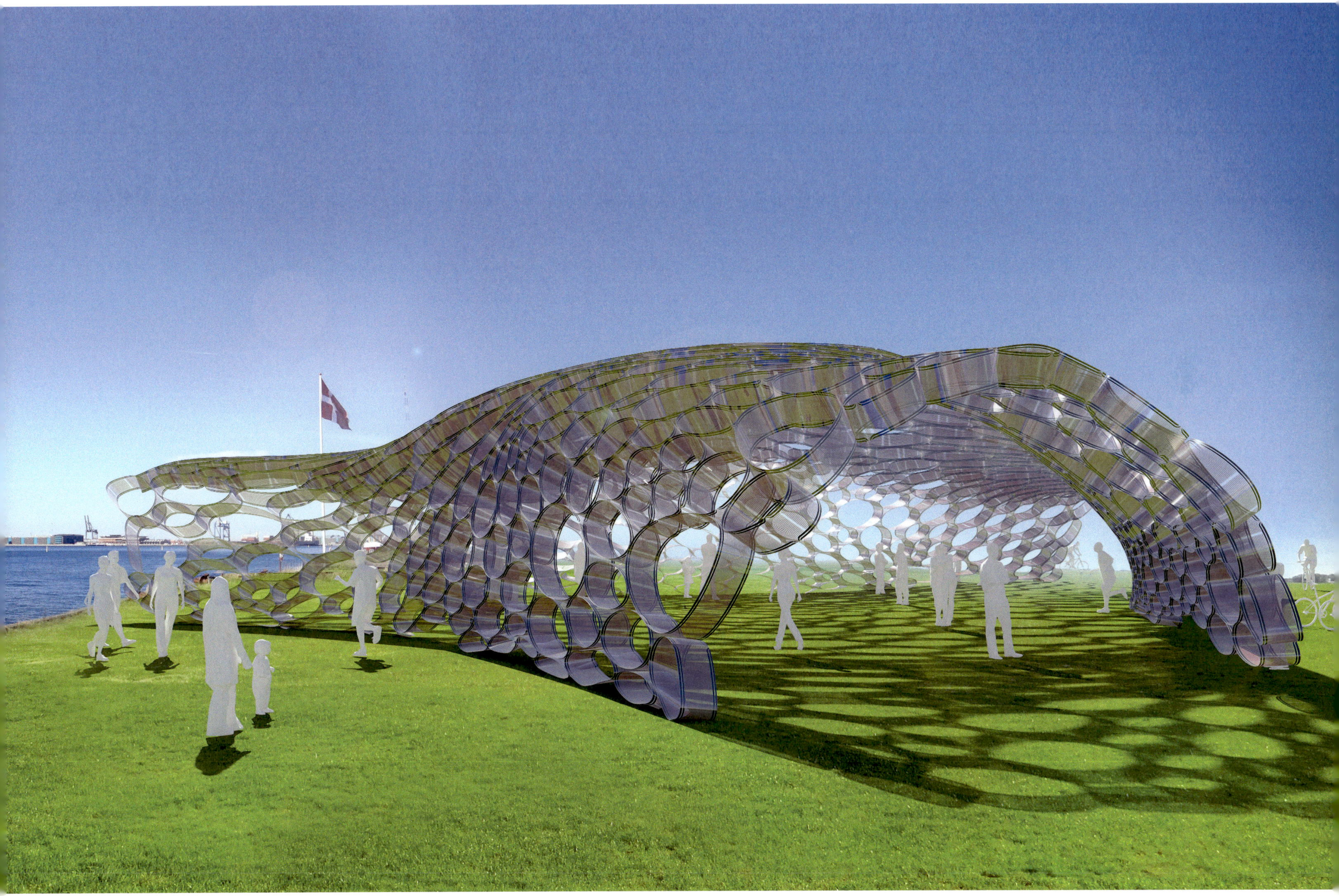

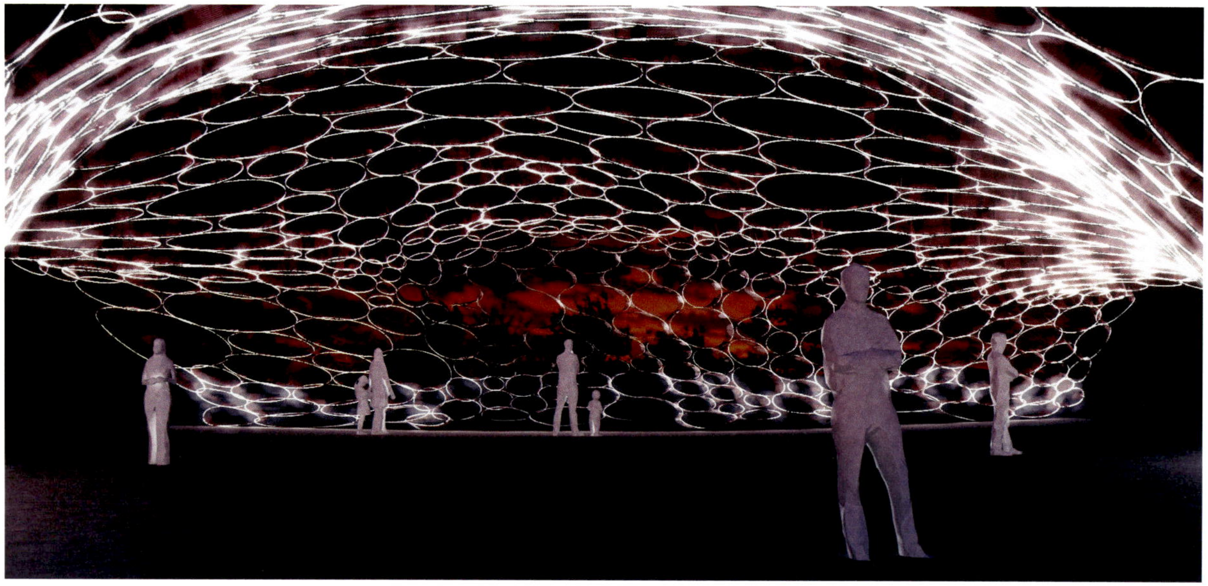

The structural skin can flex its parametric form while maintaining its structural integrity.

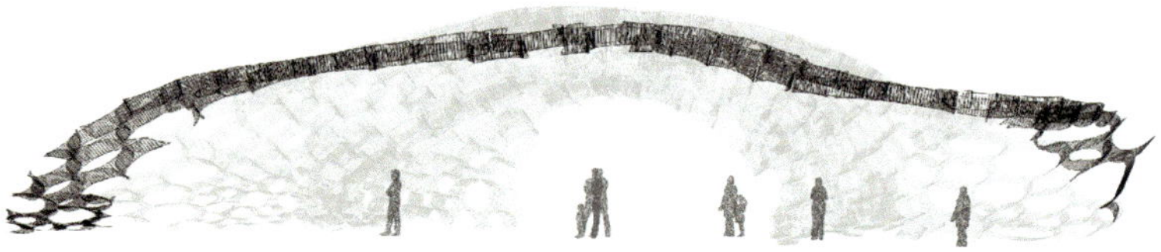

The concave skin can span over large distances and acts as a continuous, curving "multicellular" roof.

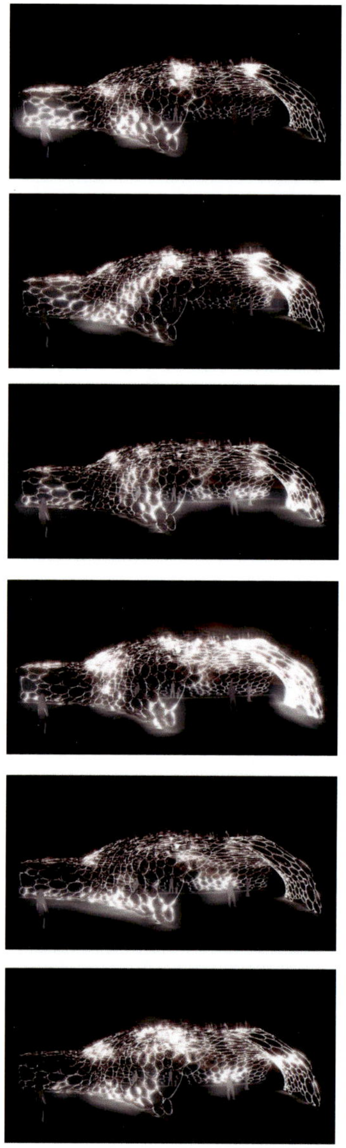

At night, *LightSponge* comes to life with a LED light performance employing a portion of the power it has absorbed throughout the day.

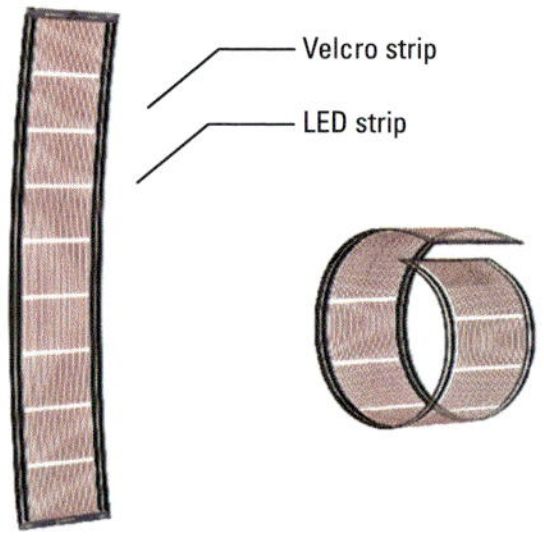

OPV plastic ring

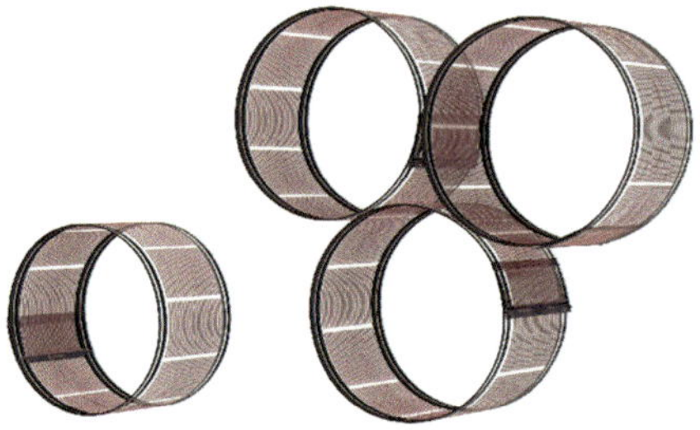

Assembly method

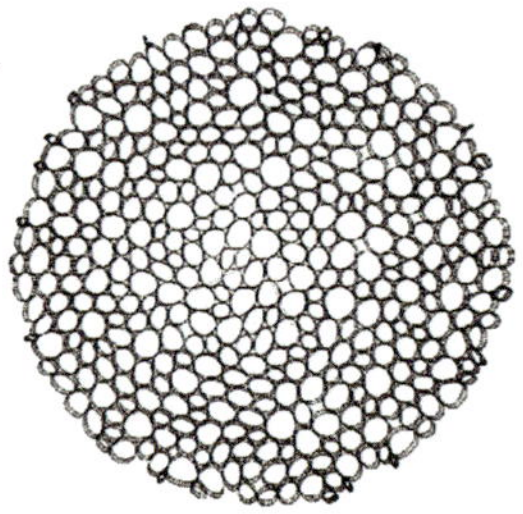

Packing circles

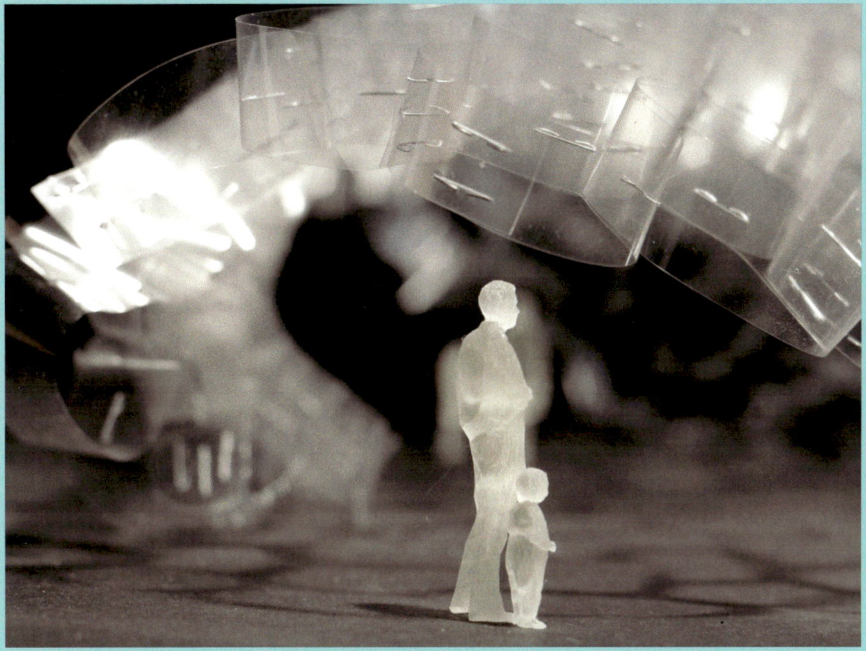

Structural study

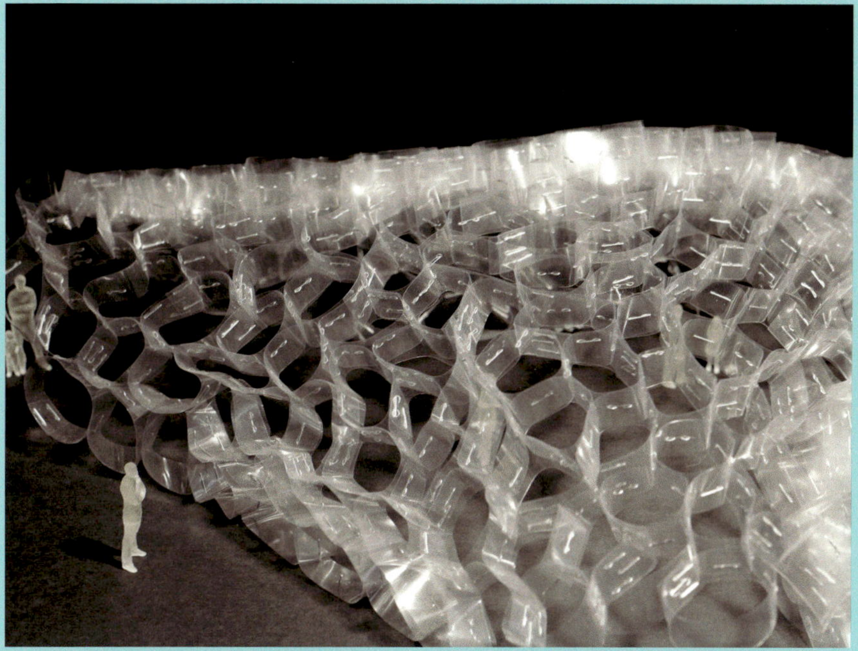

Circle packing is the study of the arrangement of circles on a given surface such that no overlapping occurs and all circles touch another.

Aetherius

Aetherius expresses the poetic potential of the natural resource of wind, as well as its potential to generate energy. Although wind does not have a shape or color, it can come alive and express its powerful elegance through the use of technology. The intervention aims to transmit the beauty of wind through its material manifestation. *Aetherius* enables the contemplation of nature as a source of beauty.

The installation moves according to the weather conditions—most specifically to the wind speed, frequency, and direction. The wind direction is not always constant. Even though the prevailing wind is westerly, there is always a microclimate created within an urban context. *Aetherius* responds to this unpredictability.

Two structures each cover a volume of 40 meters long by 45 meters wide by 30 meters high. The undulating facade is supported by two lateral walls that act as an enclosing architectural feature facilitating the organization of public events, such as concerts and markets.

The wind actuates a series of ultra-light wings attached to a pivot that can rotate within limits, adjusting for possible proximity to visitors. This initiates an interaction between the visitors and the structure.

The top of each wing is linked to a supporting beam with springs that have different levels of elasticity. Therefore each wing moves in a different way than its neighbor wing. This results in an ever-changing facade that responds to surrounding environmental conditions.

TEAM:
Marilu Valente

TEAM LOCATION:
London, UK

ENERGY TECHNOLOGIES:
wind-driven hydraulic cylinder generators

ANNUAL CAPACITY:
4,200 MWh

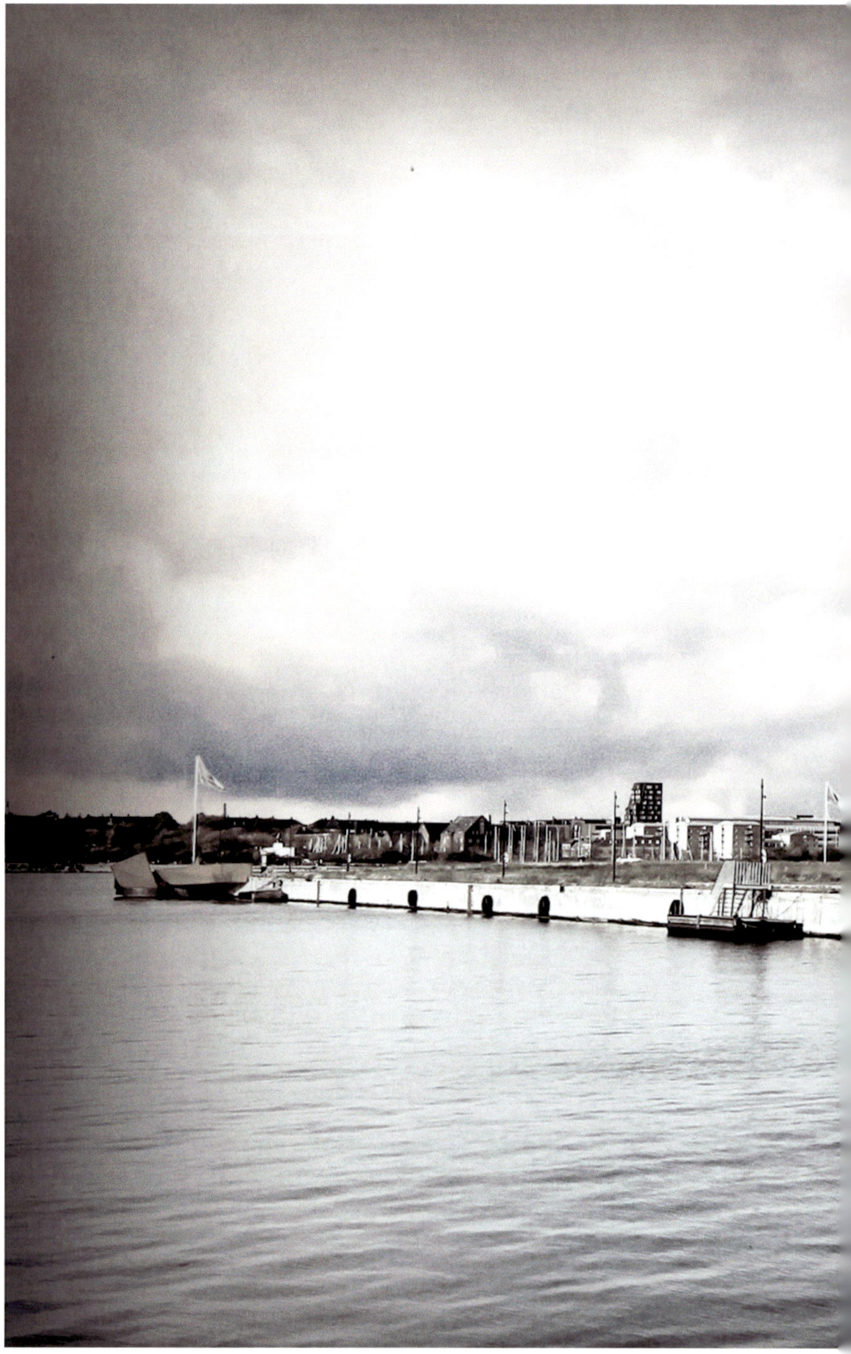

The efforts to reimagine the design site are targeted to attract people and inspire them to develop a more sustainable lifestyle.

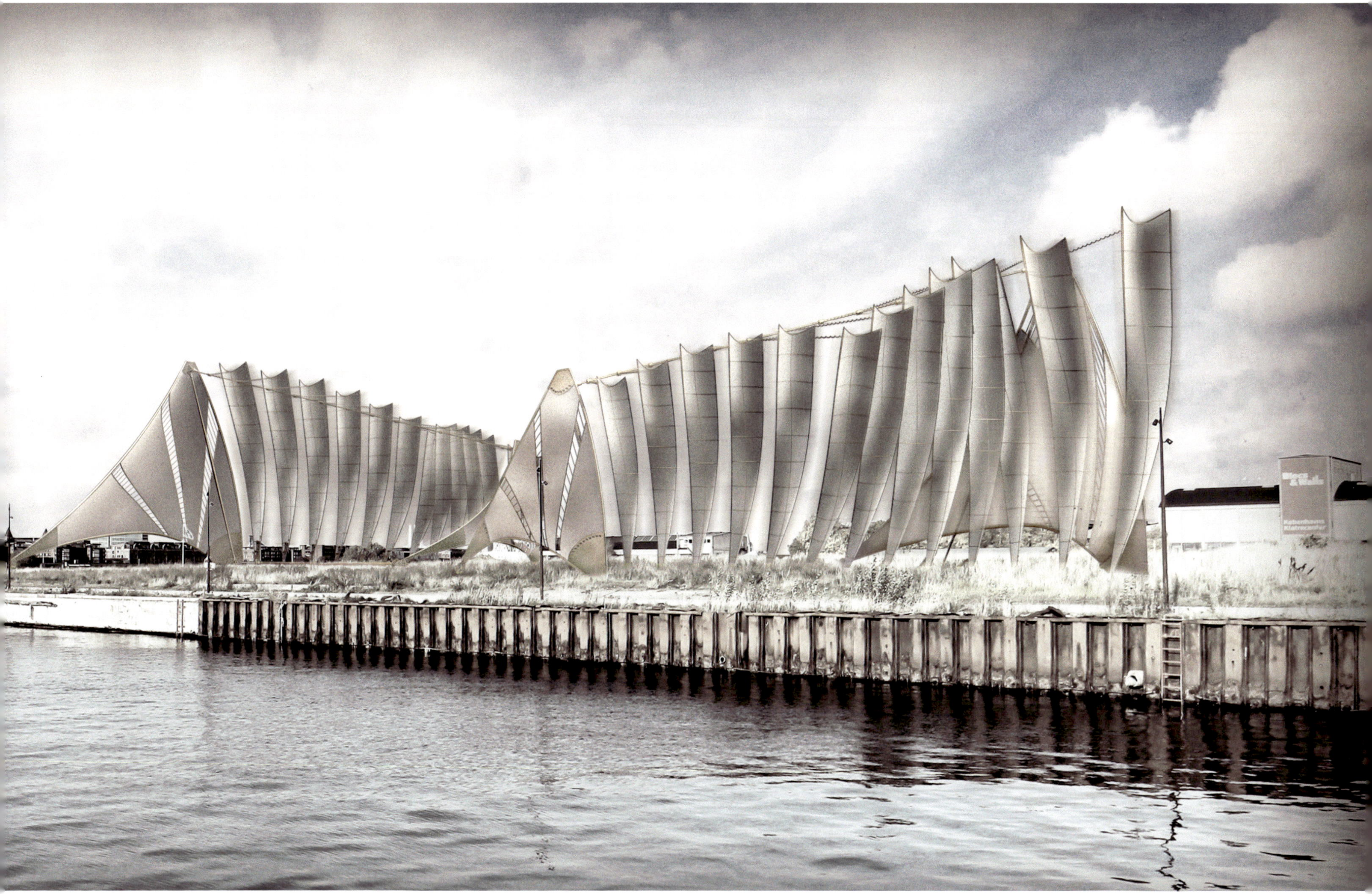

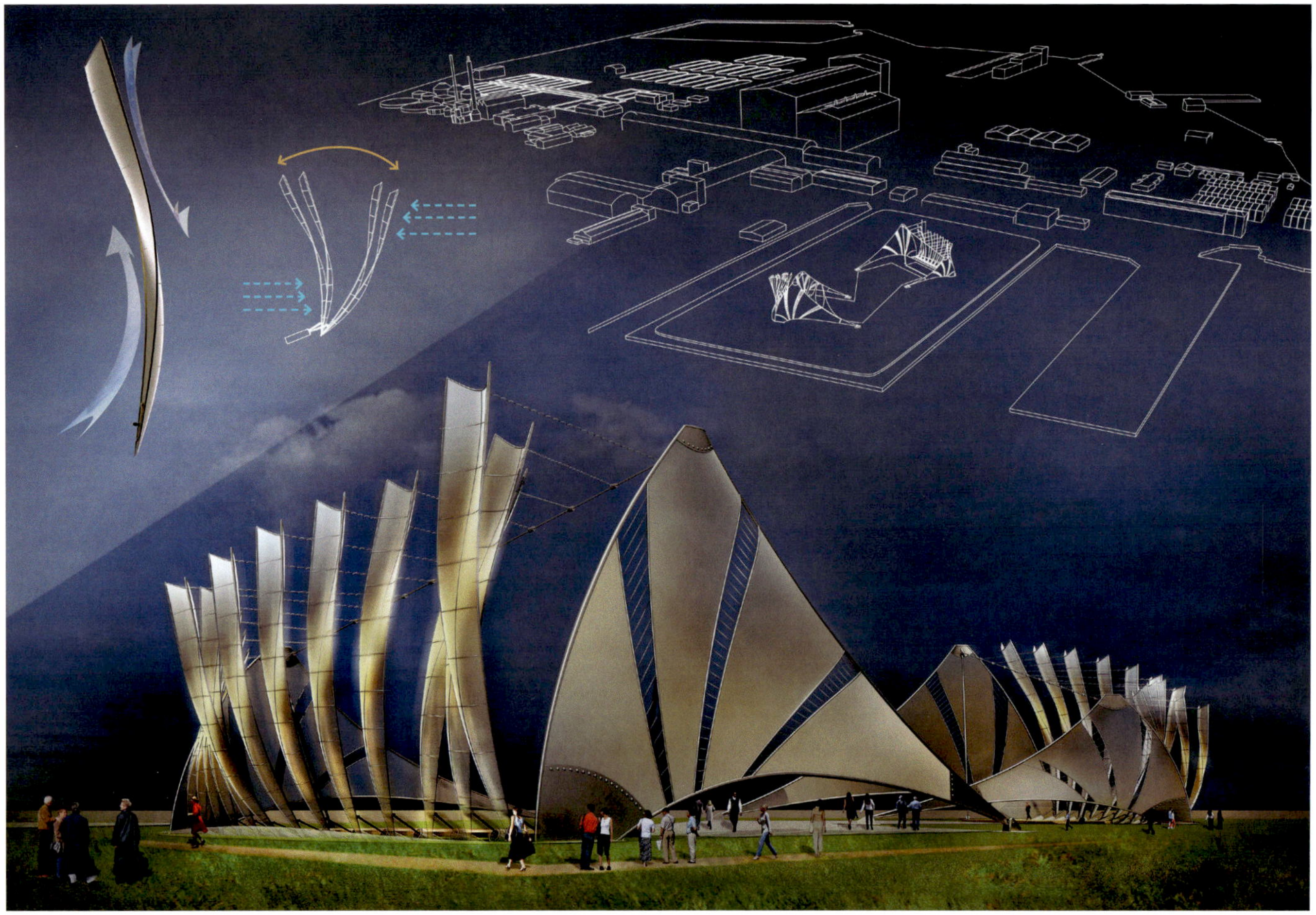

Aetherius expresses the poetic potential of wind as well as its potential to generate energy. The installation moves according to the weather conditions and more specifically to the wind speed, frequency, and direction.

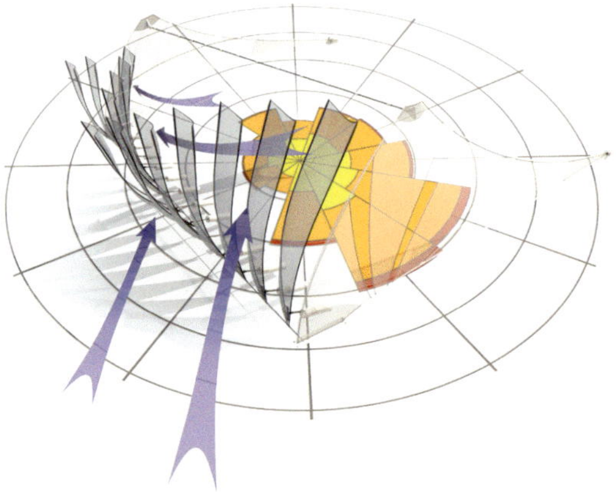

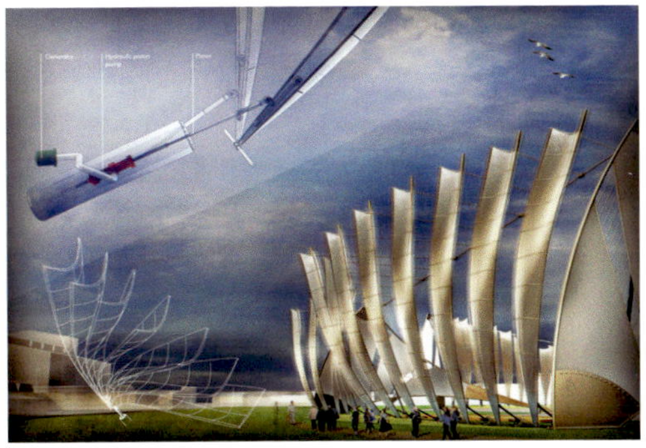

A series of wings are attached to horizontal pivots, which rotate and actuate hydraulic piston pumps. This alternating linear movement is then transformed into electricity by a generator.

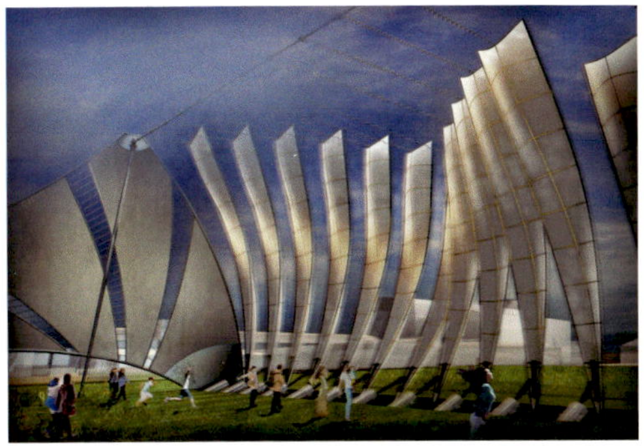

The top of each wing is linked to a supporting beam with springs that have different levels of elasticity. Thus each wing moves in a different way than its neighbor wing. This results in an ever-changing facade that responds to its surrounding environmental conditions.

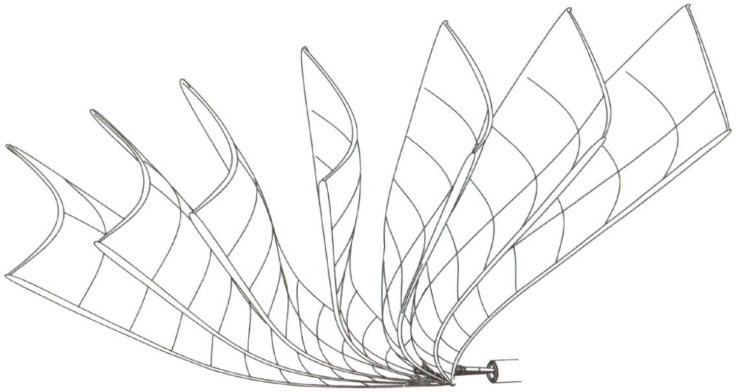

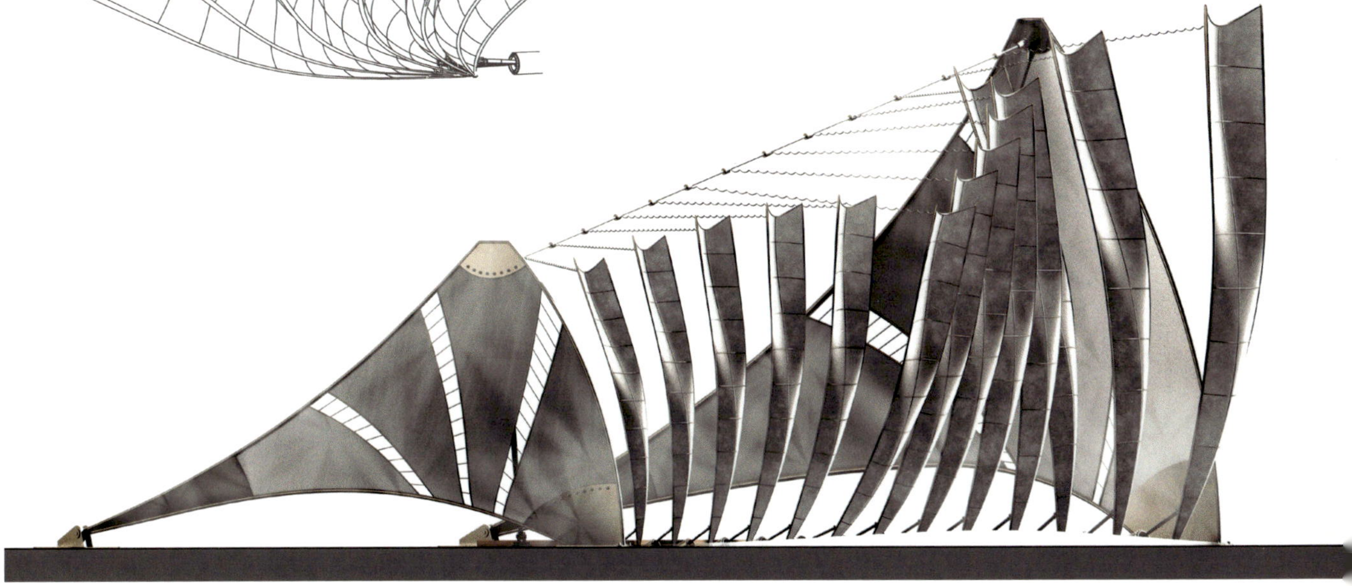

Elevation

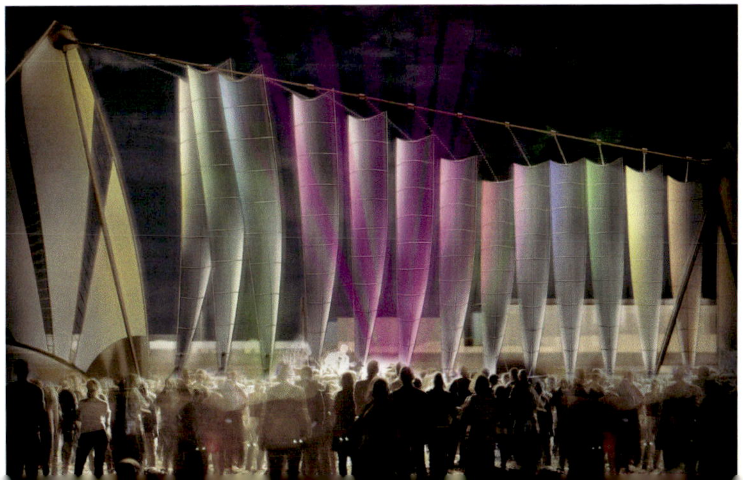

Nighttime illuminations

Wind Decoder

TEAM:
António Pliz, Cláudia Branco

TEAM LOCATION:
Borba, Portugal

ENERGY TECHNOLOGIES:
aeroelastic flutter (Windbelt™)

ANNUAL CAPACITY:
710 MWh

Wind Decoder is a climate-responsive artistic work that promotes an understanding of climate conditions and the technologies available to harvest their energies. The work is similar to a hypercube comprised of three cubes where the outer cube is 30 meters high, and the two inner cubes are 26 meters and 22 meters high.

Old constructions found on the site, including steel and a wooden amphitheater, give dialectic depth to *Wind Decoder's* materials. These physical links contrast old and new materials, such as wood and aluminum. The orthogonal lines of the former industrial facilities help define the new form, evoking memory and connecting the sculpture to the history of the site.

The cubic form is associated with stability and balance—two ambitions for future societies as citizens' environmental consciousness grows. These ambitions are represented by the outer cube. The slightly rotated and twisted inner cubes represent the wind's dynamic force, affecting its stability—an analogy to the imbalances of the past. Lights provide a visual representation of the production of wind energy.

The vertical planes are filled with 8200 Humdinger Windcell™ panels of 1.0 m x 1.0 m x 0.05 m. Independent Windbelts™ with LEDs will perform the light effect. Despite the main wind streams travelling from west to southwest, the cubic form is designed to react to the movement of the wind from every quadrant.

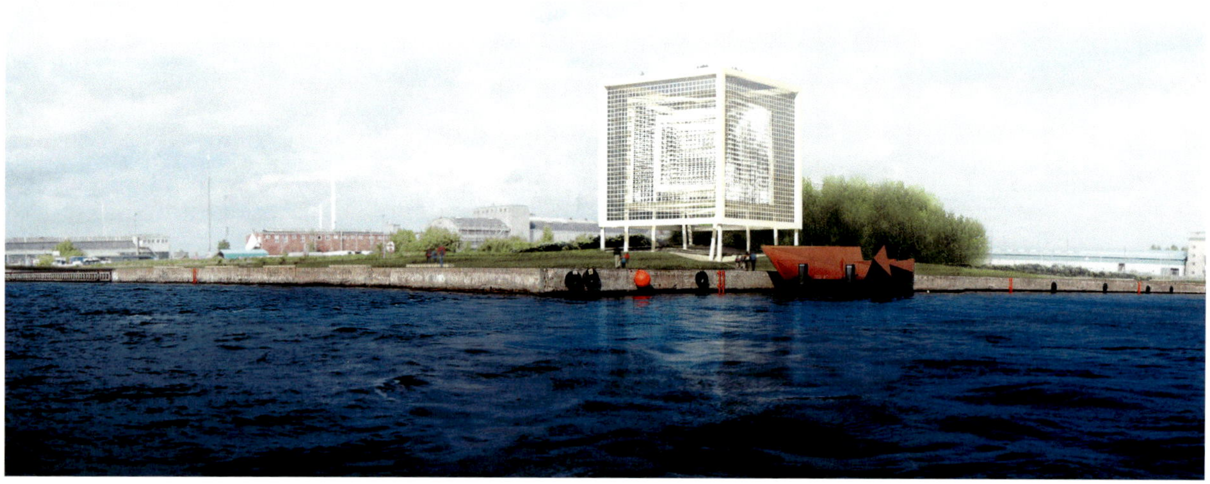

View from the Copenhagen harbor bus

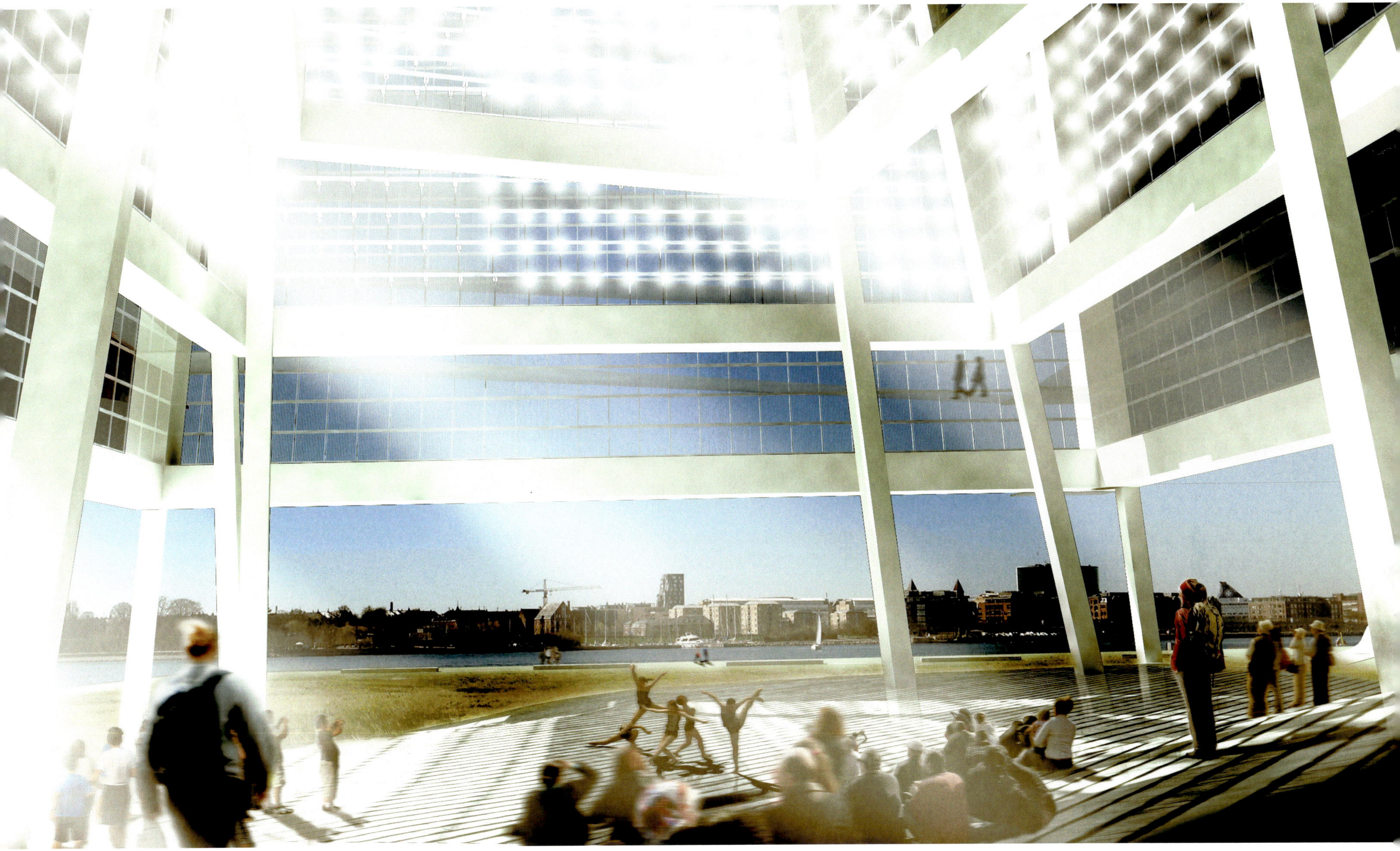

View from inside the central square

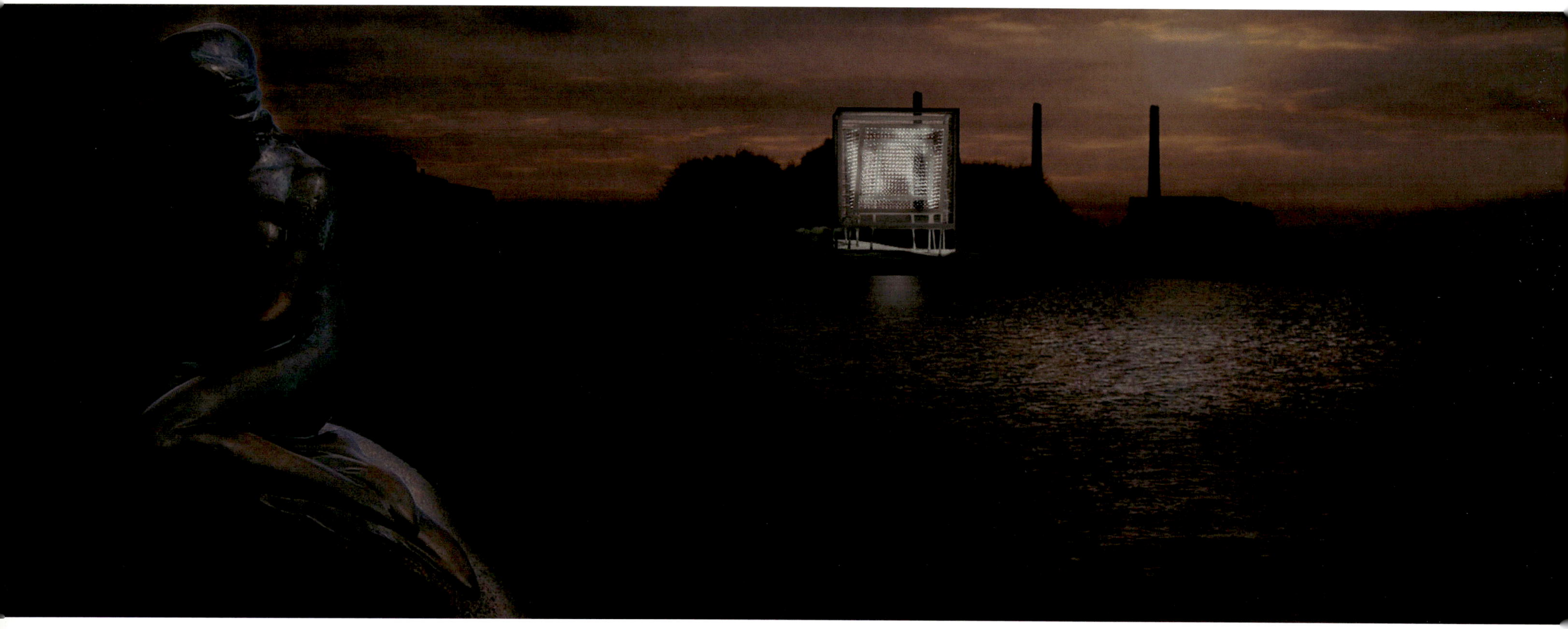

Night view from across the harbor

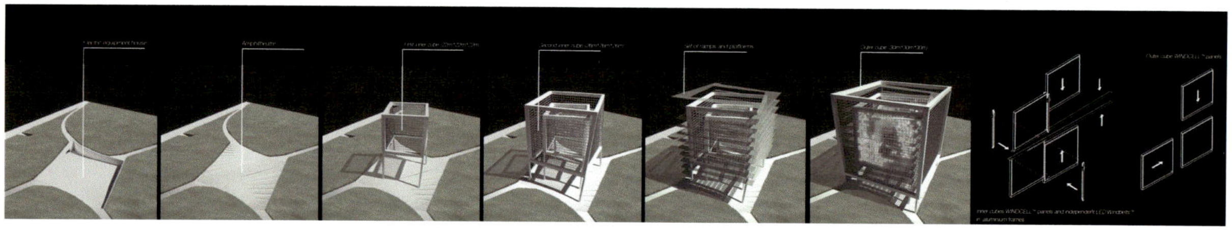

Six perspectives show how the structure is organized.

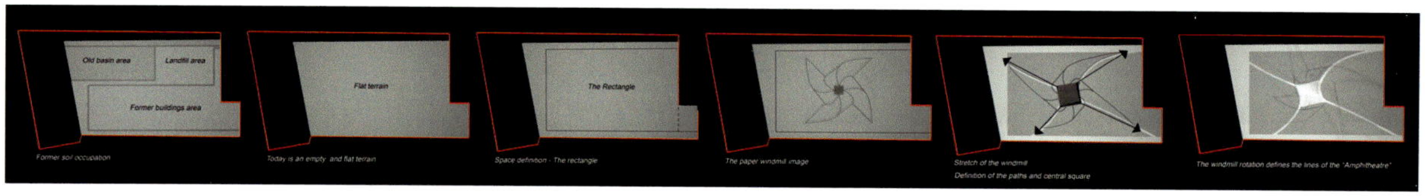

Just as the paper windmill is formed through folding and stretching a simple square, the site begins as a flat terrain, where paths are then shaped by stretching the edges towards the rectangle corners.

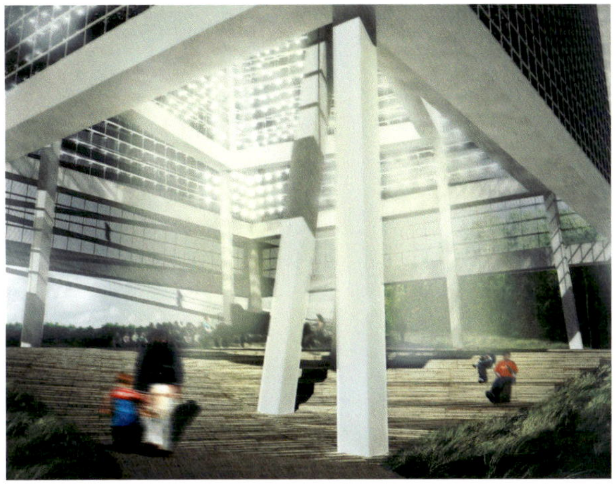

Below the hypercube

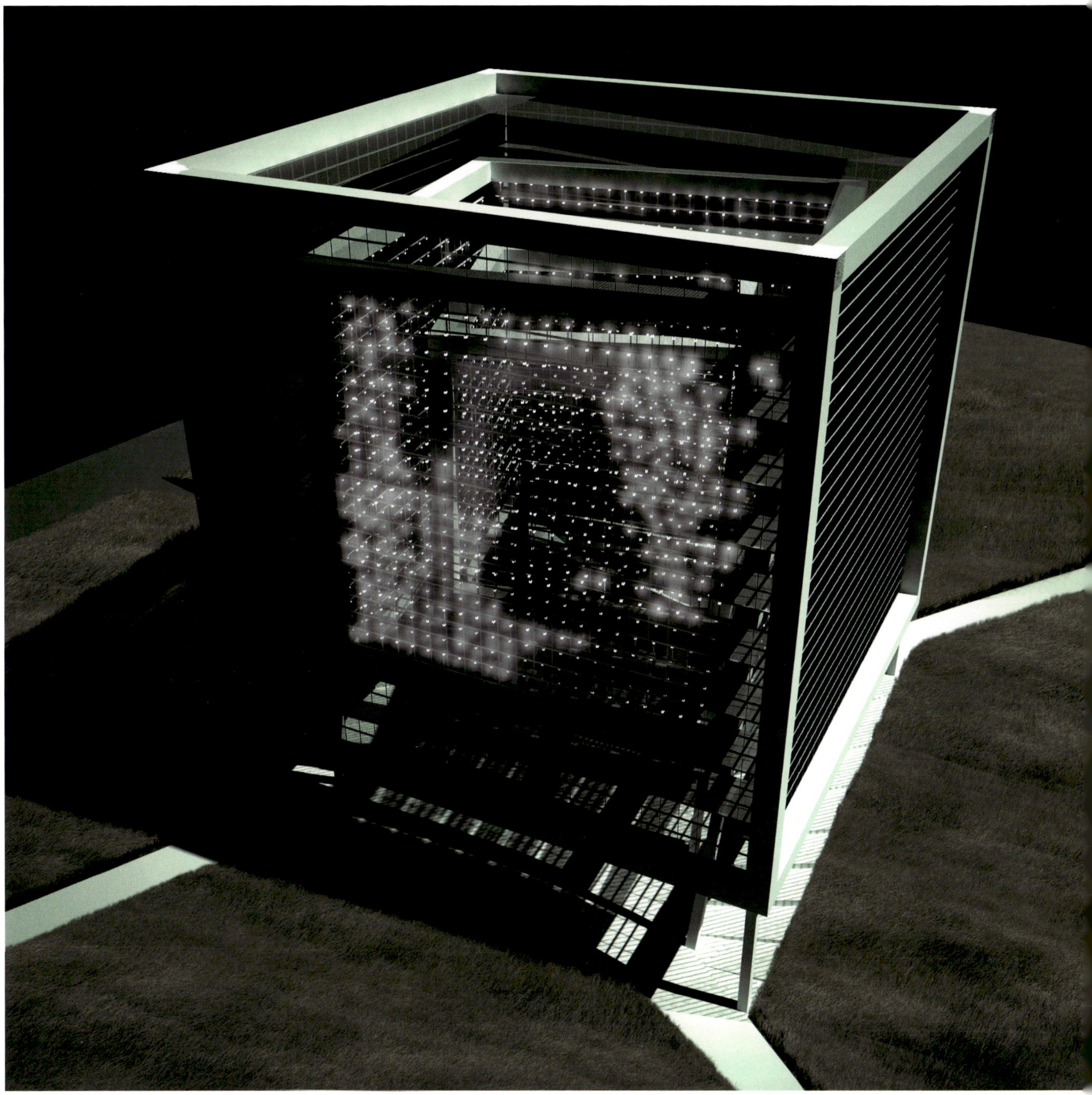

The central square marks the *Wind Decoder* location, a point of tension where social energy and natural energy meet. A set of ramps lead to the top of the structure, allowing visitors to experience the inner spaces while viewing the city and water from a series of vantage points along the way.

The Sound of Denmark

The force of the wind has carried the development of Denmark since ancient times. Wind strength assisted the Vikings as they colonized new lands, and now wind power provides renewable energy to Denmark. Because this natural force has always been rooted in the development of the country, the wind is the sound of Denmark. *The Sound of Denmark* creates a space for reflection and to feel this natural force that drives Denmark. It is a space for remembering the past and envisioning the future. It is a sound landscape.

The Sound of Denmark consists of four Viking horn trios, each trio consisting of a large, medium, and small horn. The largest horn is carved with the letter of the alphabet that refers to the natural force "sun." The next in size is carved with the letter that refers to the natural force "water." And the smallest horn is carved with the letter that refers to the natural force "ice."

The technology—compact wind acceleration turbine—concentrates the wind and increases its velocity as it passes through the horns. Each Viking horn comprises four wind turbines, and the tubular shape makes the lower part of the horn function as a sounding board. Each horn is composed of materials, such as wood and metal, obtained from the hulls of decommissioned ships. In this way, all principal materials used in the project are obtained from waste so that the environmental impact is minimal. In turn, the material has a direct relationship with the industrial and maritime history of its surroundings.

TEAM:
Laura Mesa Arango, Rafael Sanchez Herrera

TEAM LOCATION:
Vienna, Austria

ENERGY TECHNOLOGIES:
compact wind acceleration turbines

ANNUAL CAPACITY:
117 MWh

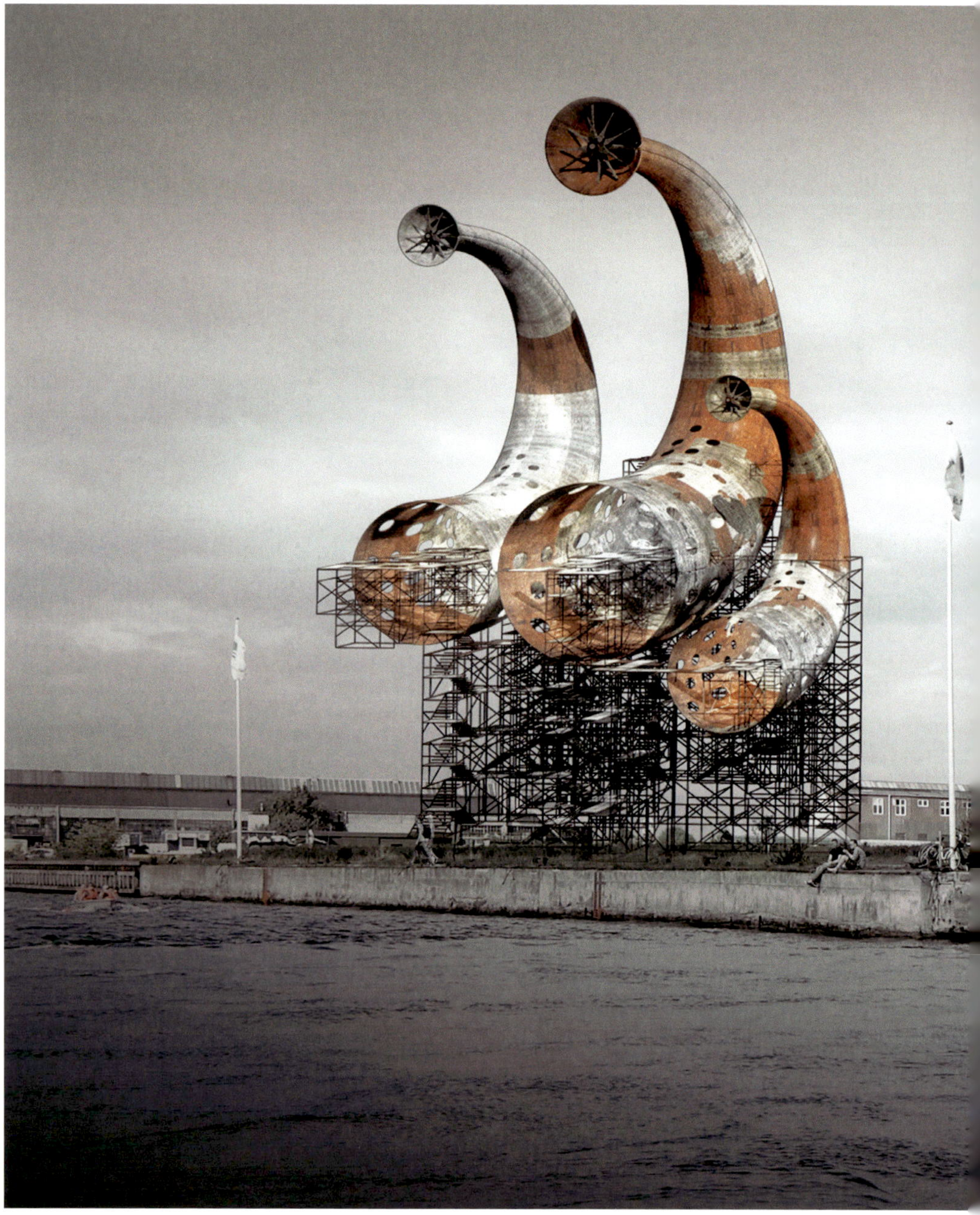

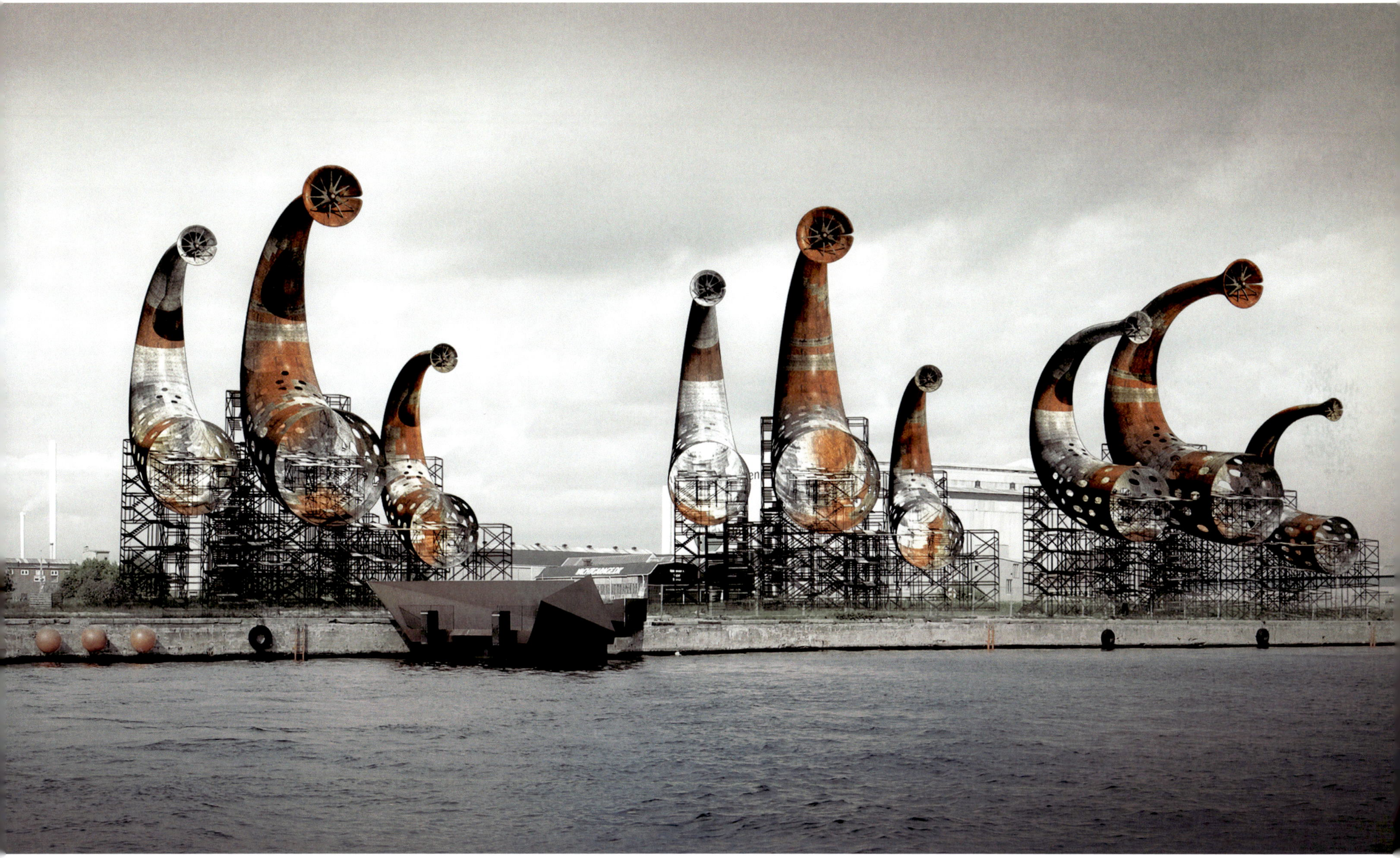

The horns open up to the west and the southwest.

The wind rose data shows that throughout the year the wind comes primarily from these directions.

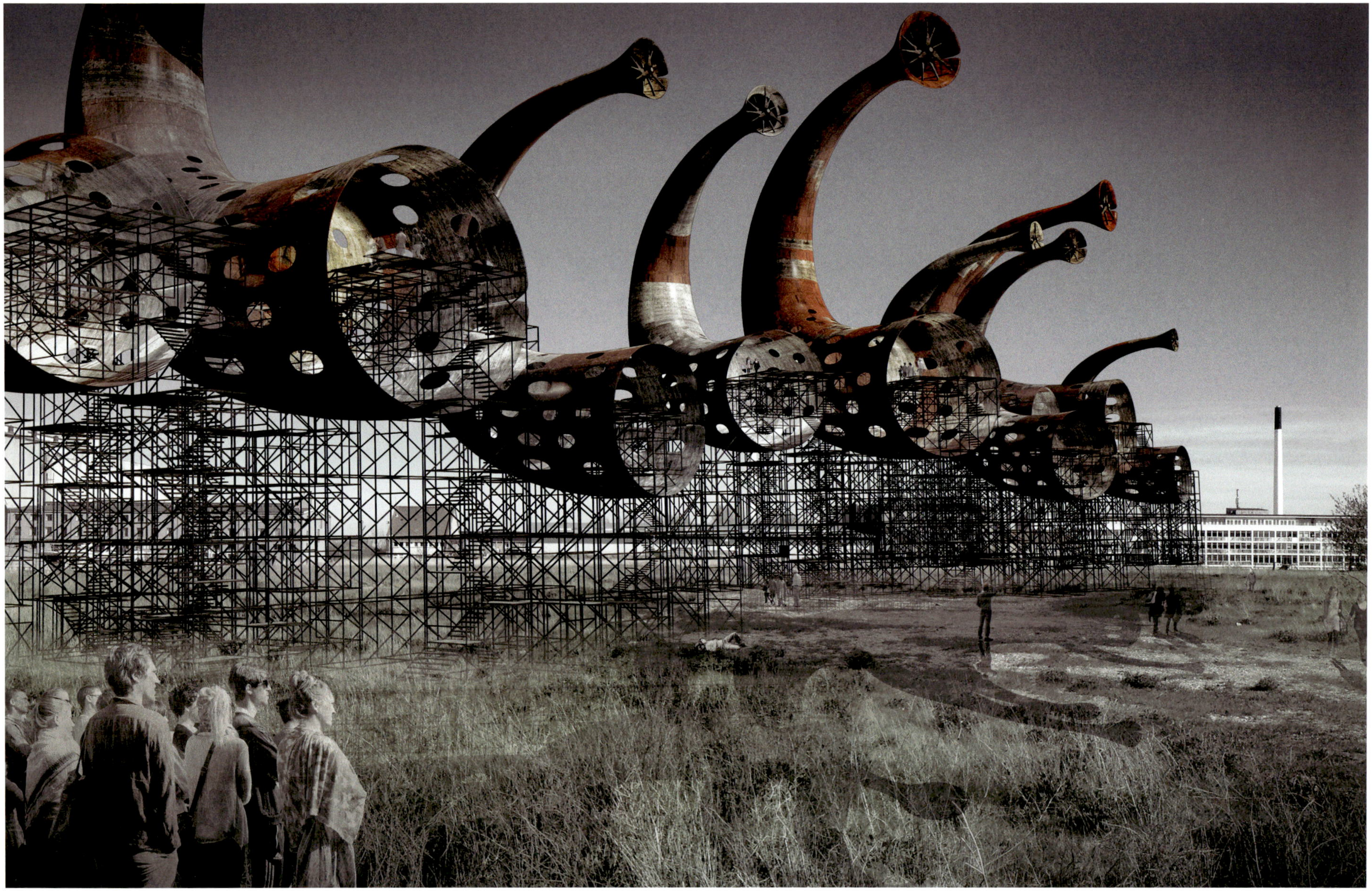

The wind enters into the upper nozzle, passing through the turbines. As it exits, it emits a beautiful sound through the lower nozzle—the sound of the wind and of renewable energy.

Compact wind acceleration turbines concentrate the wind and increase its velocity at the turbine.

for the Viking horns. It also encourages interaction between the visitors and the artwork.

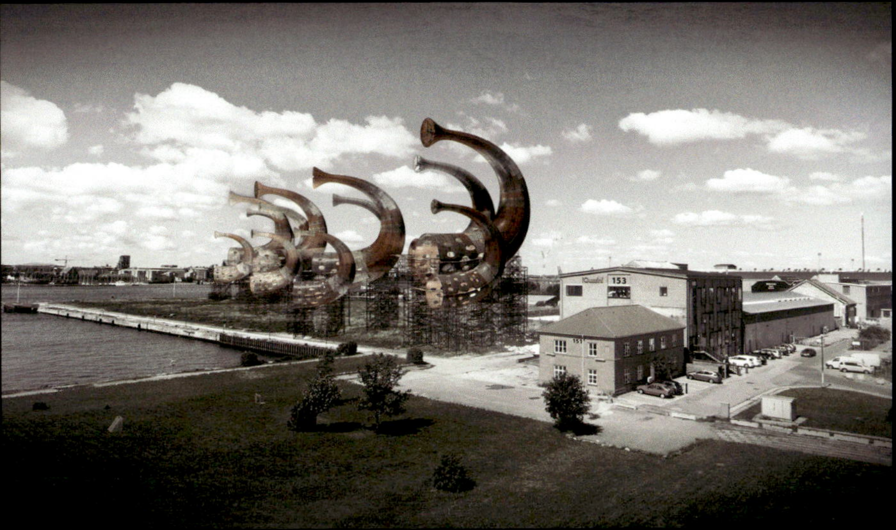

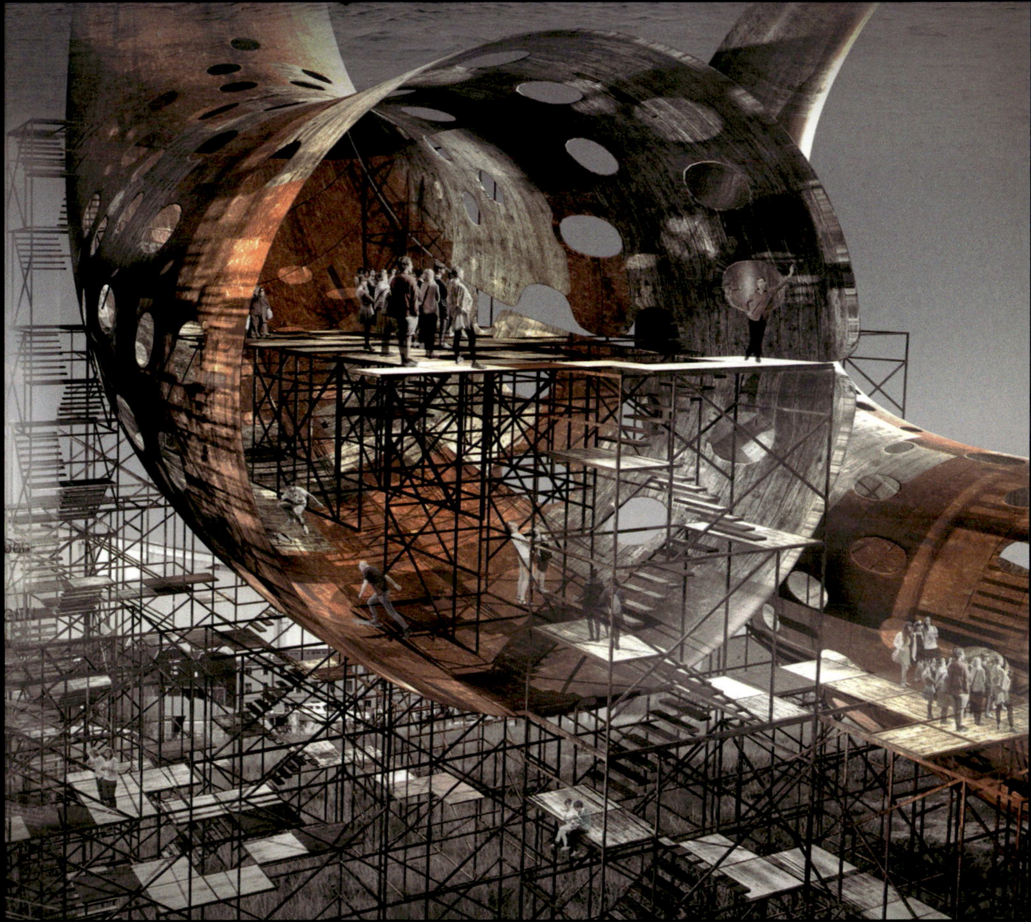

Each Viking horn comprises materials, such as wood and metal, obtained from the hulls of old ships. This gives the installation an industrial and naval aspect while also reducing its ecological impact by incorporating recycled materials.

Zephyrus Park

Zephyrus Park harnesses the power of available wind that blows across the Copenhagen harbor. Wind vector and sun angles form vaults that funnel wind into an array of wind turbines, amplifying wind speed, and increasing the efficiency with which each turbine produces energy for the City of Copenhagen. On the surface, a park provides community space along with semi-private areas created with the topography.

The vaults are set up to collect wind as much as possible throughout the year. The five intake vaults are positioned to take advantage of the abundant west wind. The funnels significantly increase the wind speed so that each turbine produces much more energy, allowing for a lower number of turbines.

The connection between surface and interior is made at the harbor bus dock, which allows a view deep into the vaults to see the turbines at work. This will be the most heavily accessed entry to the site and so will give users an idea of what is happening below as they inhabit the park above.

The form of the wind-funneling vaults produces a dynamic topography for the park, creating varied spaces from the public center of the site, to smaller, semi-private fingers that branch out on top of the vaults. This stratified landscape frames views of the changing sunset, and allows for renewed experiences of the site at different times of the day and throughout the year.

TEAM:
Mike Sollenberger

TEAM LOCATION:
Philadelphia (PA), USA

ENERGY TECHNOLOGIES:
wind turbines within acceleration vaults

ANNUAL CAPACITY:
250 MWh

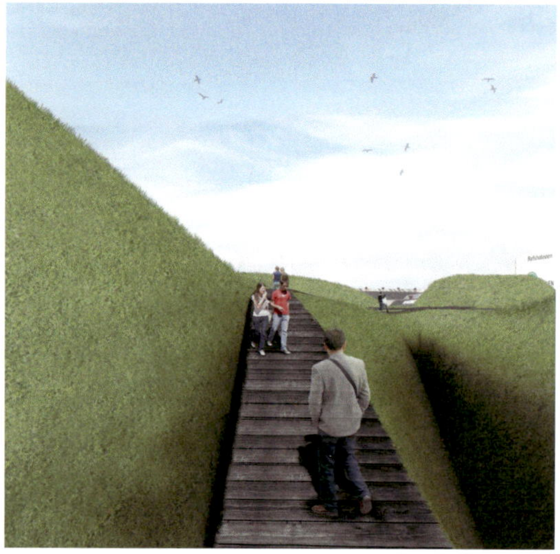

View from the bottom of one of the ramped walking paths

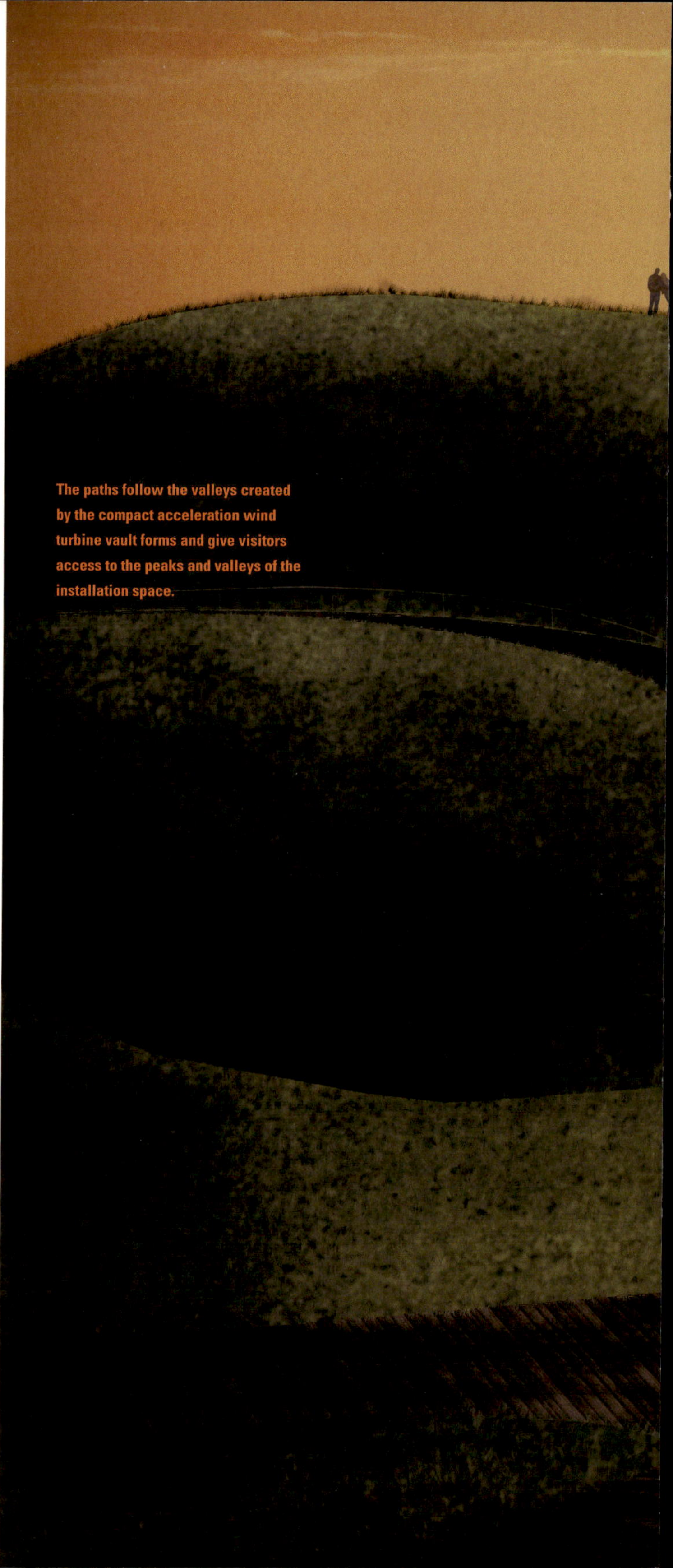

The paths follow the valleys created by the compact acceleration wind turbine vault forms and give visitors access to the peaks and valleys of the installation space.

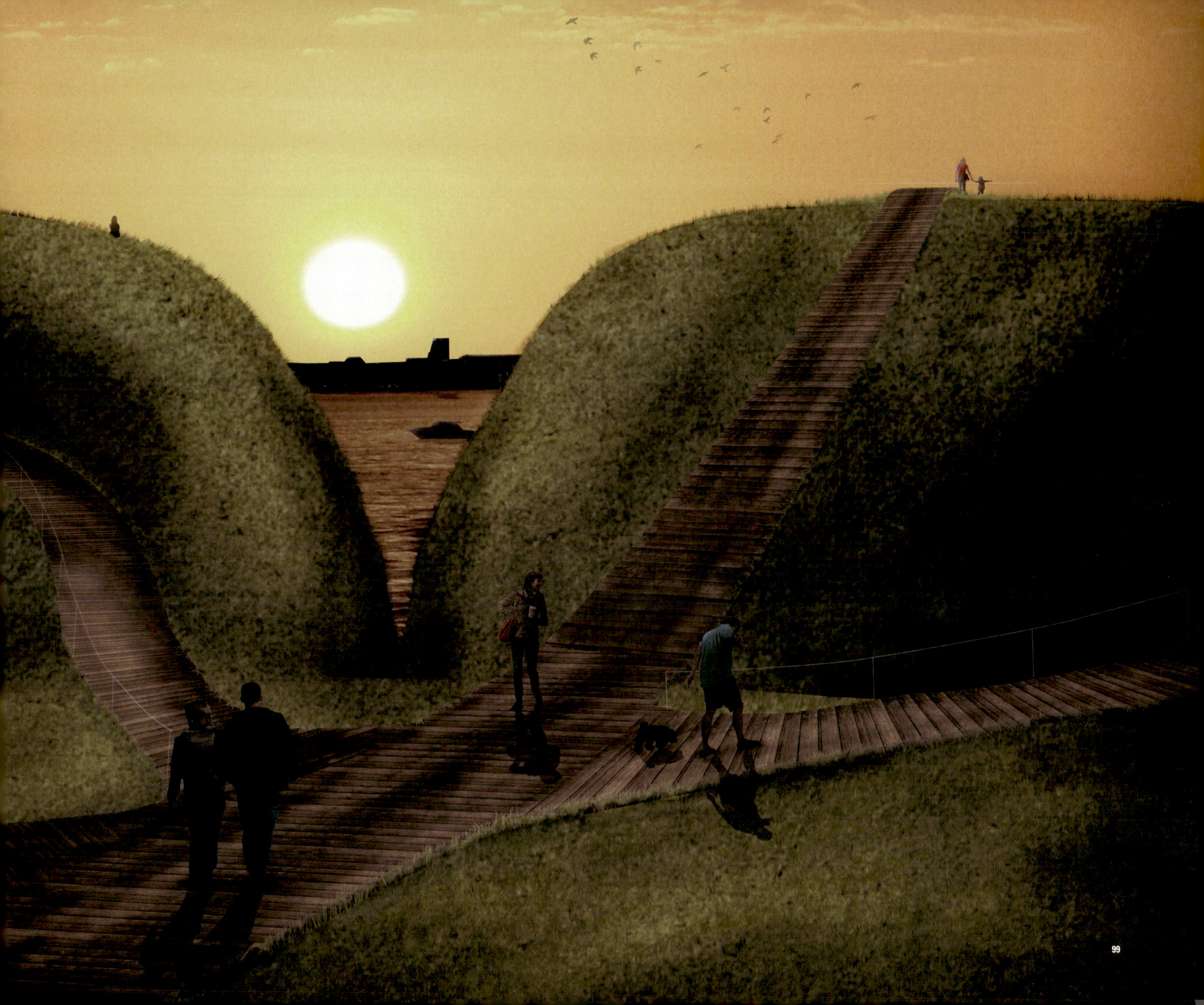

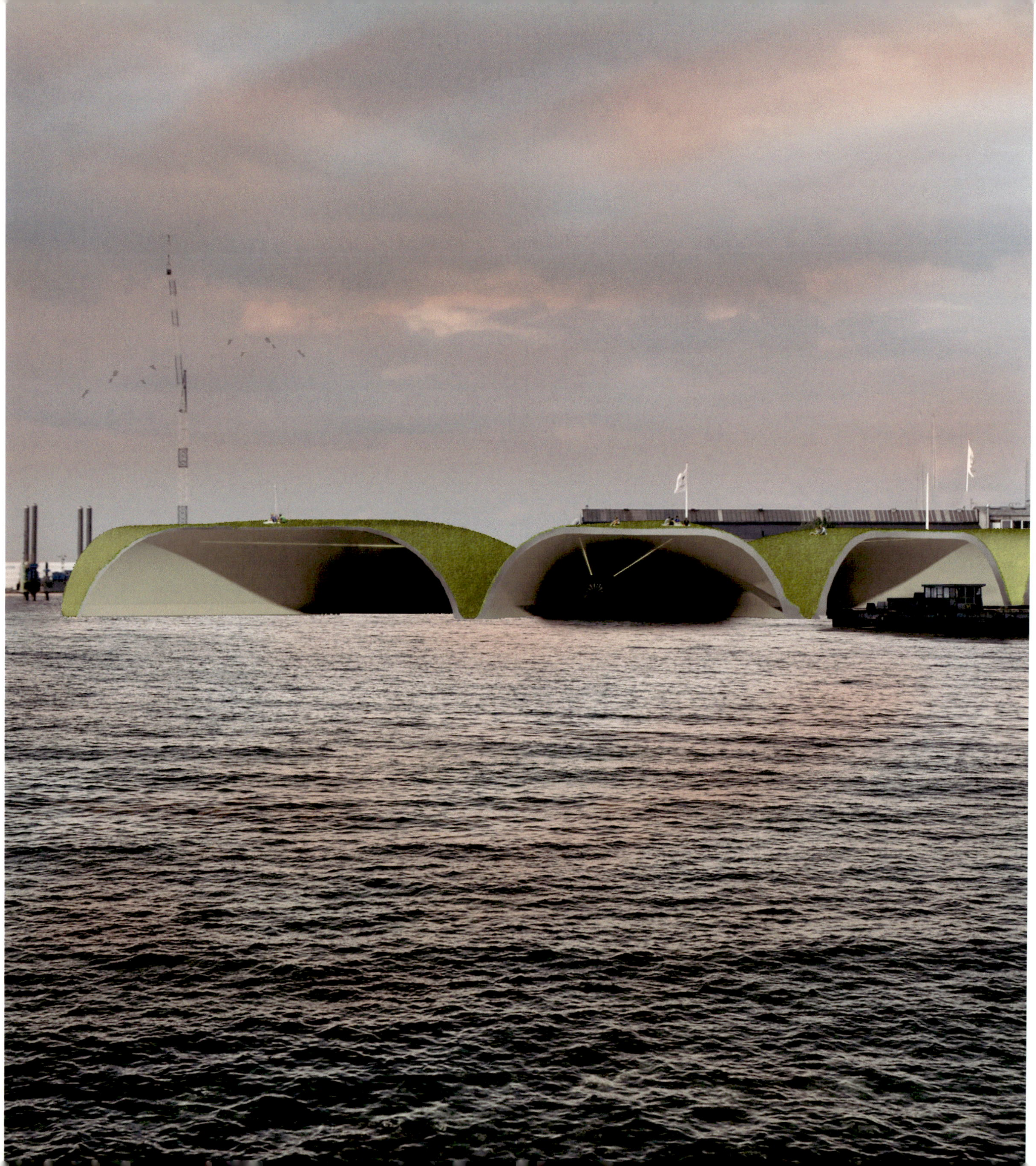

The intake vaults as seen from the harbor waters. Wind blowing across the water is funneled and accelerated towards the turbines within.

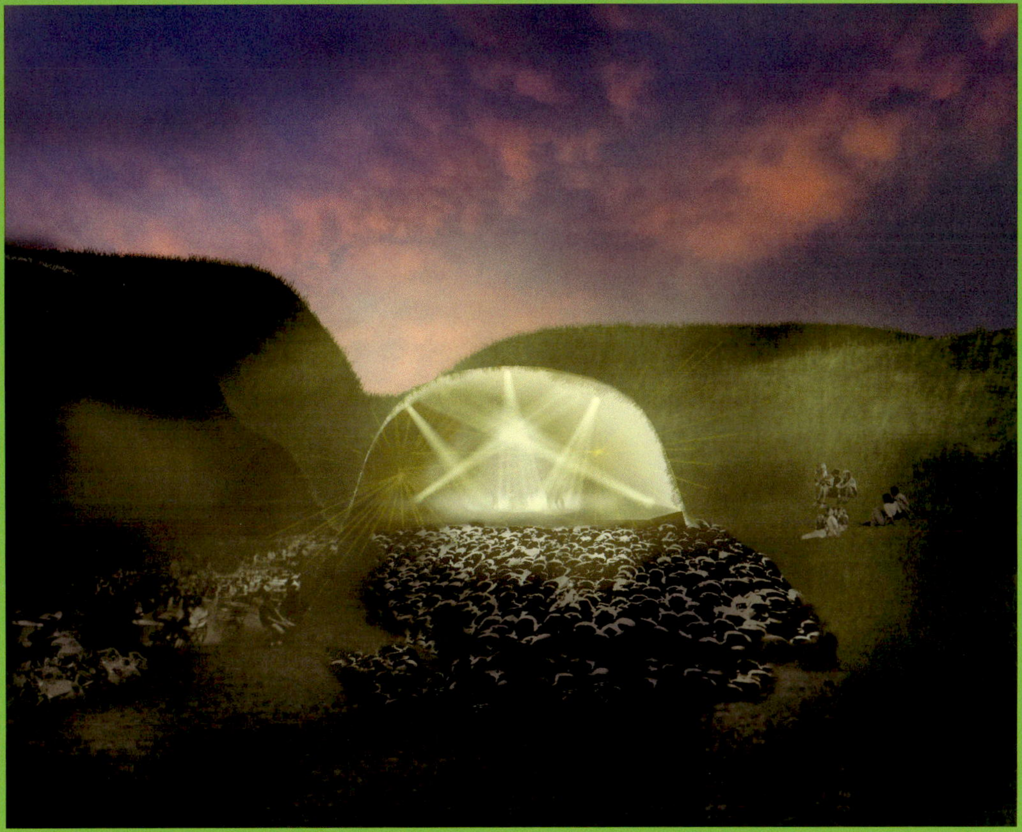

The amphitheater space on the east side of the site

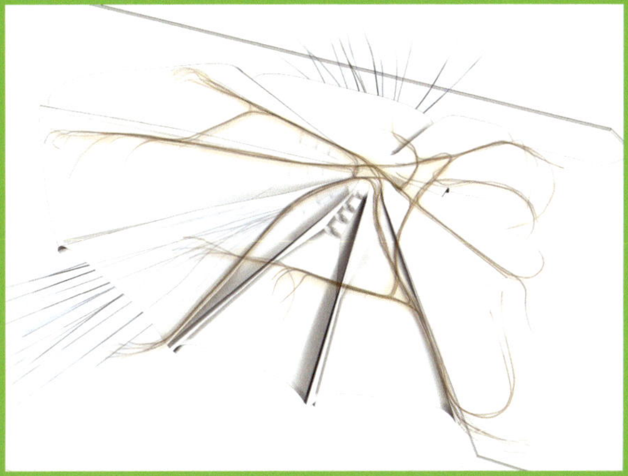

Towards the back of each wind vault, turbines convert the concentrated power of the wind into electricity and exhale the remaining breeze through to the other side of the site.

GRID Slide

TEAM:
Morten Rask Madsen, Julie Trier Brøgger, Julie Rindung, Natalia Guerrero Gutiérrez, Artis Kurps, Kevin Bailey, Søren Laurentius Nielsen, Per Møller, Jesper Ahrenfeldt, Tobias Thomsen

TEAM LOCATION:
Copenhagen, Denmark

ENERGY TECHNOLOGIES:
algae biofuel

ANNUAL CAPACITY:
100,000 MWh

GRID Slide creates renewable energy for the grid based on farming microalgae. *GRID Slide's* real potential, however, lies in creating diversity and value from waste products by returning them into circulation.

GRID Slide is also a landscape sliding between land and ocean, transformed by the changing water levels of the sea around Copenhagen. It is a recreational landscape skirting the urban and the natural, landscape, farming, and industrial production.

Three long, slim structures rise 50 meters above an engineered wetland, interacting with the industrial skyline in the background. Seen from the Little Mermaid, the sleek towers contrast with the chimneys behind. From downtown Copenhagen, the structures appear as translucent screens, in contrast to the massive walls of the old B&W assembly building.

Changes in the sea level activate floating pumps that press algae-filled water into the towers. Through the force of gravity, the colored algae liquid is distributed into the towers' bioreactors. The towers expose the algae to the sun's light and enable photosynthesis.

Microalgae are fed with nutrients in the form of wastewater and CO_2 collected locally from the Lynetten treatment plant and from Amager Power Station. The algae purifies the water while its volume increases up to six-fold during the day. The energy-rich product is collected as gasification feedstock for a 2.5 MW generator located in the bottom of the eastern tower—visible to visitors but enclosed by glass screens. Byproducts of the production cycle include phosphorus, potassium, and recalcitrant carbon, which have many commercial applications, such as fertilizer.

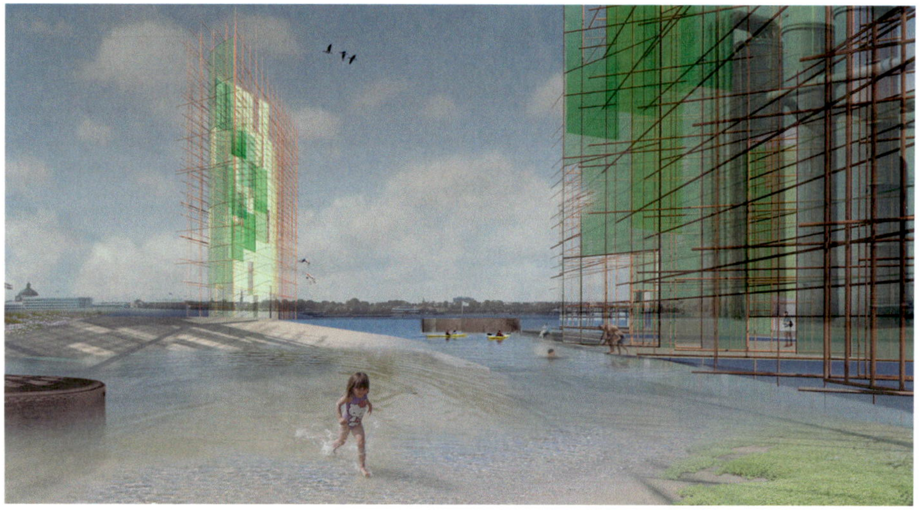

***GRID Slide* is a landscape of skewed surfaces that allow water to shape the land as it floods and drains. At the tower base, biomass is transformed into gas and a nutritious ash through a low temperature gasification process assisted by the heat of the sand.**

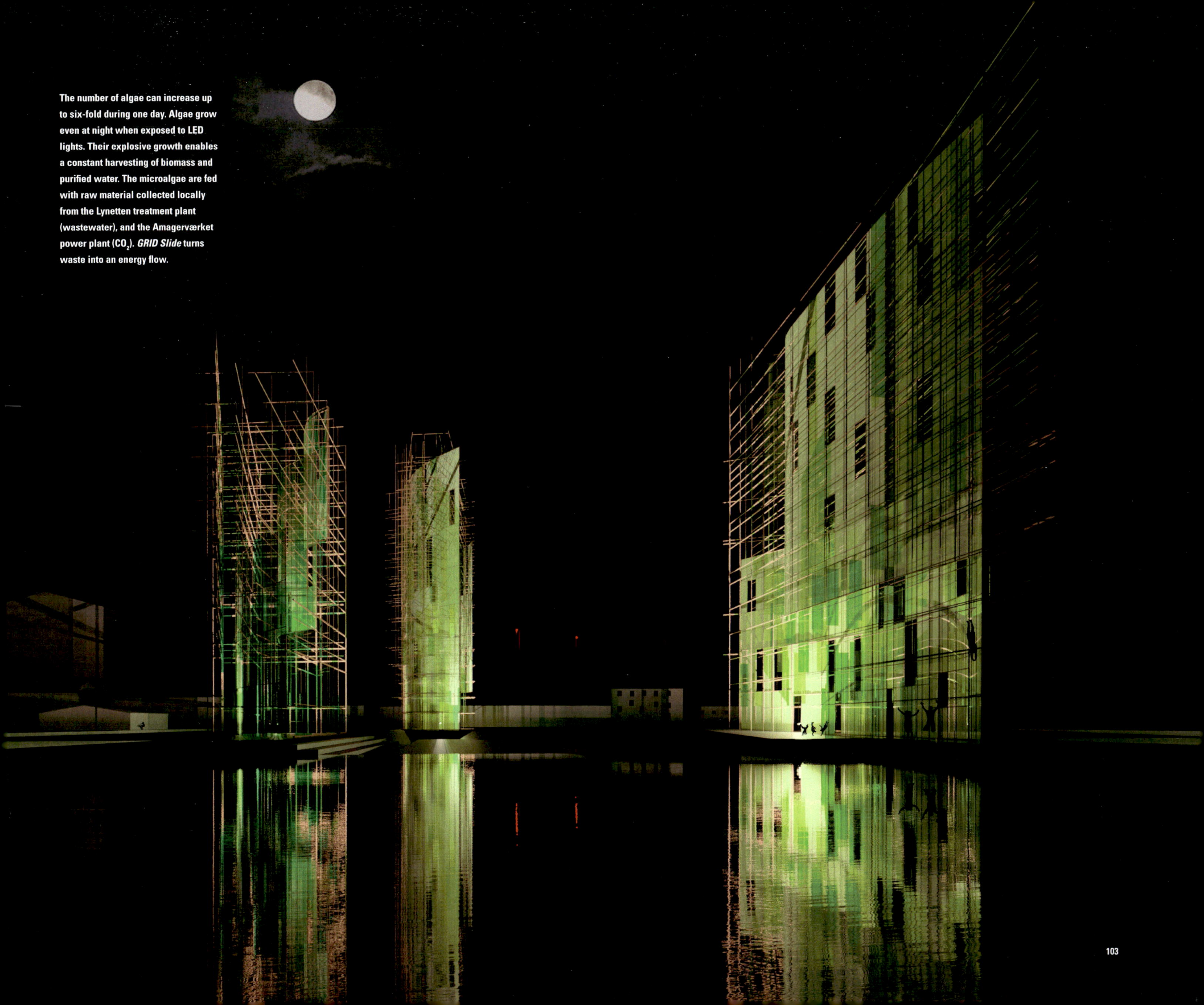

The number of algae can increase up to six-fold during one day. Algae grow even at night when exposed to LED lights. Their explosive growth enables a constant harvesting of biomass and purified water. The microalgae are fed with raw material collected locally from the Lynetten treatment plant (wastewater), and the Amagerværket power plant (CO_2). ***GRID Slide*** turns waste into an energy flow.

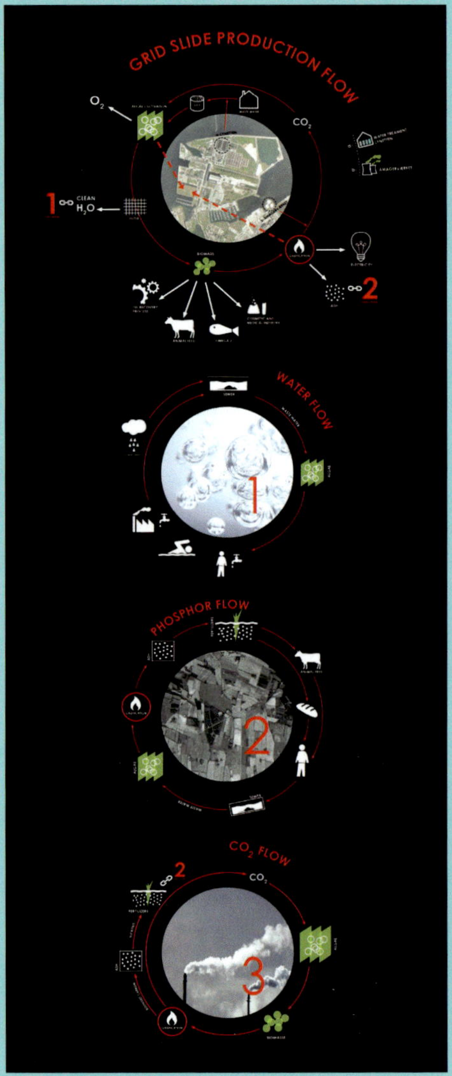

Closing the loop on energy and nutrient streams to create waste-free cycles.

1. Pumps
2. Gasifier
3. Algae towers
4. Vegetation
5. Water basin
6. Water-level dependent basins

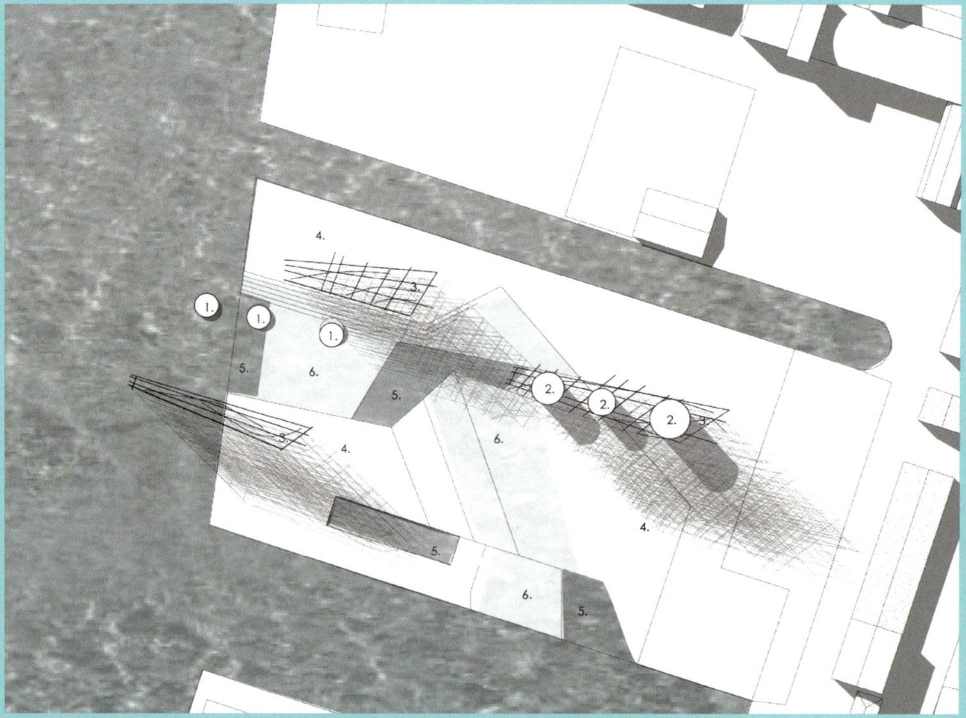

Plan diagram of *GRID Slide*

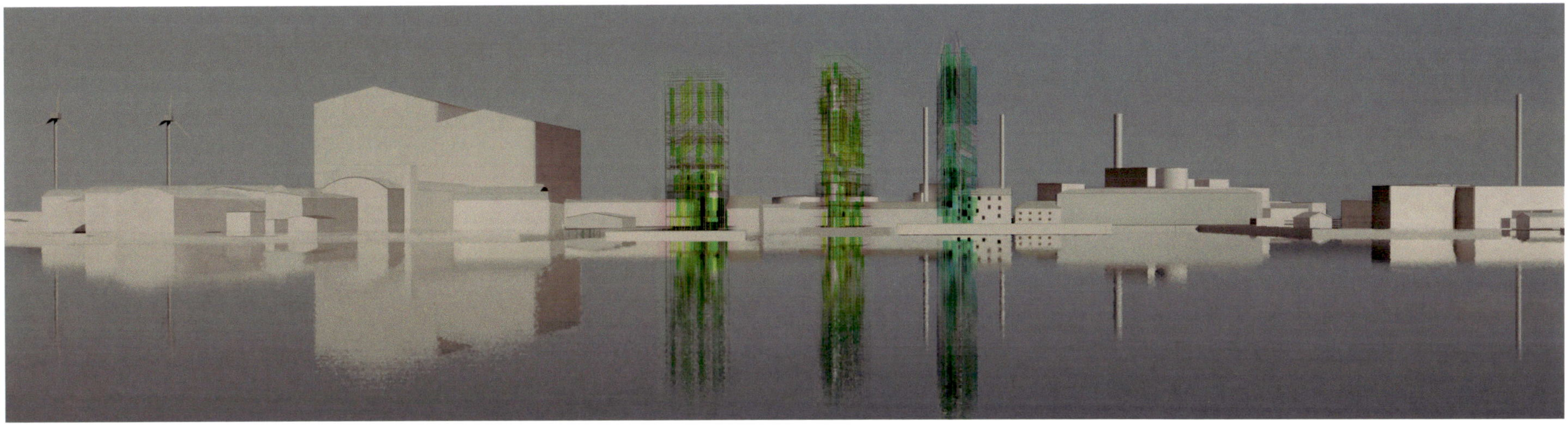

Three minimal corten steel structures rise like narrow oblong towers. The relationship between the towers and the industrial skyline behind the plot changes as visitors walk along the Copenhagen harbor quay.

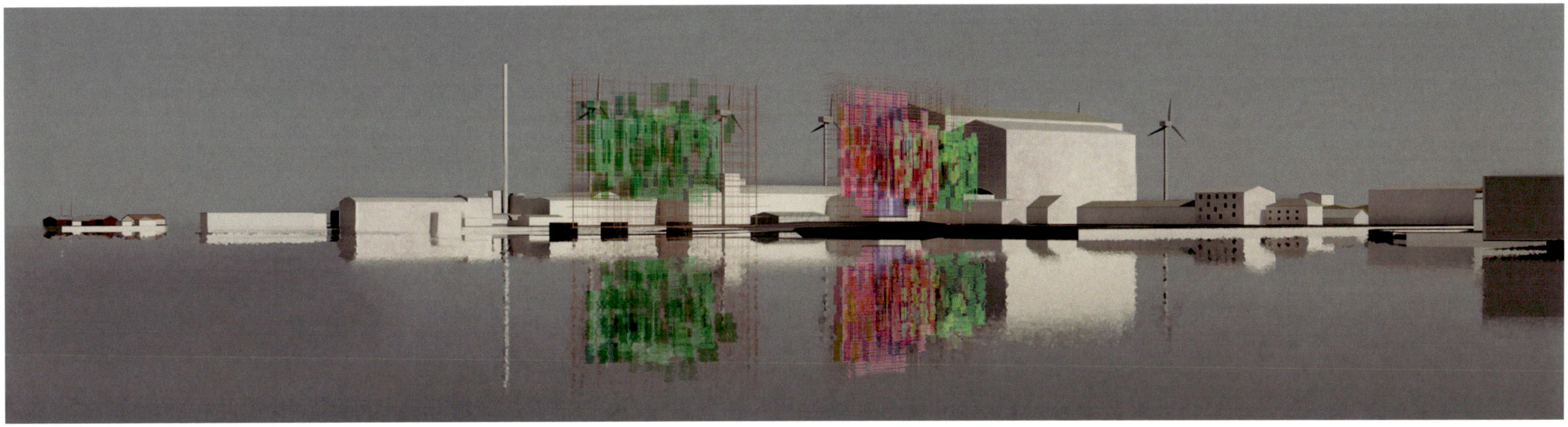

The panels turn during the day to expose the algae to light and enable photosynthesis, creating an ever-changing composition of translucent screens.

Echo of Wind:

Visualizing the Beauty of Wind Patterns

TEAM:
Mathias Bank Stigsen,
Olga Krukovskaya

TEAM LOCATION:
Copenhagen, Denmark; Moscow, Russia; Vienna, Austria

ENERGY TECHNOLOGIES:
elastomeric piezoelectric toroids

ANNUAL CAPACITY:
700 MWh

Few things are more present in Denmark than wind. It is both the driving force behind the country's many wind turbines as well as a force that has shaped its landscape. Wind by itself is invisible but, through interaction with its surroundings, its patterns and behavior can be captured, and its natural force can be harvested for electrical gain. The movement of leaves, branches, waves, and fabric are all ways in which we visually witness the wind. One of the most practical visualizations of wind movement is in sails, which have helped shape Denmark as a great seafaring nation throughout the course of its history.

Echo of Wind consists of a variety of static and dynamic springs distributed over the entire site. Whereas the static springs can host various forms of platforms, providing viewing spaces, the dynamic springs are moved by the wind. The bending of the lightweight carbon fiber springs transforms the mechanical forces of bending and motion into electrical current through piezoelectric generators. Textiles connect the dynamic springs, helping to both catch and bend the springs while creating an echo of the wind pattern. Visitors are invited to interact with this dance between wind, fabric, and electricity.

Echo of Wind is highly scalable and flexible, and can be adapted to various sizes and formats. It is laid out in a grid, but within the grid is the possibility of varied arrangements for diverse purposes. This flexibility is achieved through a shared plateau in concrete, which shelters the electrical and mechanical devices. More than a purely energy-producing intervention, *Echo of Wind* is a public space that enables a dialogue between wind and visitors, nature and humans.

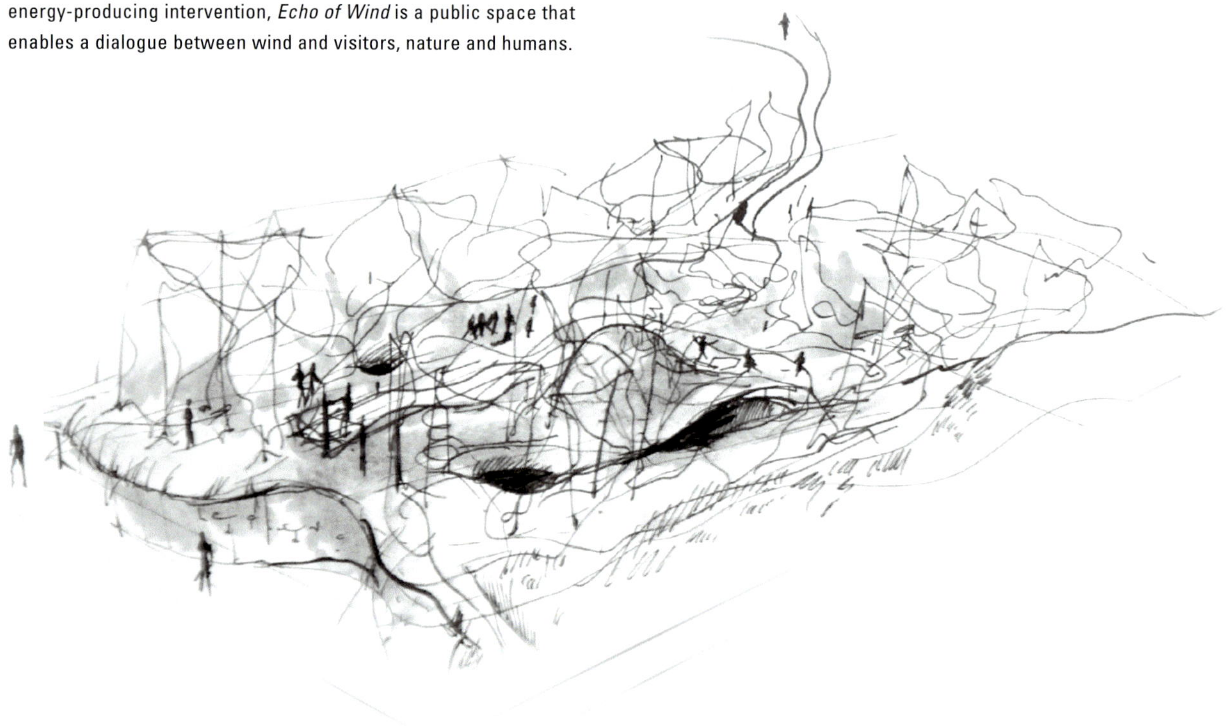

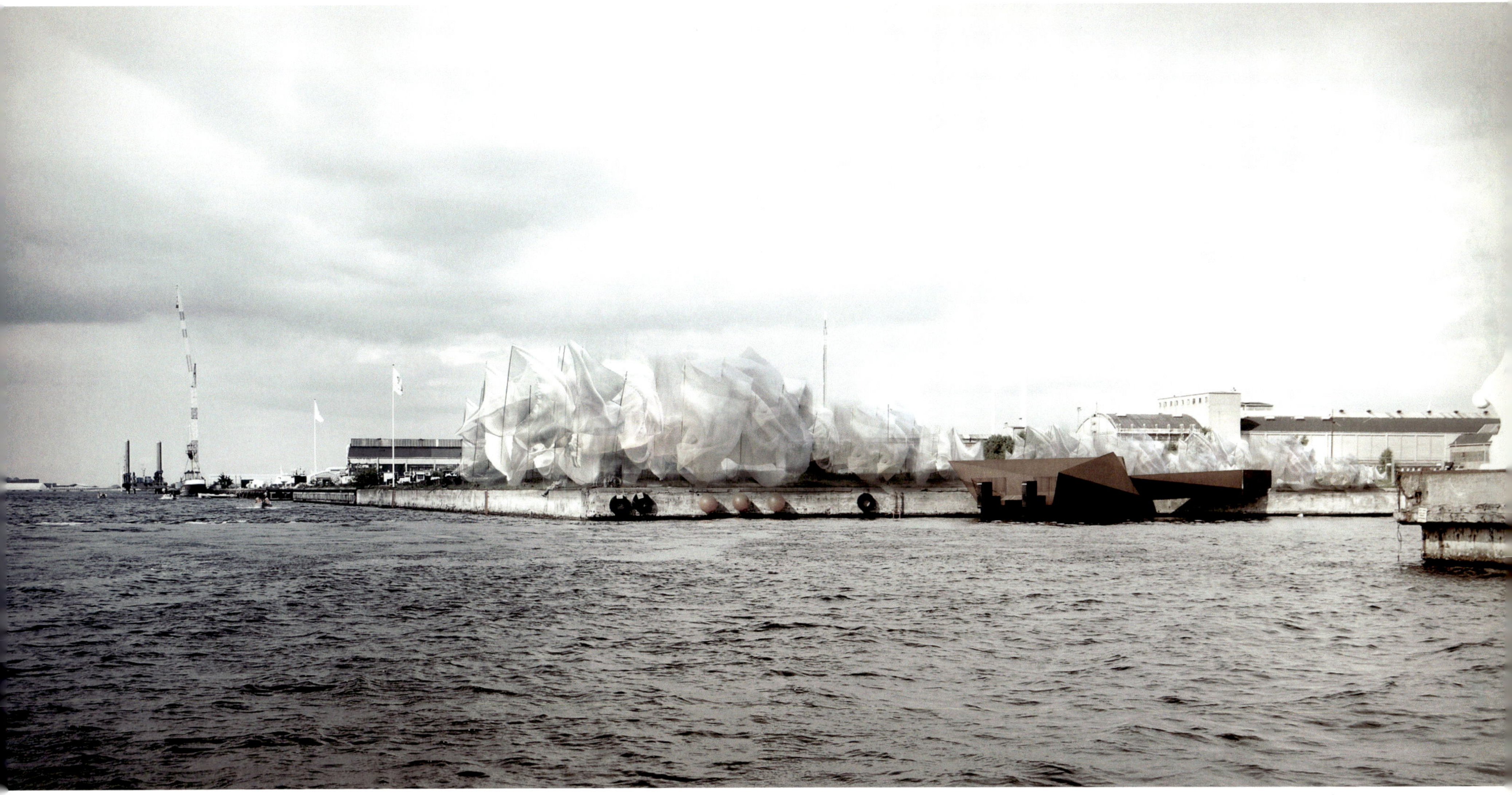

View from Copenhagen harbor bus

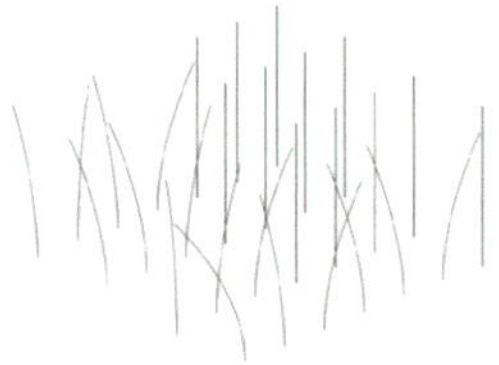

Static vs. dynamic springs

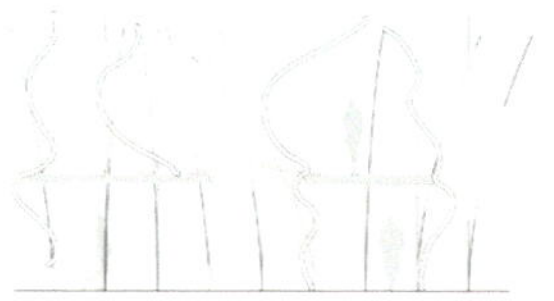

Platforms within dynamic springs

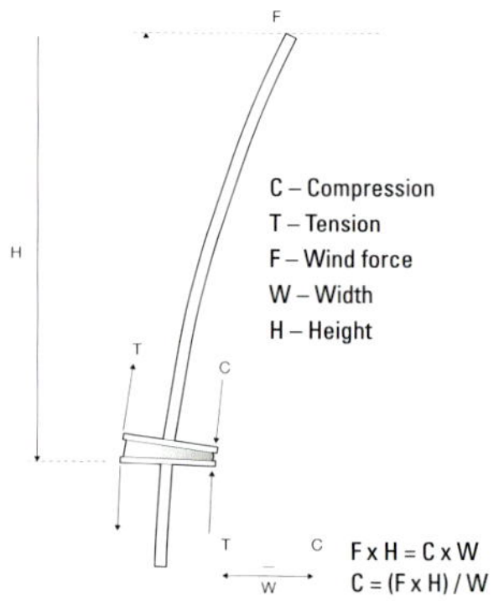

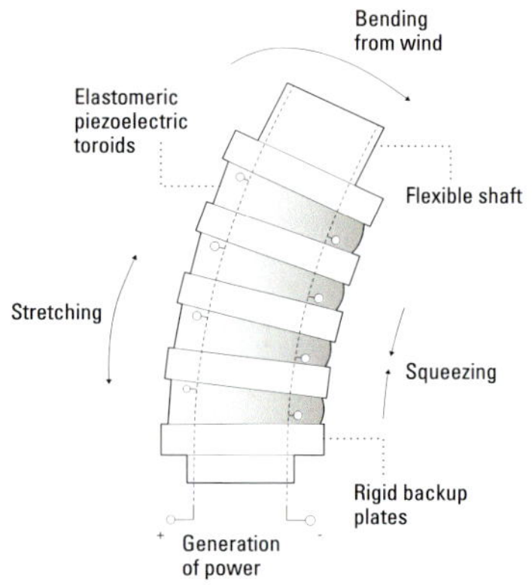

Energy concept for dynamic springs

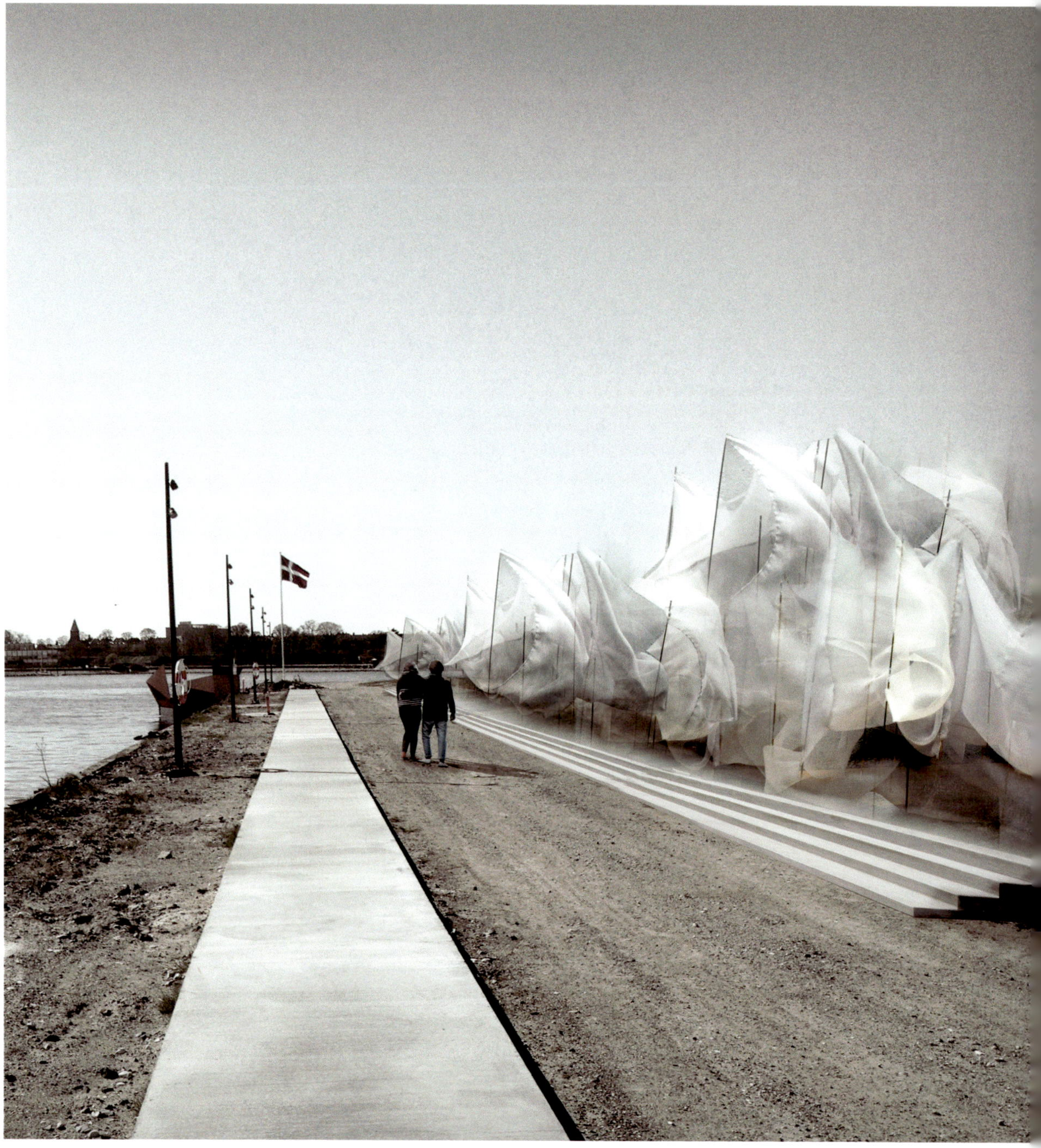

Echo of Wind embodies the transition from static to dynamic.

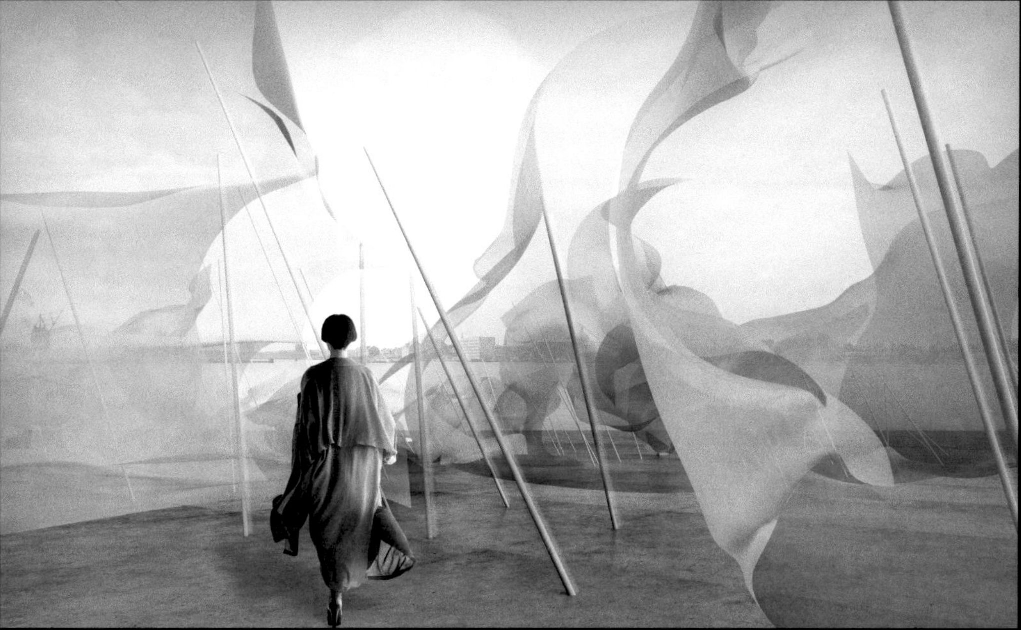

Textile has the ability to capture and visualize movement. It has a memory in its behavior that both reflects its surroundings and its constraints by which it is composed. This duality in the behavior of fabric has been an aesthetic fascination for artists and architects throughout centuries. An architectural proposal based on this aesthetics has a supple sensibility and therefore a possibility to integrate and capture literal and phenomenal motion in relation to its surroundings, its program, and visitors.

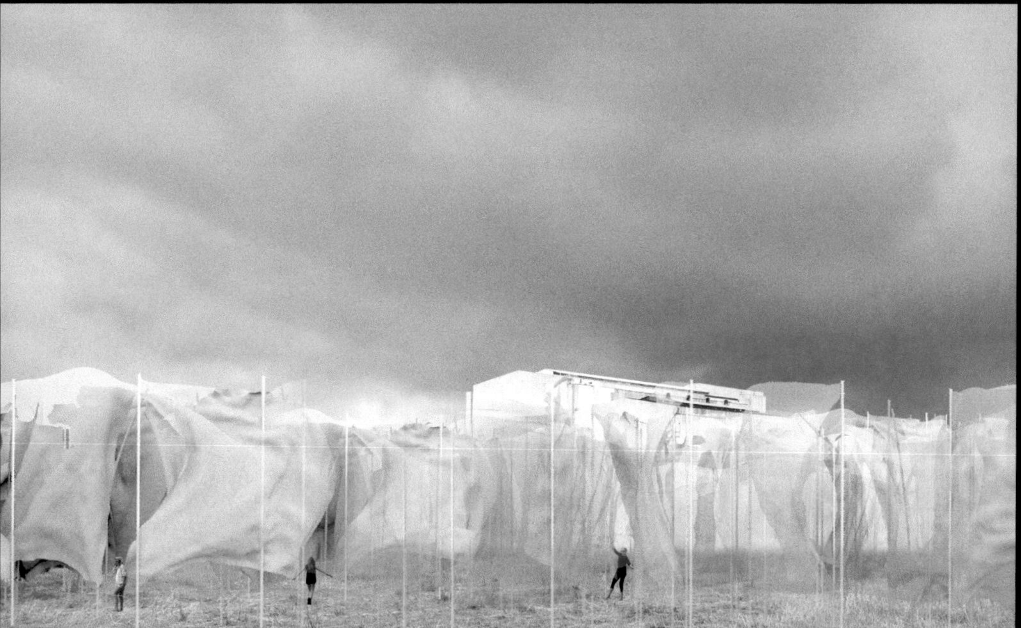

A field of springs moving in the wind

Onshore Power

TEAM:
Han Bao, Feng Xu

TEAM LOCATION:
Melbourne, Australia

ENERGY TECHNOLOGIES:
wind-driven hydraulic cylinder generators

ANNUAL CAPACITY:
3,500 MWh

It is the finest design that can change our lives, and it is the smallest changes that lead us to challenge our everyday perspectives. *Onshore Power* is situated at the interface between energy-producing infrastructures and human use. It achieves this through an integration of the ecological flows of wind, water, energy, and mobility.

A simple physics experiment can demonstrate that two strips of paper will stick together when wind blows between them due to the change in air pressure. Using this physical property, the hundreds of extremely lightweight plates that comprise *Onshore Power* move closer to each other and spring back to their original places as high winds pass through. The movement of plates transferred to the vertical shafts generates electricity.

Onshore Power is at once a generator, a recharger for boats, a dock, and a destination. It is made up of layers of flexible plates aligned to the course of the wind, with walkways and lookout points traced through its interior. As wind blows across the structures, the movement generated in the plates creates energy that is converted and stored in the microtubes that scaffold the structure. The experience within *Onshore Power* is a theatrical and powerful one, where the normally invisible strength of wind is acted out through the oscillations of the plates. The entire process of energy production, conduction, and consumption that people have come to take for granted is suddenly made visible.

Onshore Power creates possibilities for the foreseeable future, and it raises the questions: who are we, and what are we connecting with?

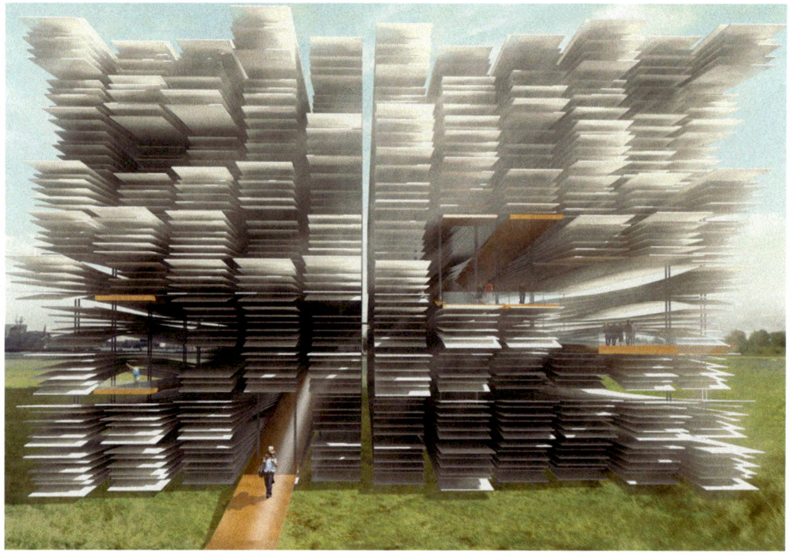

View from the vantage point of the wind as it approaches the installation

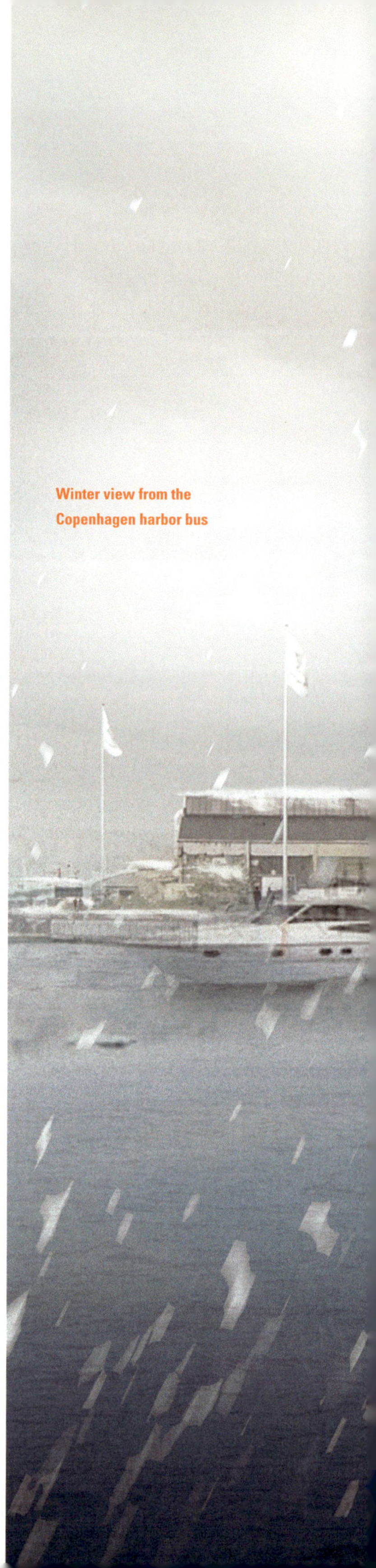

Winter view from the Copenhagen harbor bus

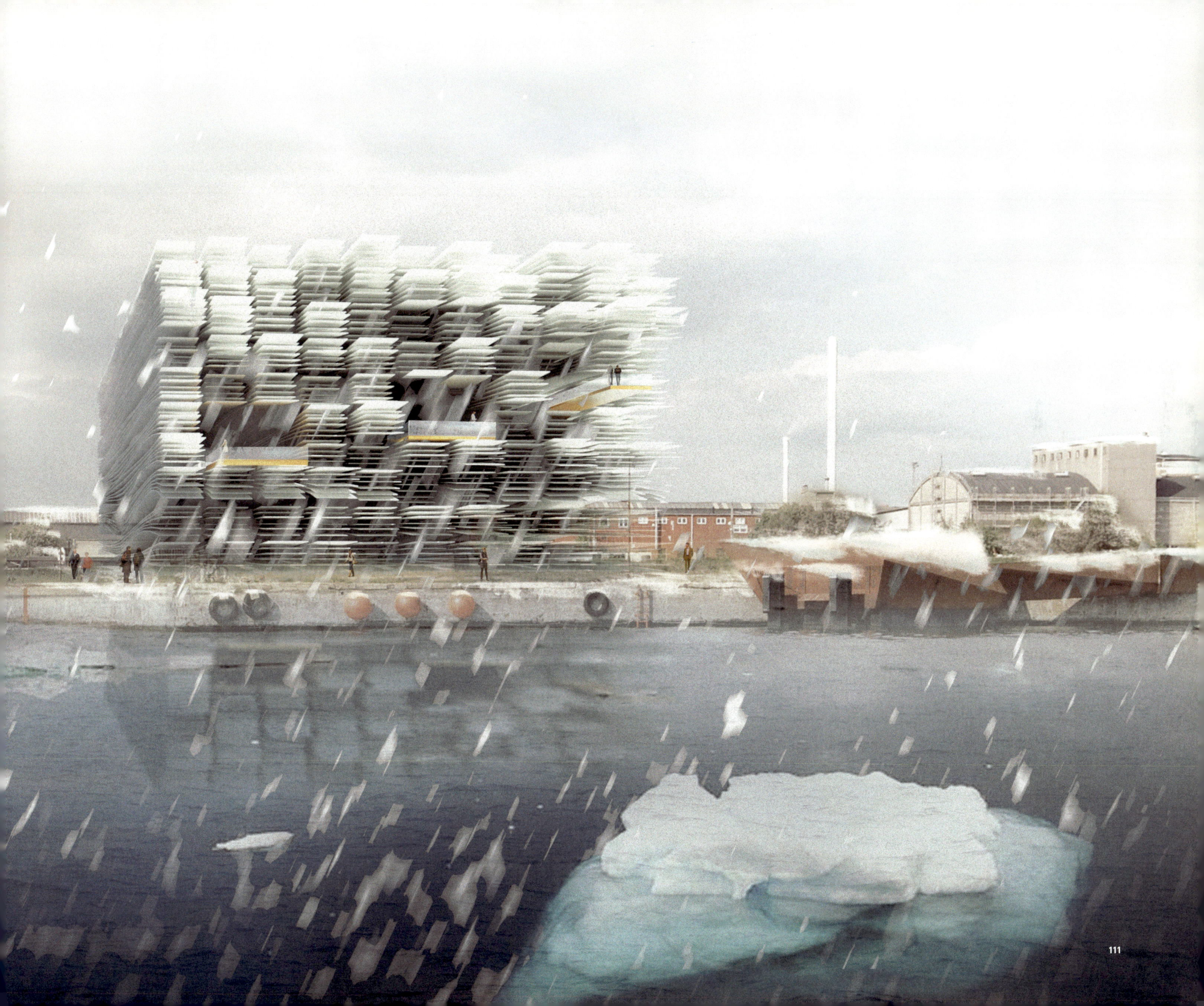

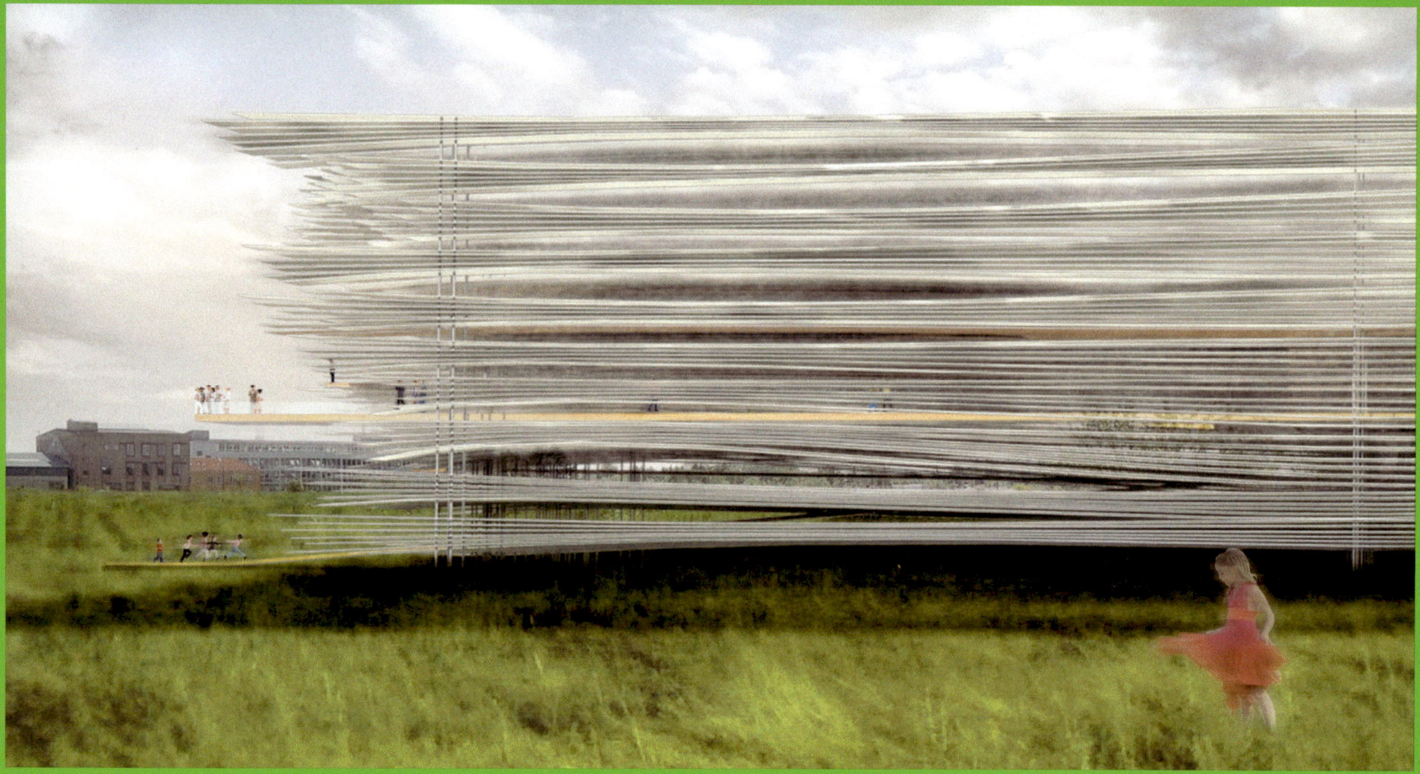

As the thin plates oscillate in the wind, the up and down movement drives hydraulic generators.

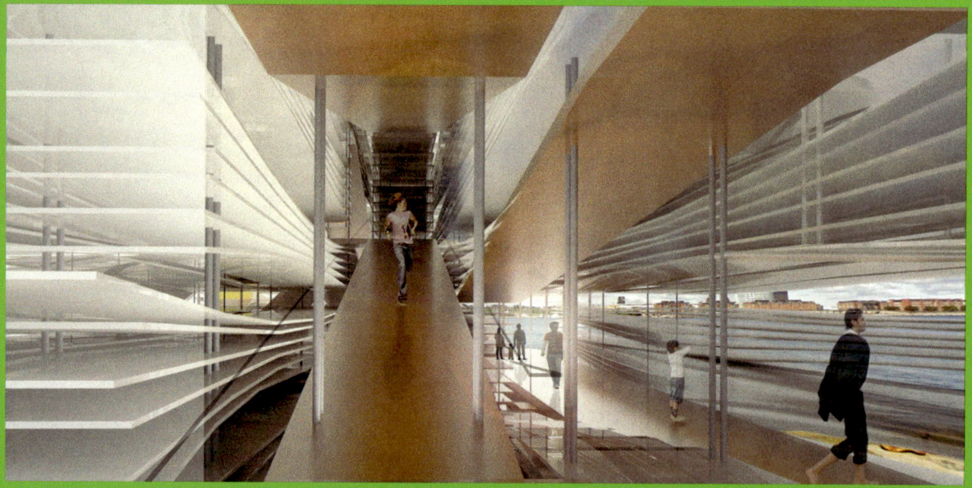

Visitors can wander into and up through the structure, stopping at various viewpoints along the way to enjoy striking views of the water and the city.

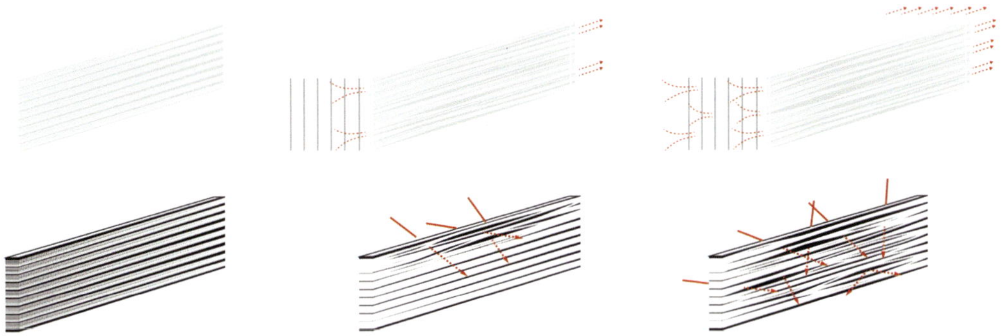

Slight shifts in the wind direction
will help to reposition the plates.

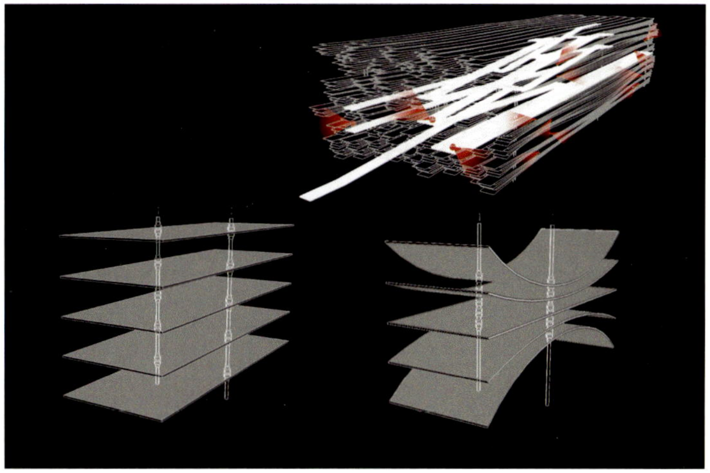

Visitor access ramps weave in
between the vertical columns that
conceal the moving pistons.

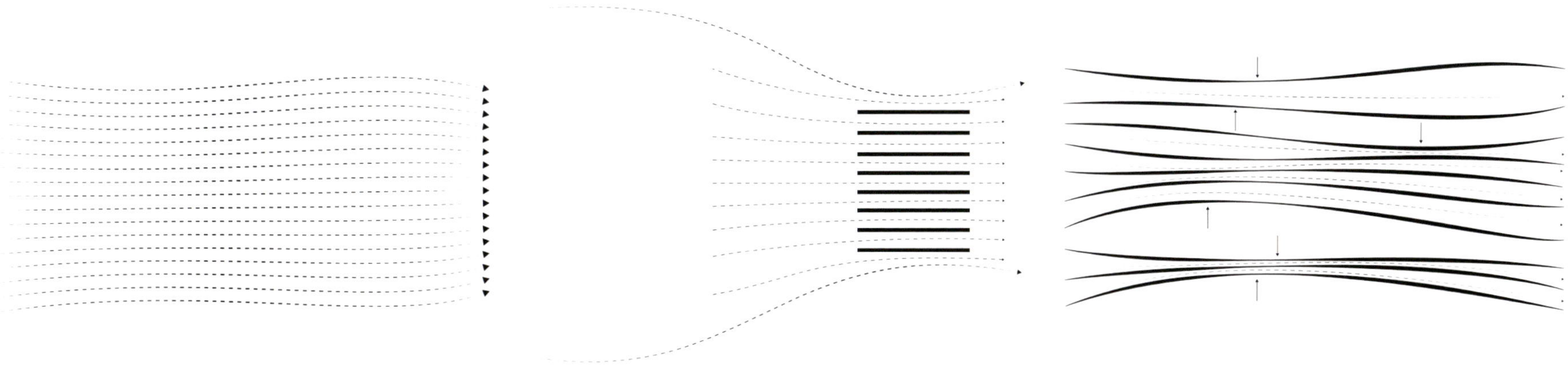

Wind creates a deformation on
the plates, which are equally spaced
when at rest.

Windshape

TEAM:
Manon Robert, Martin Le Carboulec,
Marc Antoine Galup

TEAM LOCATION:
Seoul, South Korea

ENERGY TECHNOLOGIES:
piezoelectric fabric, rotating electromagnetic generators

ANNUAL CAPACITY:
750 MWh

Wind speed is transformed into energy with *Windshape*. The 302 poles, each supporting several piezoelectric textile sails, reshape and transform the site according to the natural wind direction.

As the wind can come from any direction and change quickly, the site will never look the same from one visit to the next. To engage the public in the process, the dimensions of each sail panel are imagined at human scale. Visitors will enjoy getting lost in the changing spaces and observing others as they pass between the sails.

Some sails are intentionally fixed in order to create a passage through the site to the entrance and water taxi terminal. The others are completely free, but can be temporarily fixed by users who would like to modulate the space. These outdoor rooms created by the sail rotations can be both intimate and public.

Several factors inform the design of the sail. The main inspiration was a study of windsurfing sails and their fabrication. A succession of different types of layers on the surface of the sails (aramid fiber scrim and mylar protection) provides tensile strength and protects the piezoelectric fabric layer.

The fabric provides an estimated output power density in the range of 1.10–5.10$\mu W/cm^2$ at applied wind pressures in the range of 0.02–0.10 MPa. In addition, the dynamo system at the base converts rotational energy with an estimated peak output power of 3.5 W. LED lights located in the poles softly illuminate the area at night, making it visible from the opposite shore as a reminder of the power of Copenhagen wind.

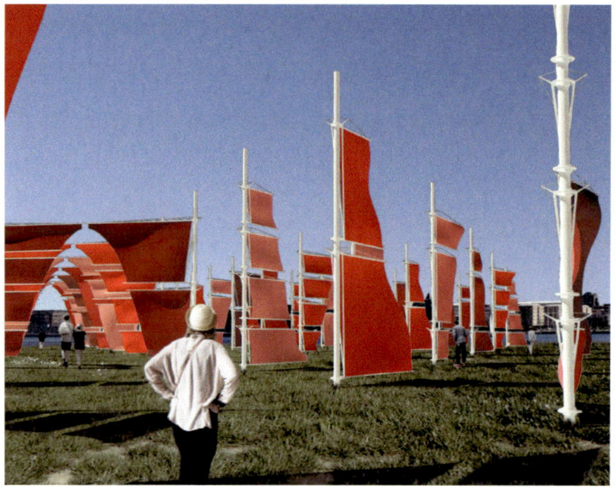

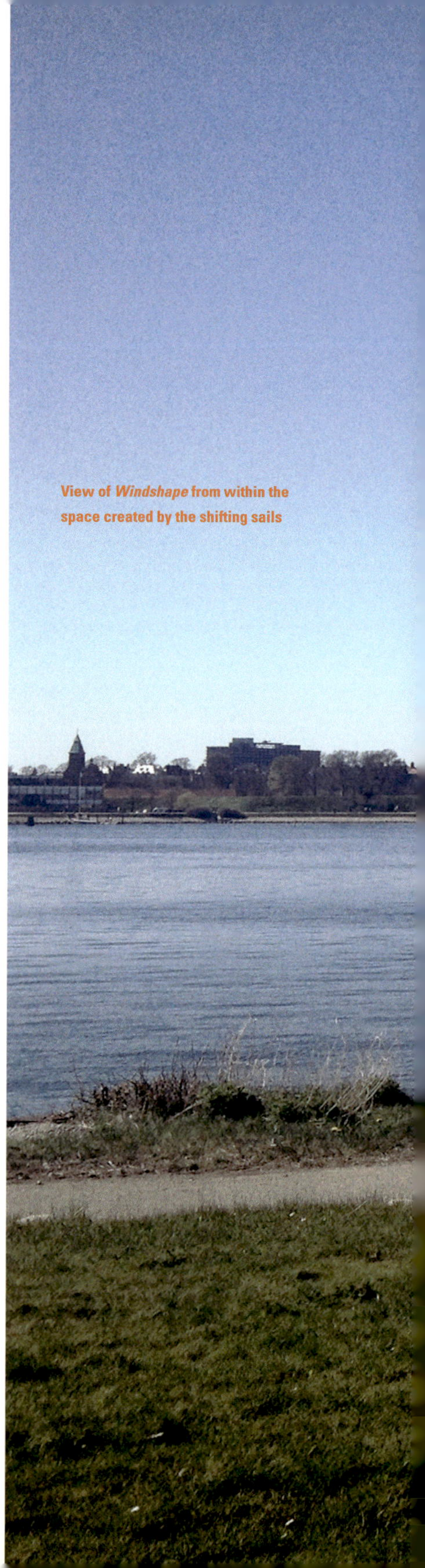

View of *Windshape* from within the space created by the shifting sails

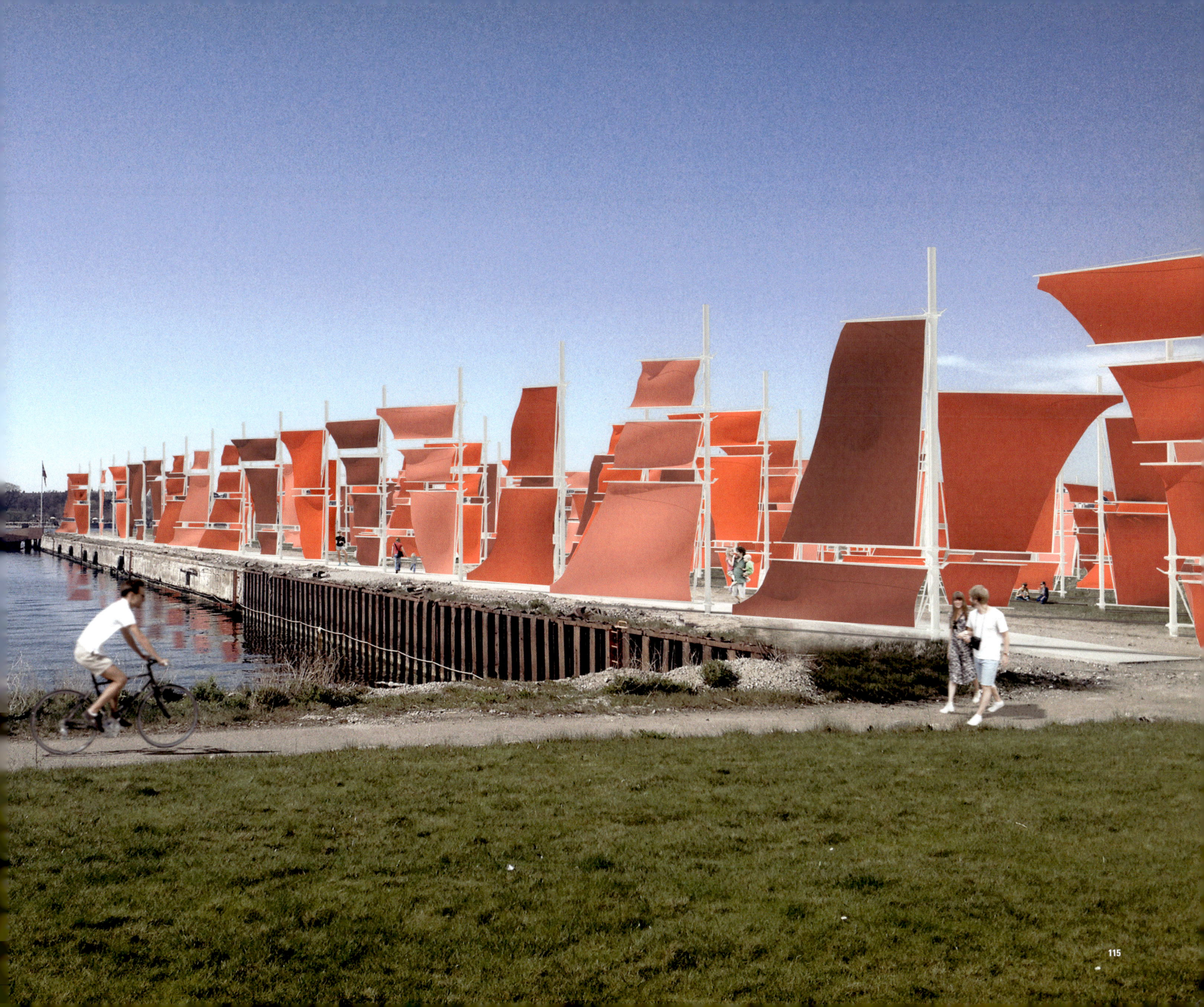

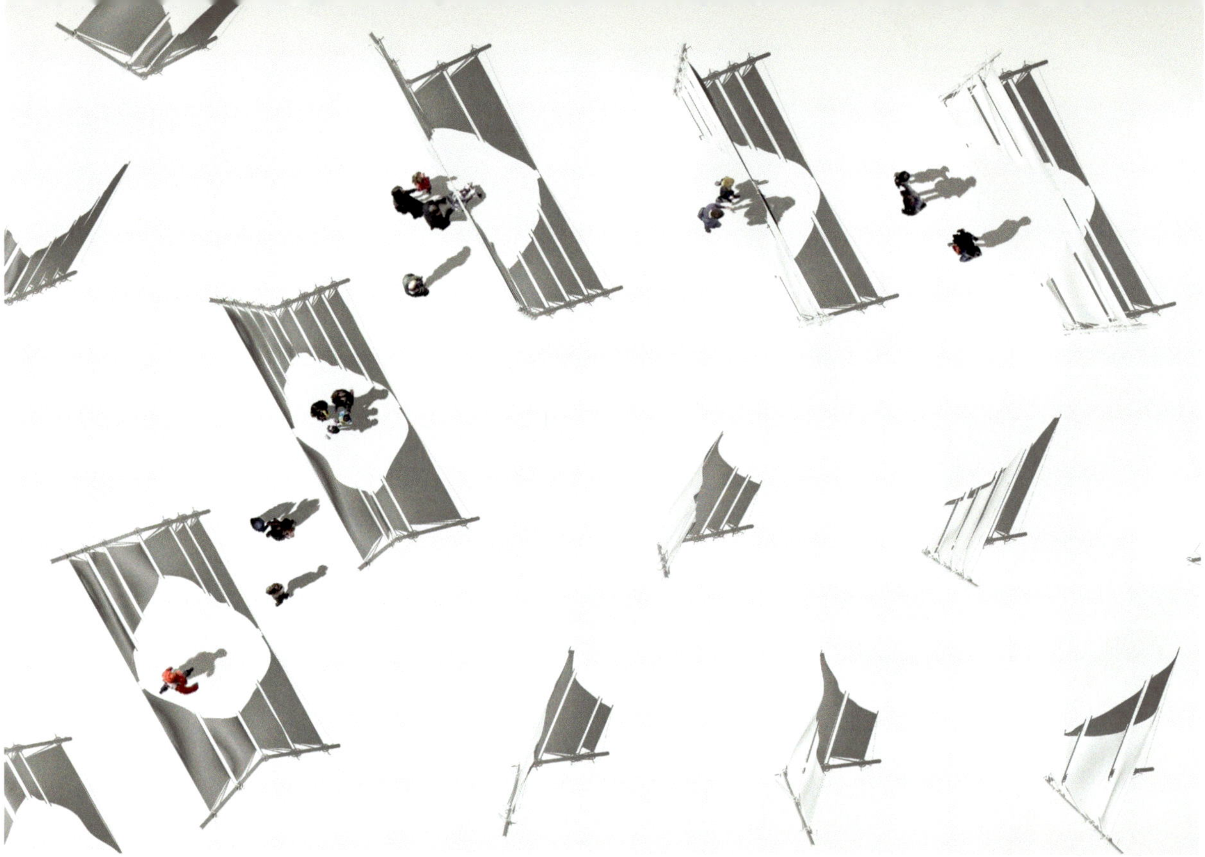

As the sails shift position, the changing spaces between them are a playful experience for all ages. These ever-changing outdoor rooms can be used for many different purposes.

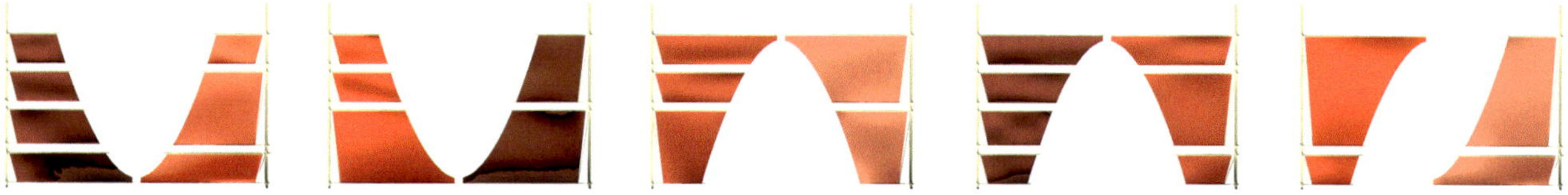

Multiple variations of the sail pattern

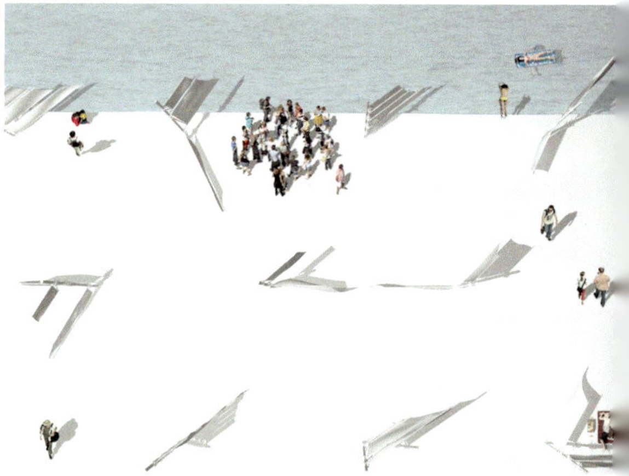

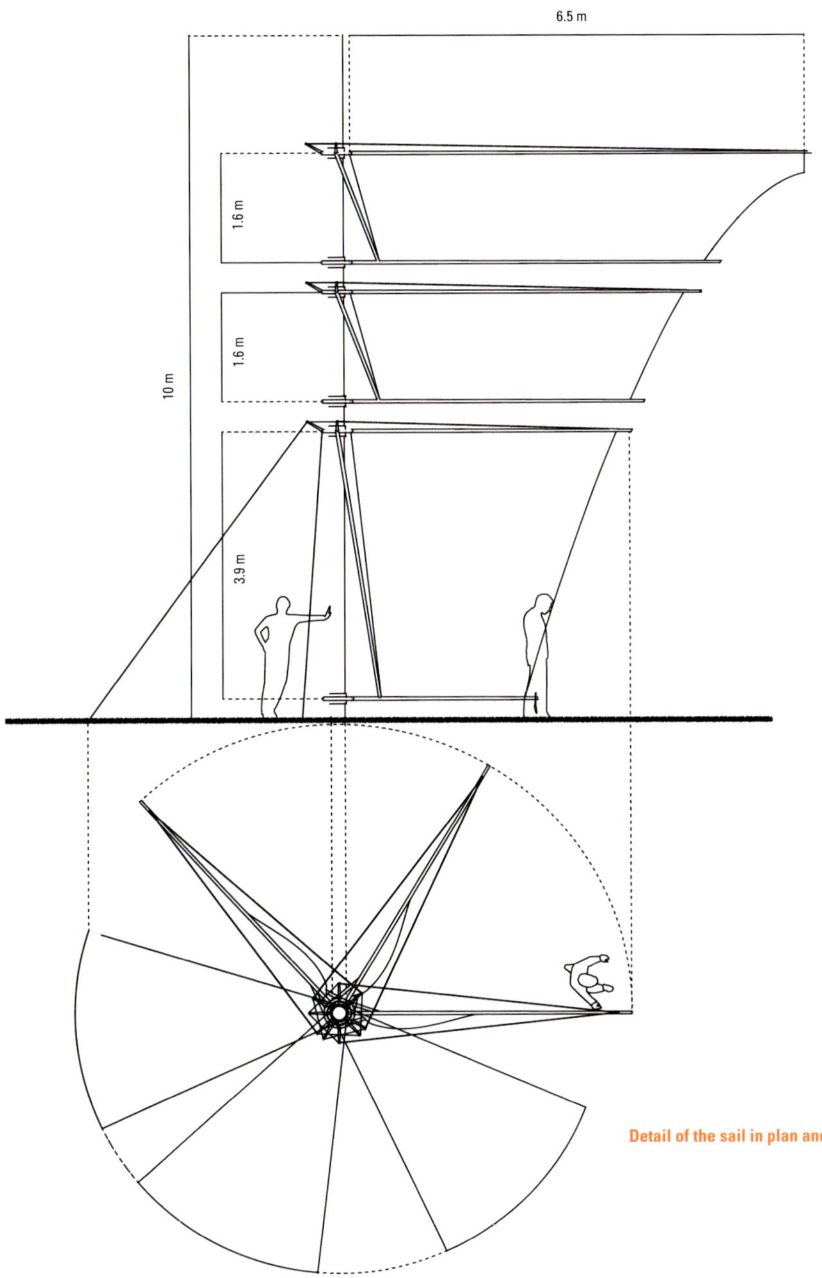

Detail of the sail in plan and elevation

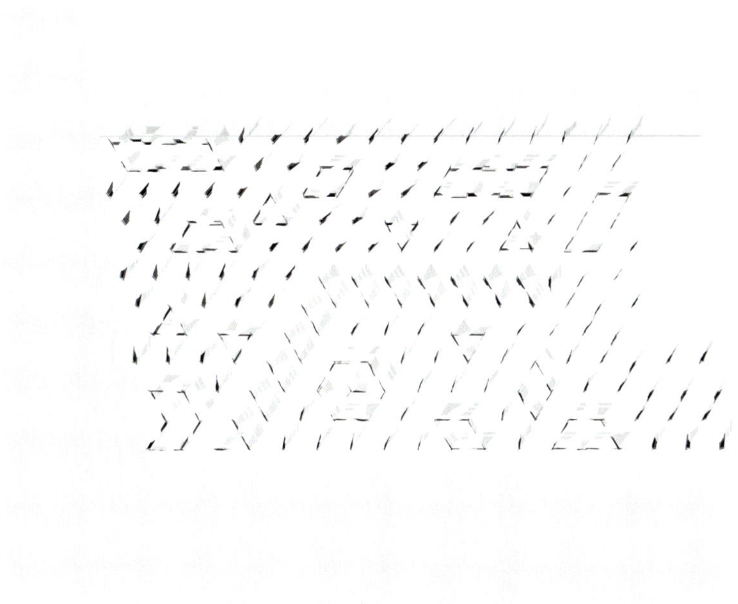

Site plan

The Ephemeral Machine

The Ephemeral Machine combines an image of the historic steam engines produced by Burmeister & Wain (characterized by massive volumes with pistons moving vertically) with the contemporary vertical elements of the site (chimneys, cranes, silos, and antennas), resulting in an evocative energy installation and landscape artwork.

The movement, and energy produced, is the substantial element of this installation, while the form and the materials of its components are nothing more than ephemeral temporary conditions. This 120-meter high installation occupies the sea portion in front of the former shipyard and is composed of more than 550 vertical elements. Each element is divided into an underwater structure affixed to the sea floor, a floating buoy-like raft, and a transparent vertical helium balloon connected by a steel wire to the raft. The entire system is easily set in motion by the waves and the action of the wind on the balloons. Rafts on the surface of the water are accessible to visitors, whose actions contribute another element of motion and energy production.

The Ephemeral Machine utilizes existing technologies: buoy-type wave converters and semitransparent photovoltaic film. The balloons and the rafts are conceived to counterbalance each other in order to generate soft and gentle movements.

From a landscape point of view, the artwork creates a dense volume—suggesting the shape of engines but also the voluminous encumbrance of present-day cargoes and ocean liners. However, the form dematerializes due to the lightness and transparency of the balloons, which are the only visible elements from urban-scale distances.

TEAM:
Alfio Faro, Denise Leclerc

TEAM LOCATION:
Paris, France

ENERGY TECHNOLOGIES:
thin-film photovoltaic (balloons), APB 350 (Autonomous Power Buoy® by OPT)

ANNUAL CAPACITY:
1,000 MWh

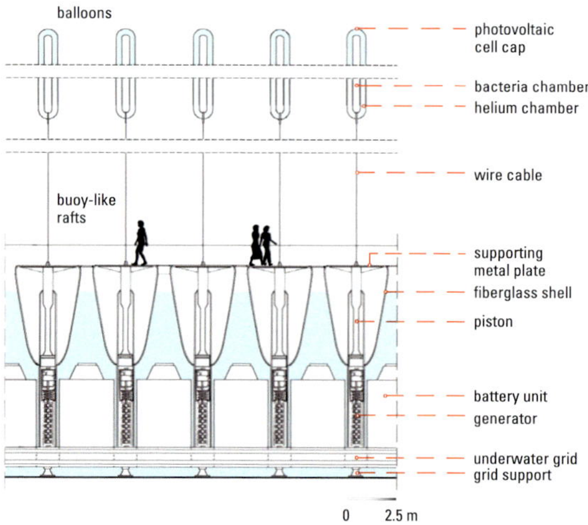

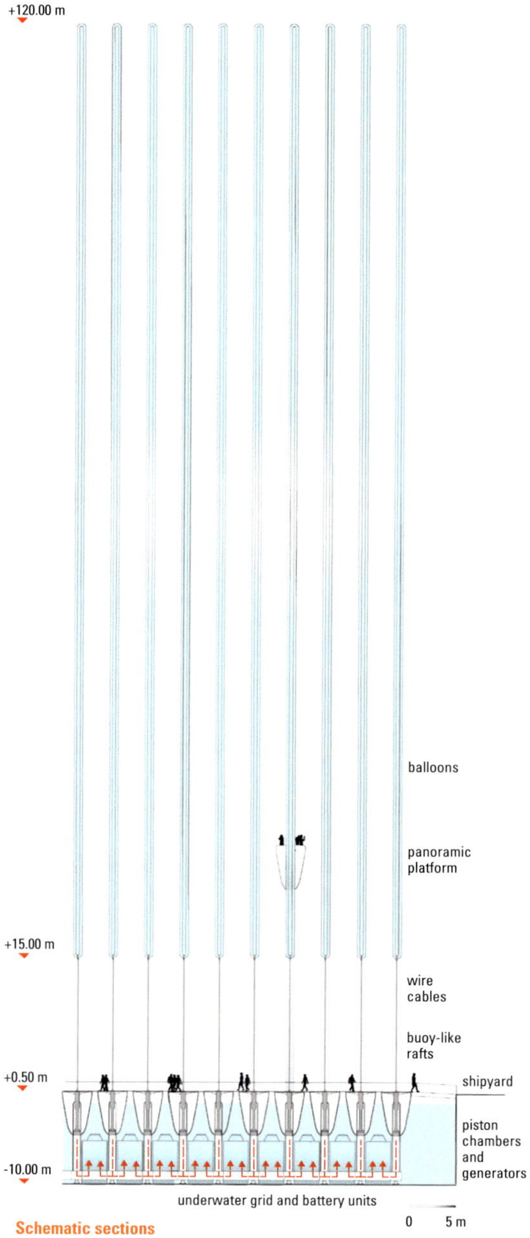

Schematic sections

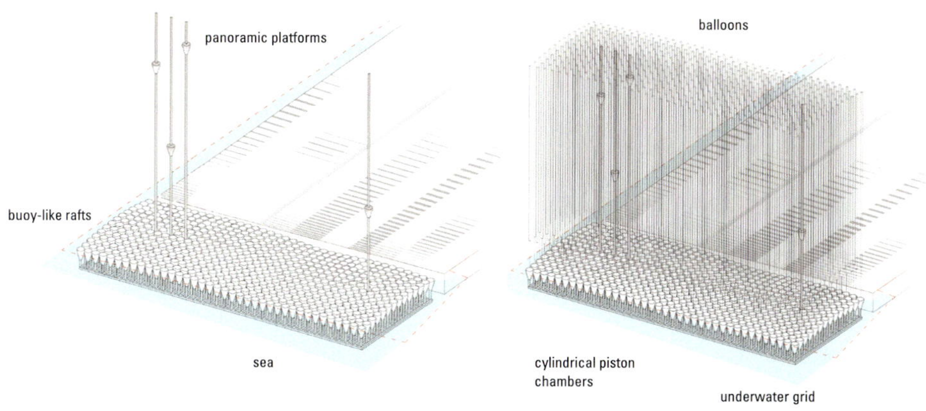

Winter view from the harbor

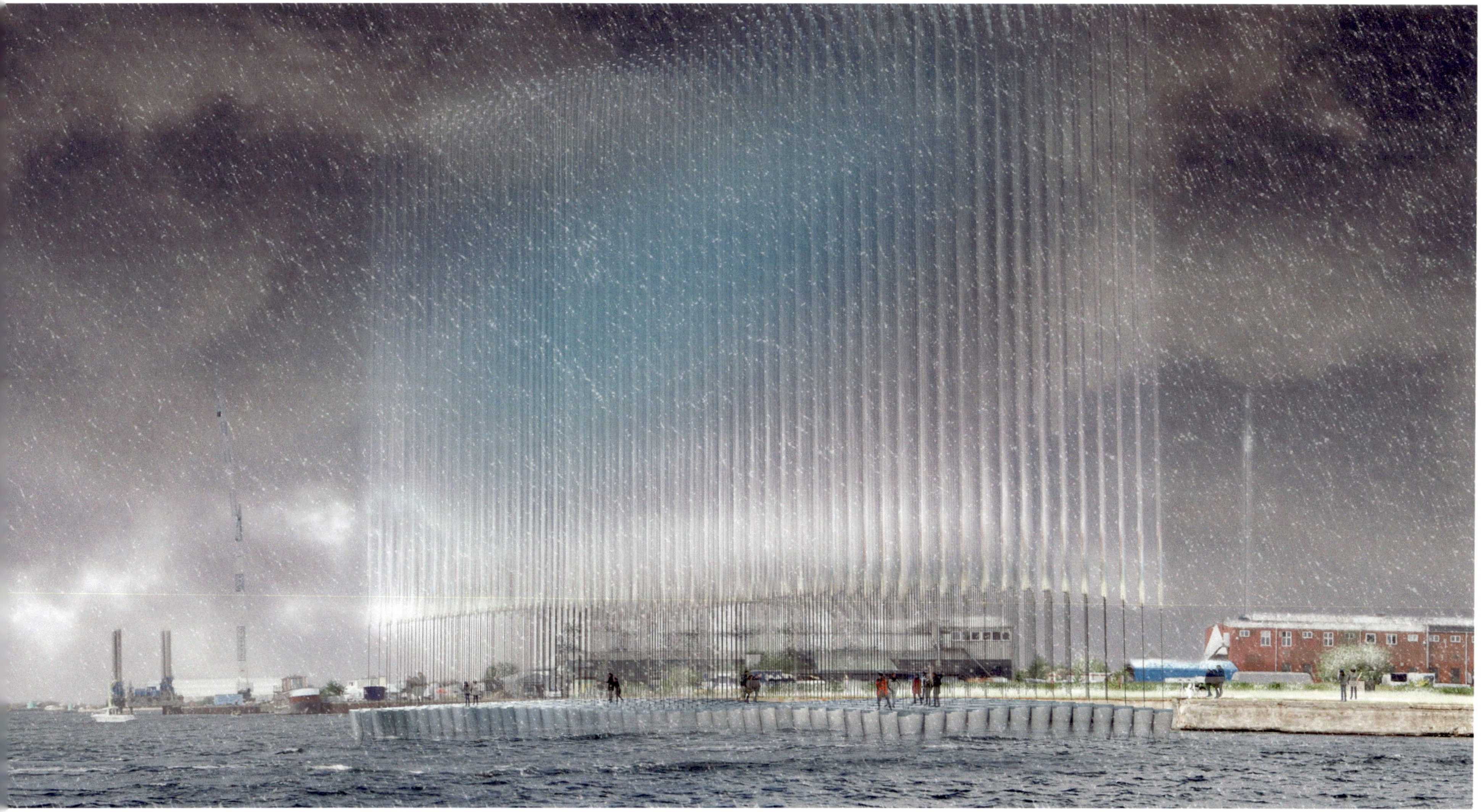

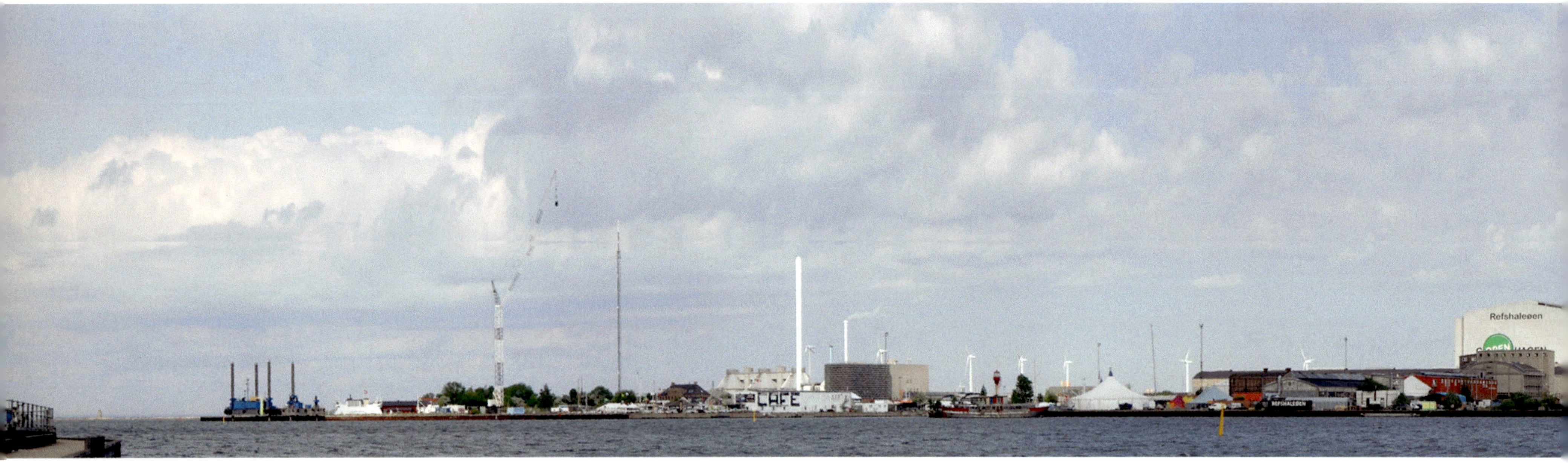

During the day, the semi-transparent thin-film solar balloons seem to almost disappear in the sky.

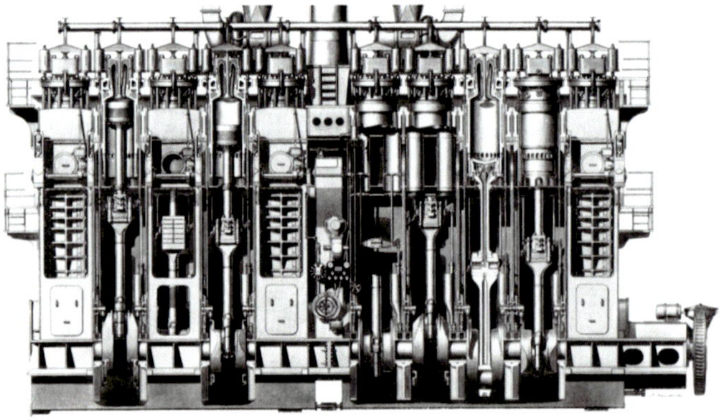

Burmeister & Wain steam engine provides the formal inspiration

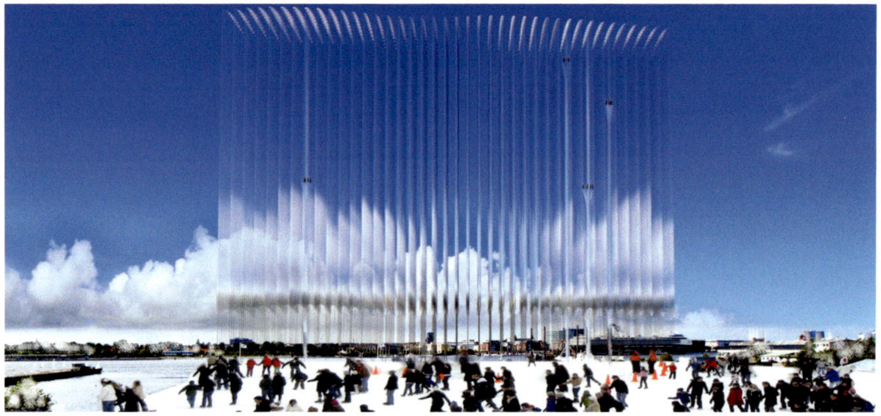

View from Sønder Hoved

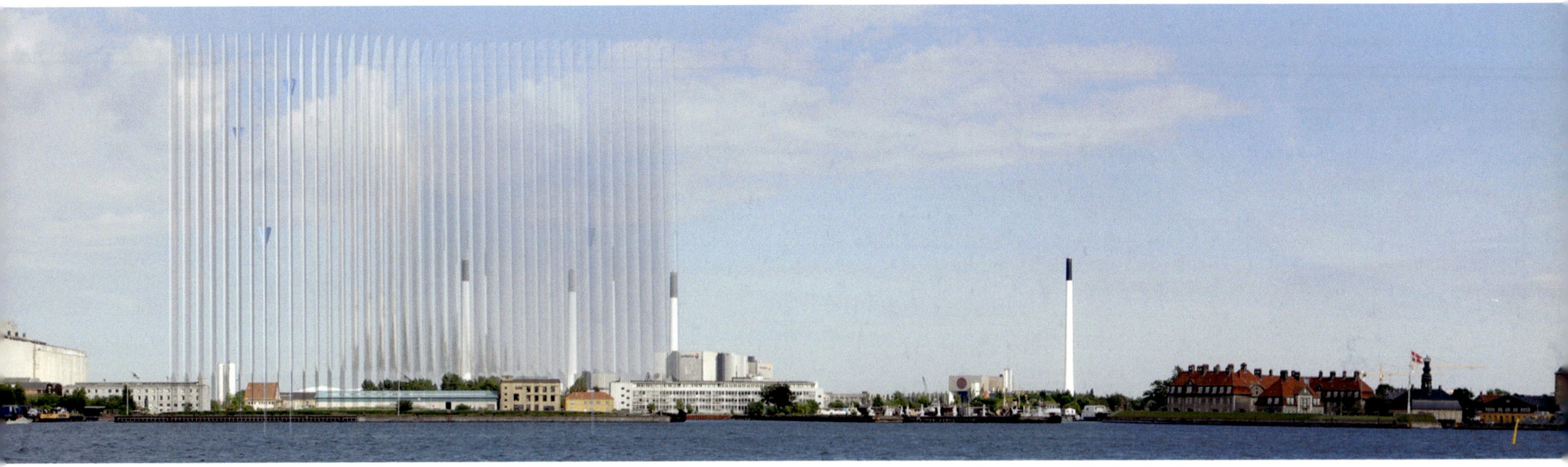

The Ephemeral Machine

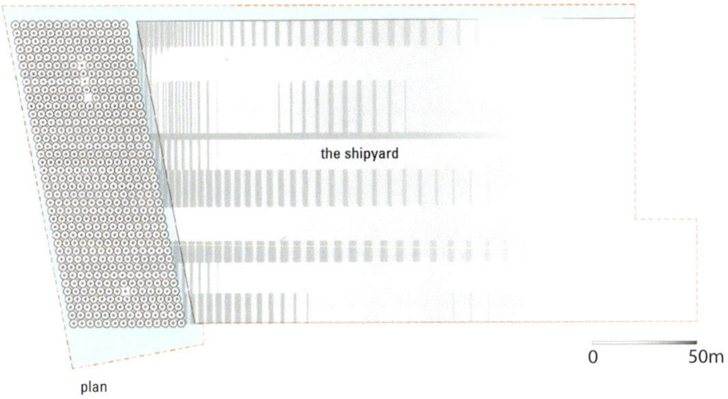

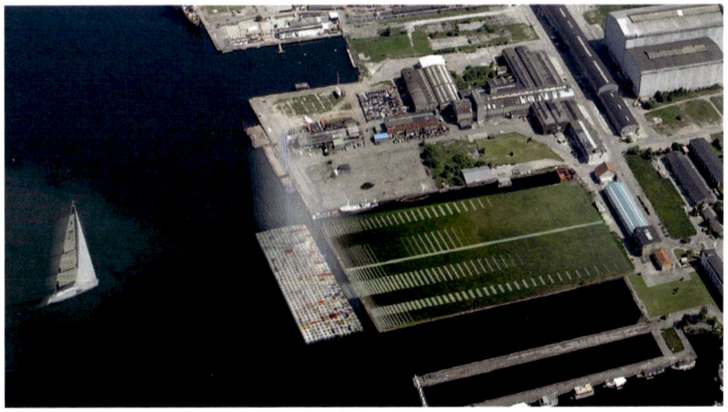

The machine occupies the harbor in front of the old shipyard.

SunBath

TEAM:
Jacob Boswell, David Shimmel, Ian Mackay

TEAM LOCATION:
Columbus (OH), USA

ENERGY TECHNOLOGIES:
solar pond, thermoelectric generators

ANNUAL CAPACITY:
750 MWh

SunBath offers a series of large, public, indoor, and outdoor baths, both hot and cold, encircling and actively participating in a working power plant. The site thus becomes a type of liminal space, blurring the distinction between land and water while also challenging the false dichotomy of a productive vs. a pleasure landscape.

Energy is generated at *SunBath* through the interactions between five key elements:

The Solar Pond

A 14,700 m^2 salt gradient solar pond is covered by a greenhouse to prevent disturbance of the water and to aid in the maintenance of its internal temperature.

The Catalyzing Pool

In a typical solar pond installation, heat from the pond is used as part of a district heating operation or passed through a standard heat engine to produce electricity. At *SunBath*, heat from the solar pond is transferred to a catalyzing pool via a series of heat pipes.

The Minto Wheel

Cold water from the harbor is pumped into the catalyzing pool via the Minto wheel and used to continuously cool the condensing end of the heat pipes. It is at this threshold between the hot pipes and the cold harbor water that electricity is generated. This is accomplished via a series of Seebeck Thermo Electric Generators (TEGs).

The Attenuating Pool

Partially warmed water flows from the catalyzing pool to the attenuating pool, the heated water of which powers the Minto wheel. This cycle creates a self-regulating system.

The Harbor

The harbor and the consistently cold water fed to it by the Øresund are the true source of power on the site. The mechanism for producing electricity at *SunBath* is not purely solar heat, but rather the difference in temperature between the warm side of the thermoelectric generator (the solar pond) and the cold side (the harbor water). The colder the harbor is, the more electricity the *SunBath* generates.

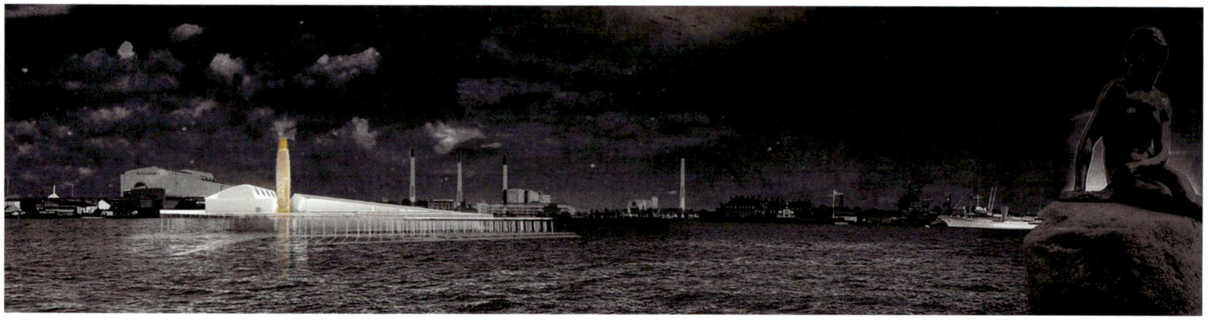

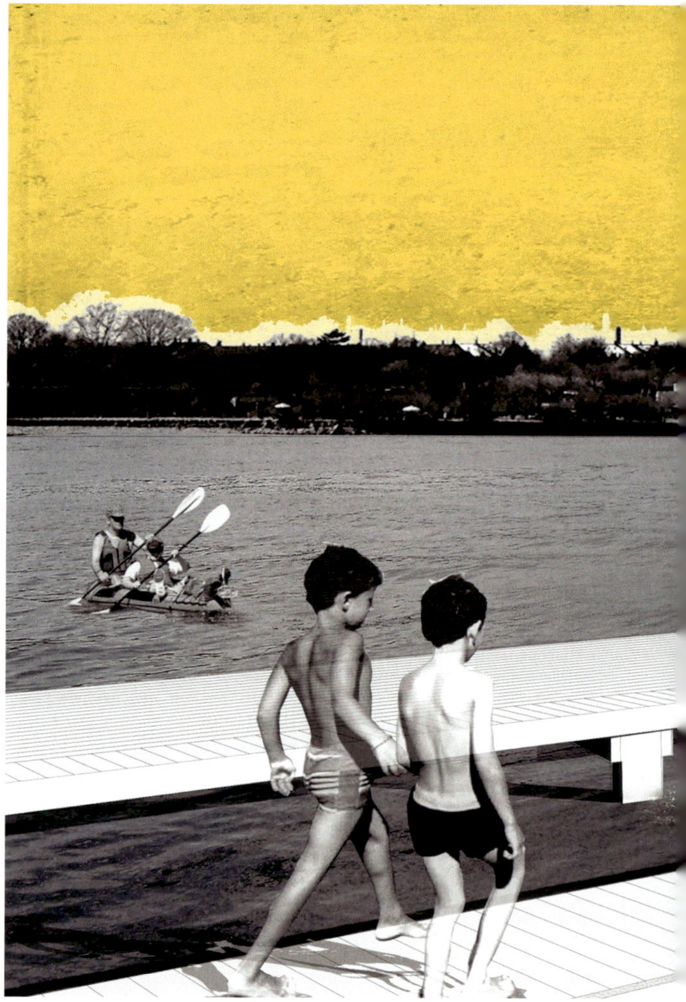

View toward the City Center from the Harbor Bath. The Harbor Bath is articulated in order to provide platforms and spaces for a variety of swimmers and swimming activities. Spaces are scaled for both children and adults and designed to host everyone from casual bathers to professional swimmers.

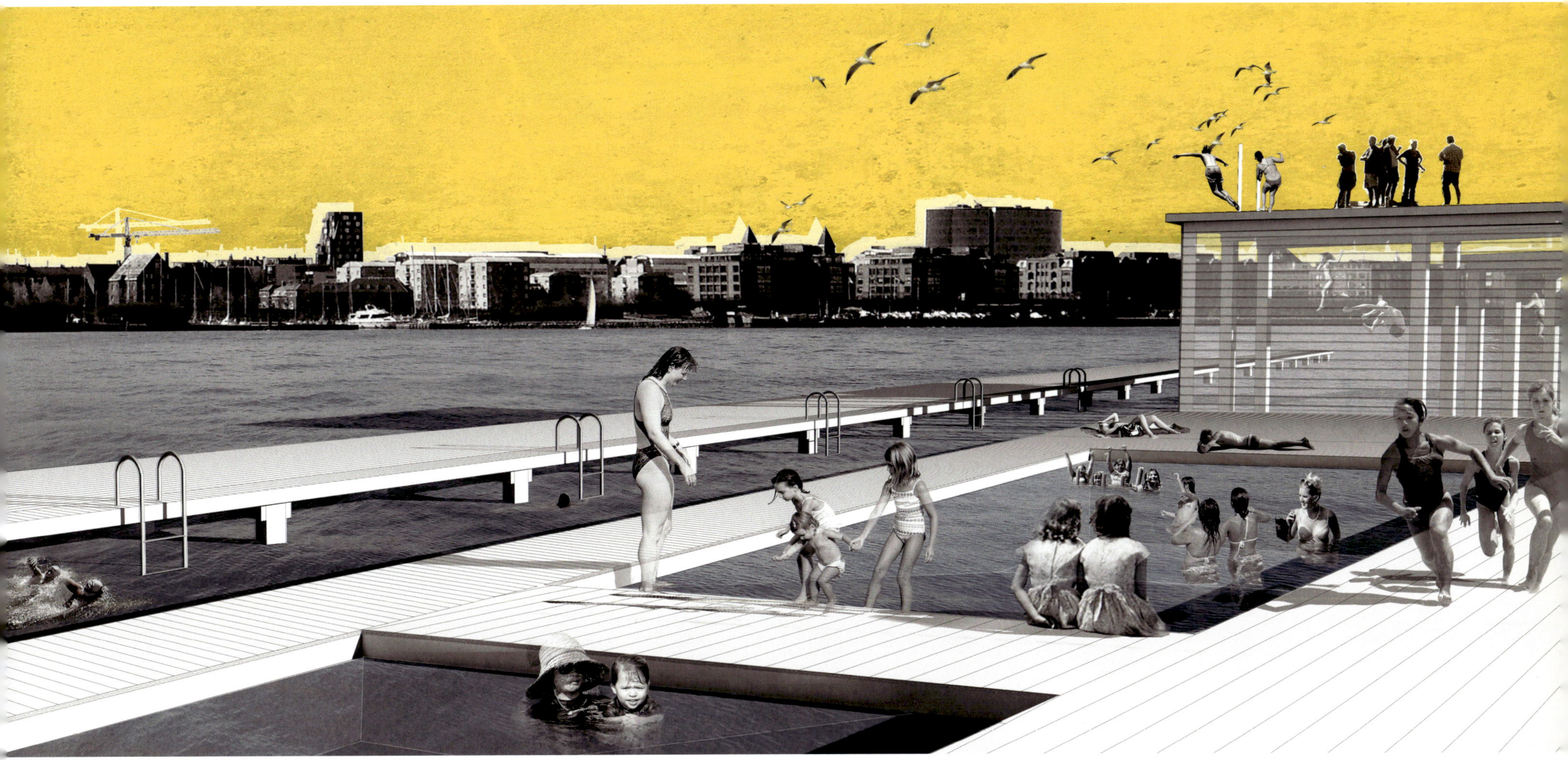

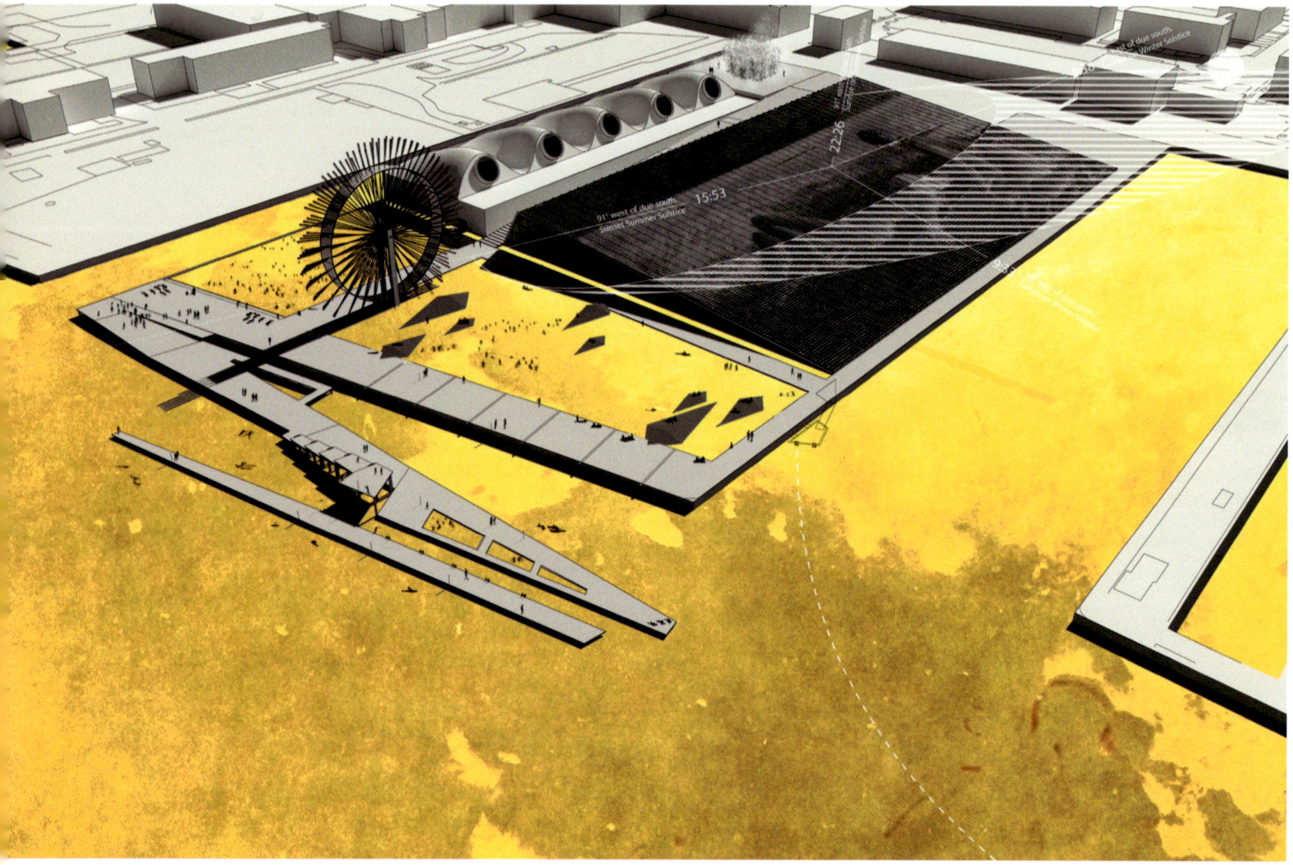

Aerial view of *SunBath*. The solar arc is shown over the solar pond, indicating the arc of the sun at both the summer and winter solstice.

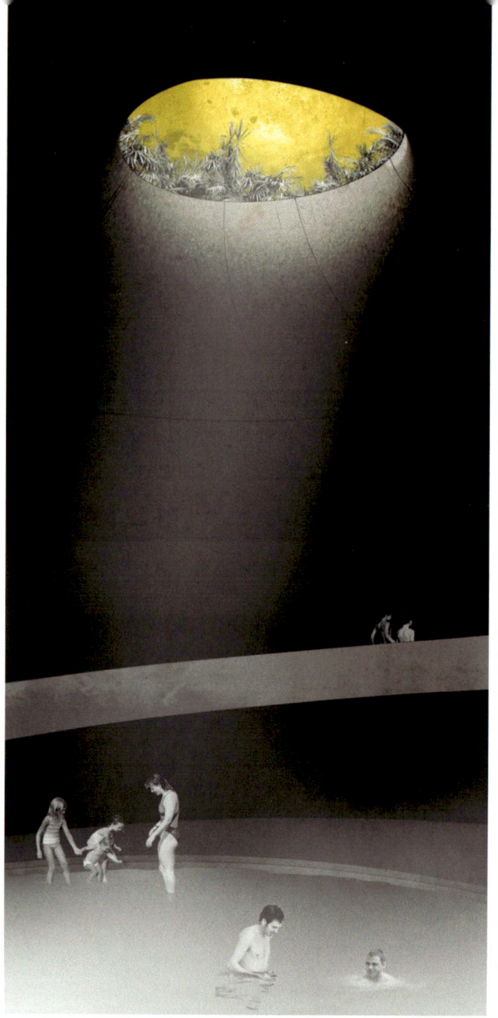

View from inside the bathhouse

The Minto wheel is a form of heat engine which, like a heat pipe, relies on a working fluid with a low boiling point (in this case Isobutane, C4H10). As the tube containing the working fluid comes into contact with the warm water of the attenuating pool, the working fluid vaporizes. The vaporized fluid rises up the tube and condenses on the opposite (cooler) end. As the vaporized fluid collects at the top of the tube, it unbalances the wheel causing a slow rotation with high torque.

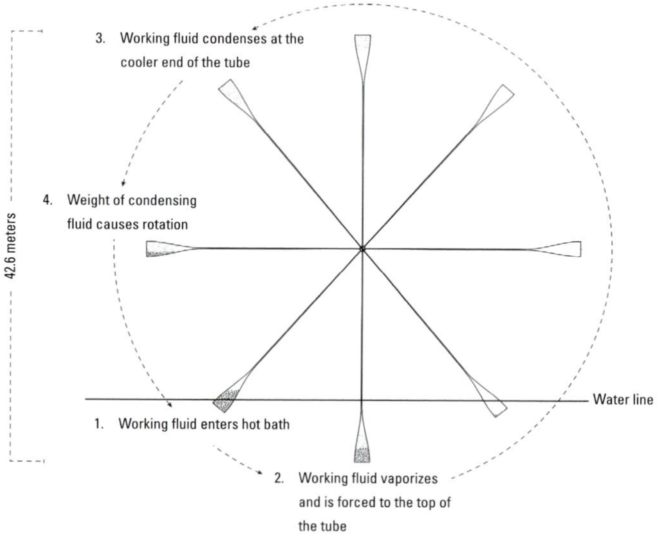

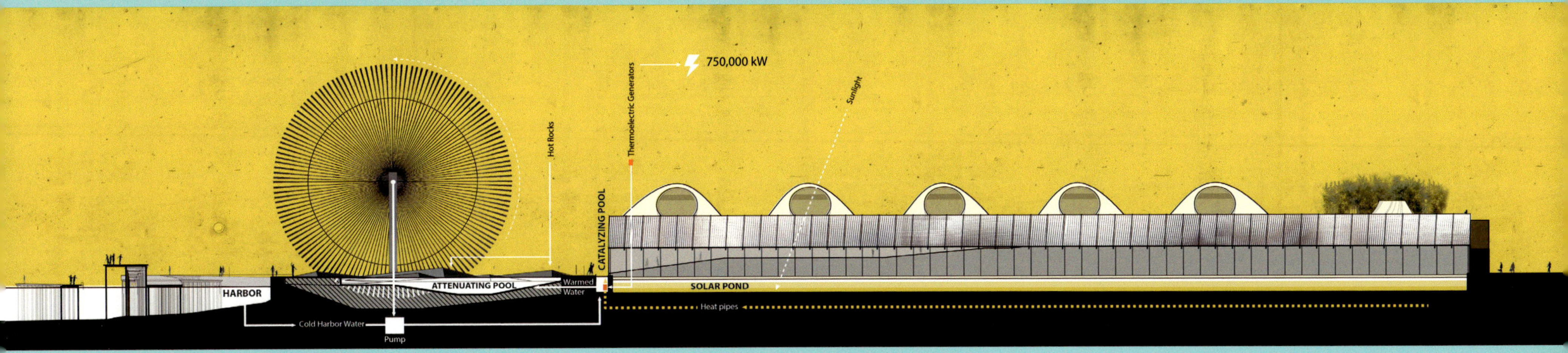

Site section looking northeast

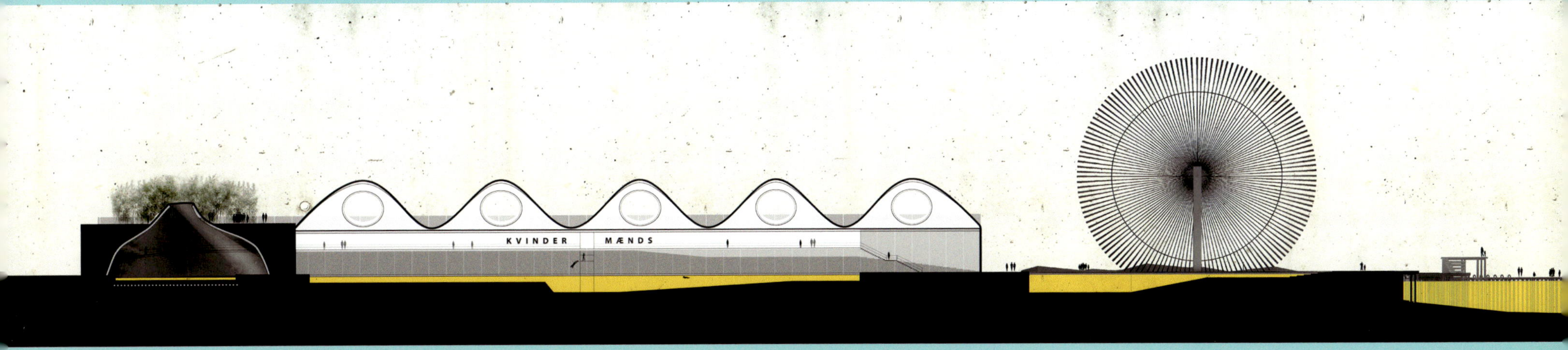

Section through the caldarium and bathhouse looking southwest. This section through the site shows the interior of the caldarium and bathhouse in relation to the site's other main features: the Minto wheel, the cinema pool, and the Harbor Bath. The large pool can be accessed via the ground floor entrance. Men's and women's changing rooms, showers, and bathrooms are accessible from inside the bathhouse and via the exterior ramp that runs between the solar pond and the bathhouse. A second floor catwalk provides access to the caldarium, allowing one to enter and descend the exterior wall of the oculus. The catwalk also grants opportunities for playful engagement with the large pool, providing access to diving platforms, fire poles, and trapeze bars suspended over the deep pool below. Clerestory oculi direct light to the interior of the bathhouse.

The Cloud

TEAM:
Tim Thikaj

TEAM LOCATION:
Copenhagen, Denmark

ENERGY TECHNOLOGIES:
wind microturbines

ANNUAL CAPACITY:
2,016 MWh

A cloud is a visible mass of condensed water vapor floating in the atmosphere, typically high above the ground. It describes an indistinct mass, especially of smoke or dust, a large number of insects, or birds moving together within a certain dynamic. With regard to fractals in mathematics, a cloud can refer to a set of multidimensional points. *The Cloud* exists as a visual metaphor—the image of a fractal that is both illustrative and abstract. It can be described as being both solid and hollow. It is mathematically predictable, while it is also ever-changing.

The Cloud is conceived as a social gathering point, where activities, such as swimming and enjoying the weather, can take place. It is an alternative, self-powered pool area, which is accessible both summer and winter.

The Cloud uses the geometry of the sphere in its facade. The circular wings of the turbines reflect natural and artificial light and, while moving, visually interact with the surroundings and the inside space. When each turbine reaches a certain speed, the edges of the blades become invisible to the eyes and transform into a three-dimensional shape with a diffused, misty, white surface. *The Cloud* measures 28 meters at its highest point. Though the cloud is a structure of symbolic status, the height is low and similar to the buildings found in the City of Copenhagen to better integrate it into the landscape.

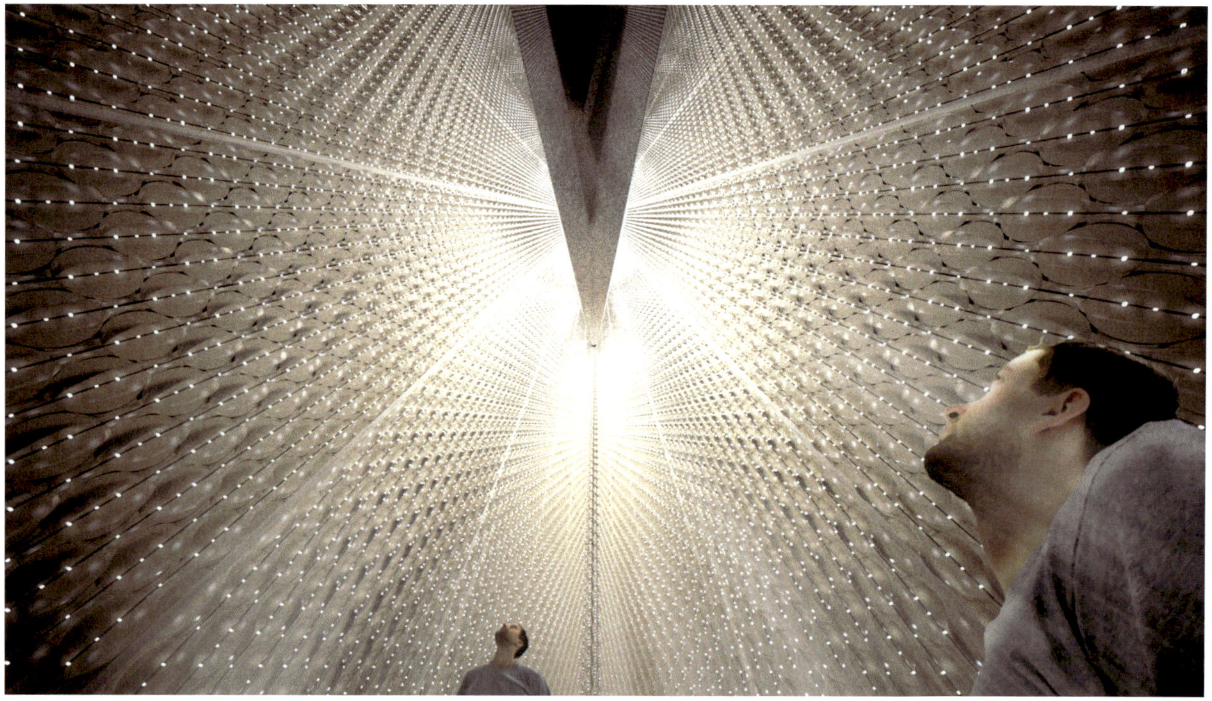

The facade consists of 40,319 miniature sphere-shaped turbines each measuring 20 cm x 20 cm in diameter. Each turbine rotates on a vertical axis. There are 436 vertical "wires" stretching between the two primary frames supported by columns—one frame outlining the roof and the other frame placed on the ground. The turbines are made of white, semi-translucent, bent acrylic material, which hangs from the vertical wires like pearls on a string, transforming *The Cloud* into a gleaming jewel when moving.

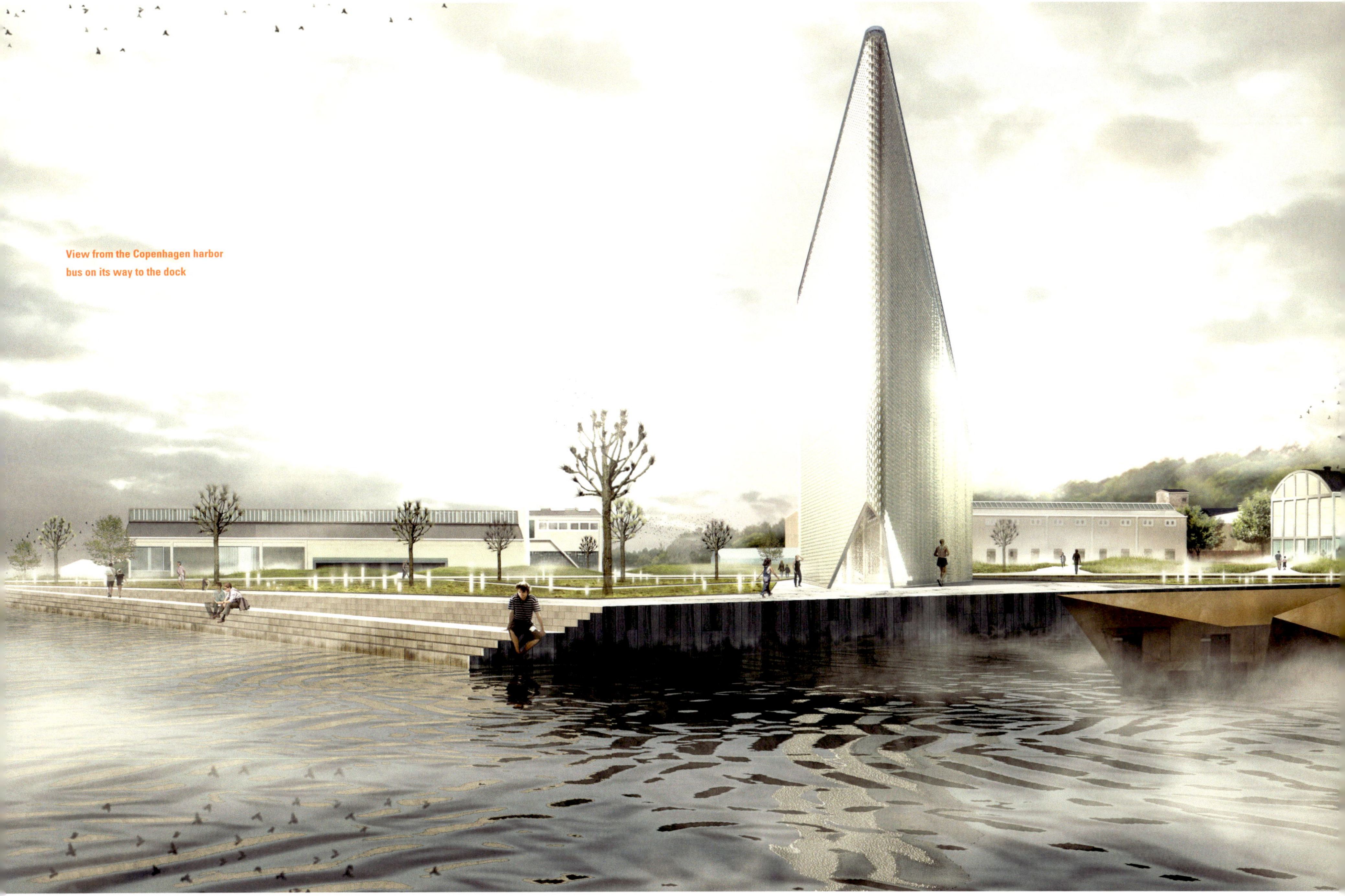

View from the Copenhagen harbor bus on its way to the dock

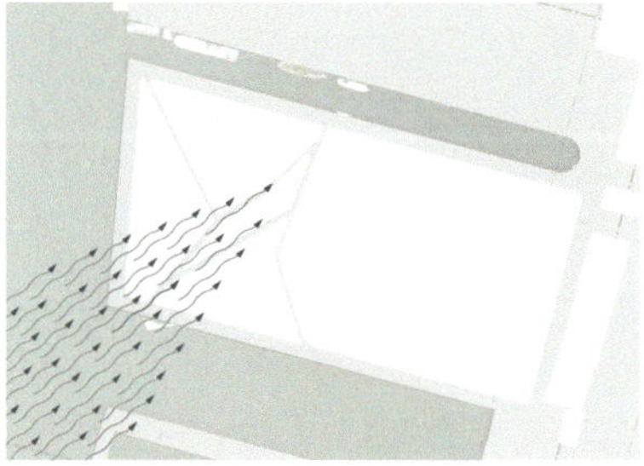
Frequent wind direction

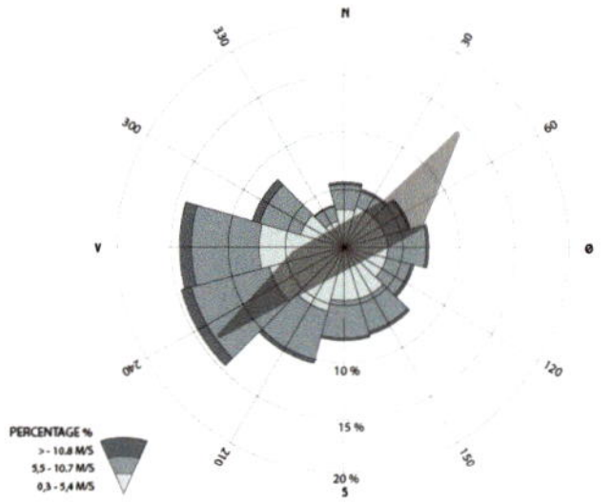

Wind rose analysis

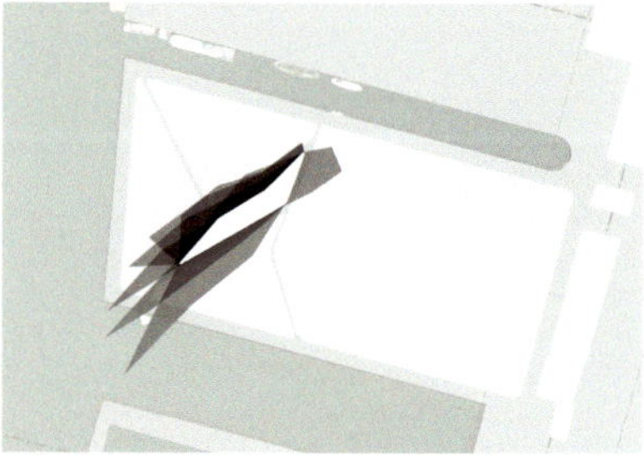
Shadow diagram (June)

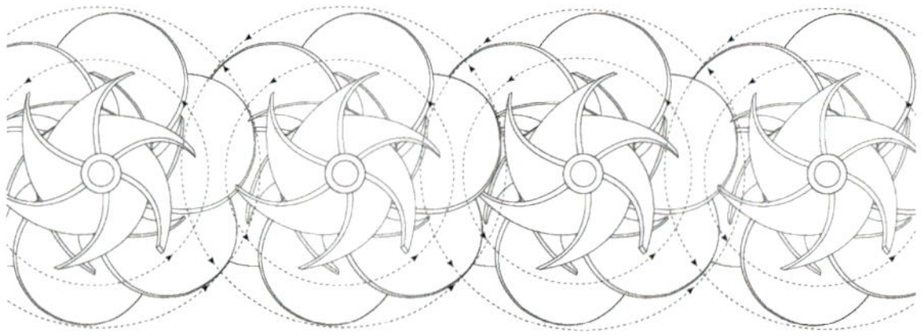
Turbines displayed as plan section

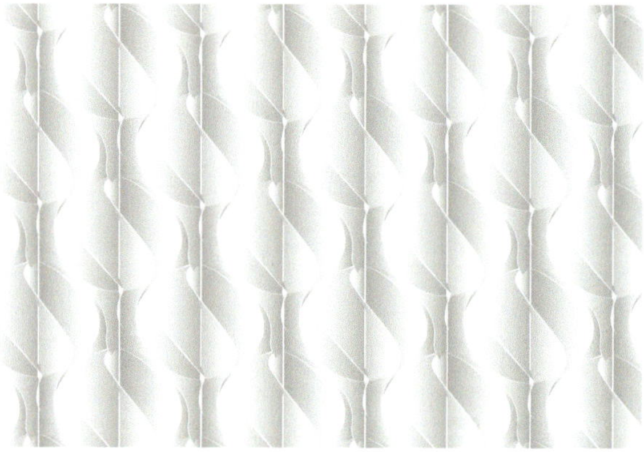
Turbines displayed as facade

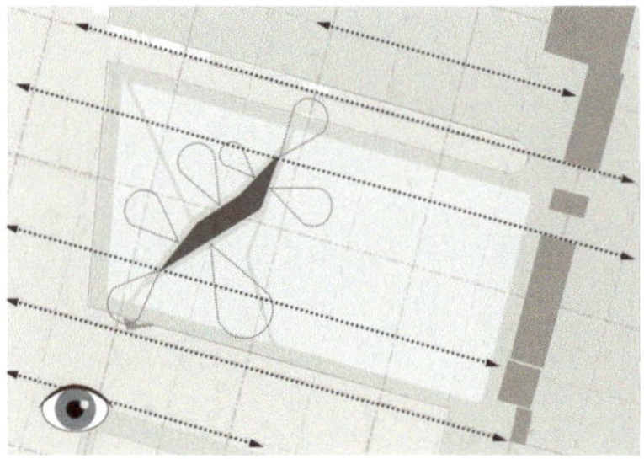
View and directions

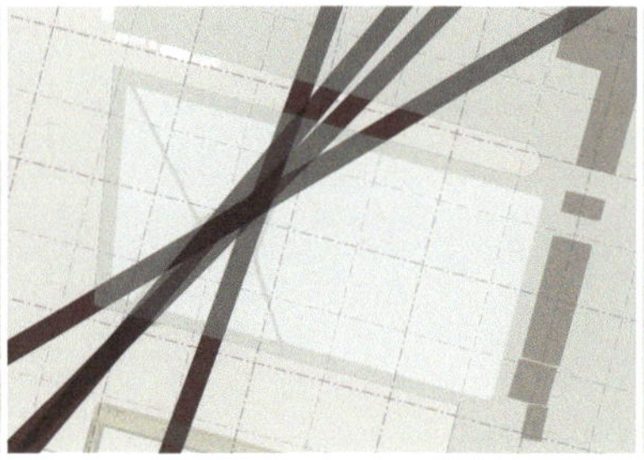
Grid structure

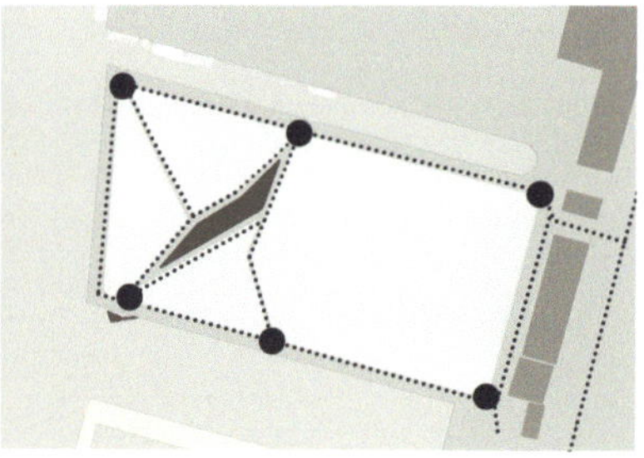
Focal points

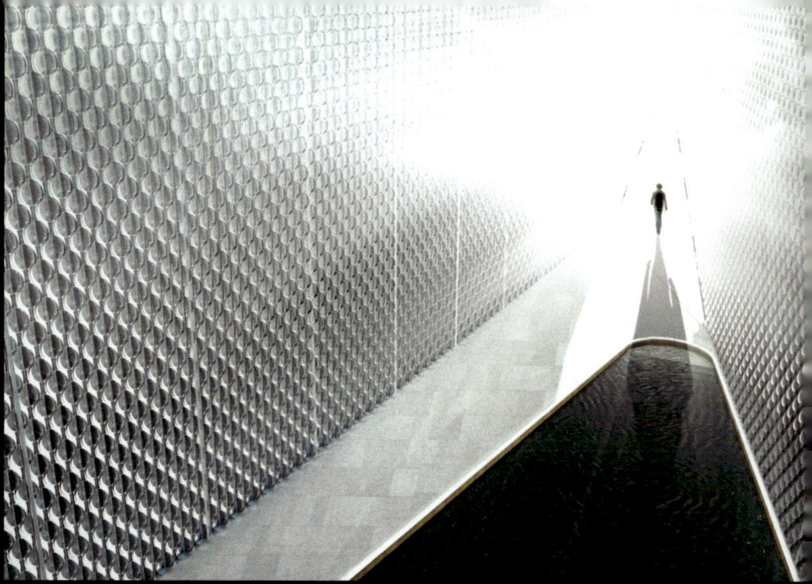

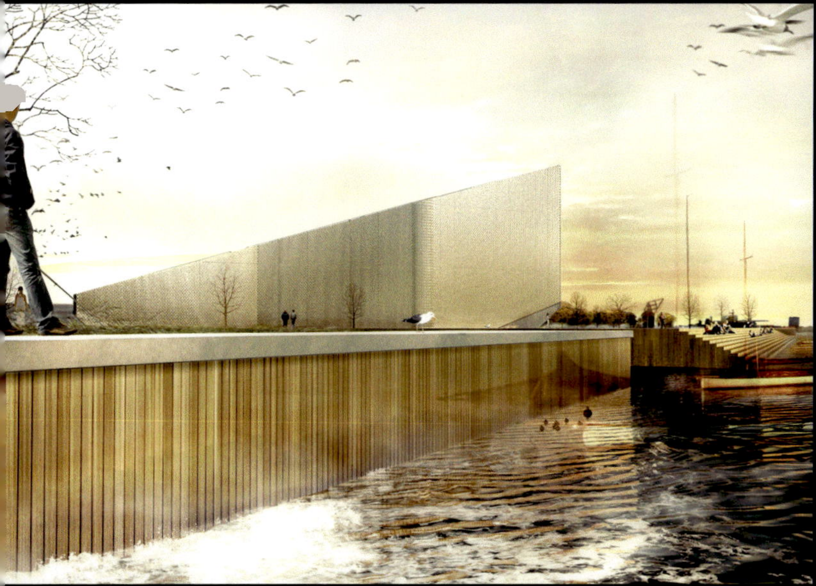

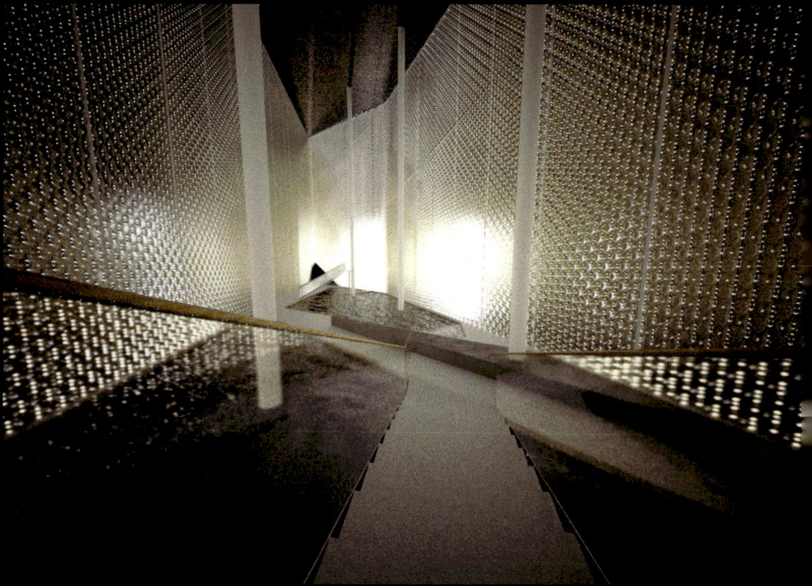

he Cloud measures 28 meters at its ighest point. Although *The Cloud* s a structure of symbolic status, the height has been designed to be similar to the buildings found in the City of Copenhagen.

Balance | Imbalance

Balance / Imbalance generates energy from wave, wind, and human activities. The design site is interpreted as an intersecting area between land and ocean, humanity and nature. The main concept of this kinetic sculpture is to demonstrate the beauty of balance and imbalance between contrasting entities.

The artwork has two arms with buoys suspended on the seaward side. The movement of the attached weight elongates and contracts Electroactive Polymer Artificial Muscle (EPAM). The expansion and contraction of this artificial muscle generates electricity, reversing the process through which natural muscles expand and contract when stimulated by electricity. Spherical solar devices are installed at the landward side as an integral part of the counterweight system.

In addition to these two primary devices, piezoelectric generators are installed at each joint within the system to convert the internal forces of compression and tension into electricity as the balance sways in conjunction with wave and wind motion. The same effect is also produced on the landward side, where visitors can activate this energy-harvesting system by making use of the attached furniture, such as seesaws and hammocks. The more interaction there is between people and the artwork, the greater the amount of electricity that is produced.

A small portion of the energy produced goes toward lighting the site at night. The intensity of the LED lighting will shift according to wave motion. This luminous wave line will act as an iconic spectacle and an impressive backdrop when viewed from the opposite shore.

TEAM:
Hideaki Nishimura

TEAM LOCATION:
Tokyo, Japan

ENERGY TECHNOLOGIES:
buoy-type wave converter with Electroactive Polymer Artificial Muscle (EPAM by SRI International and Bayer Material Science), Sphelar® photovoltaic, piezoelectric film

ANNUAL CAPACITY:
720 MWh

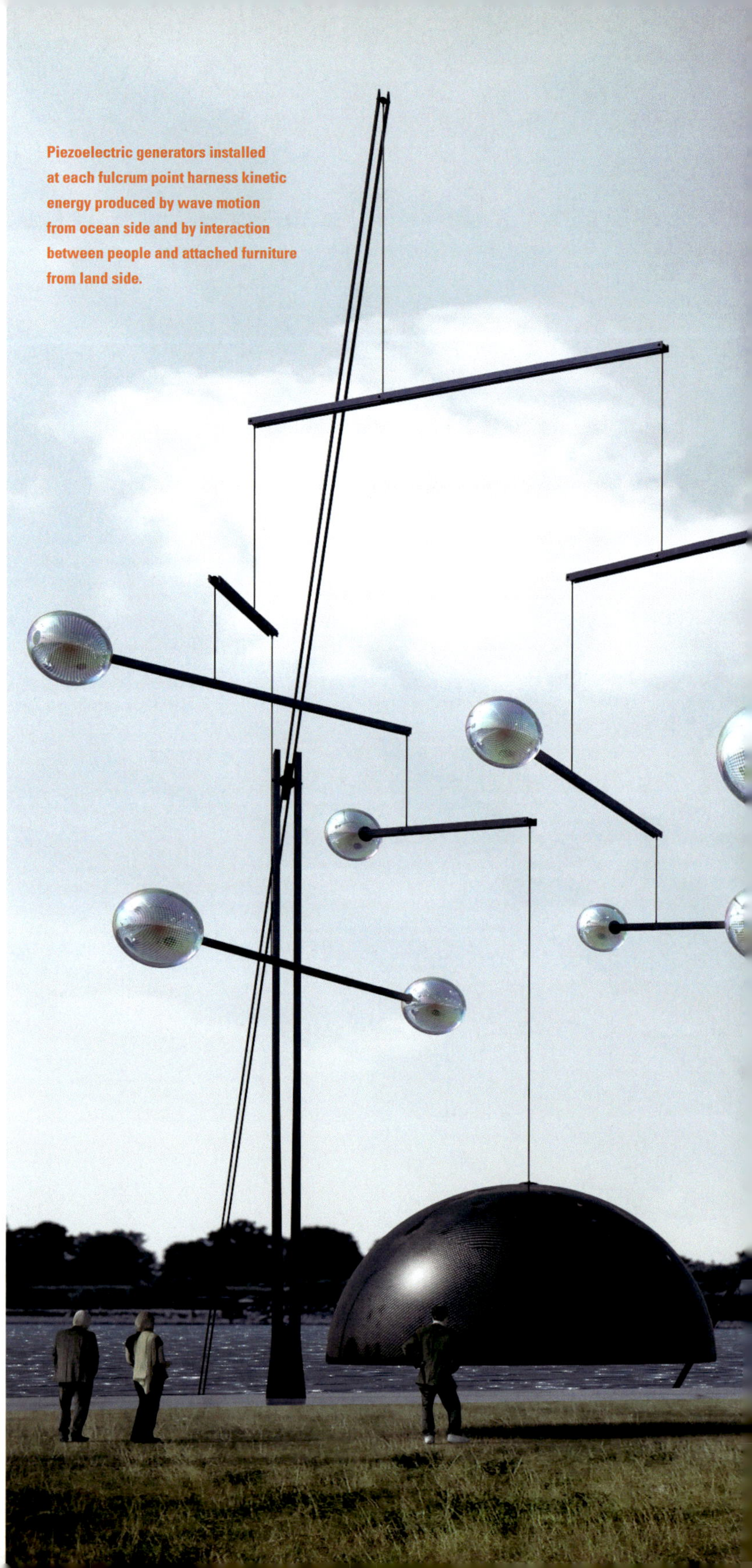

Piezoelectric generators installed at each fulcrum point harness kinetic energy produced by wave motion from ocean side and by interaction between people and attached furniture from land side.

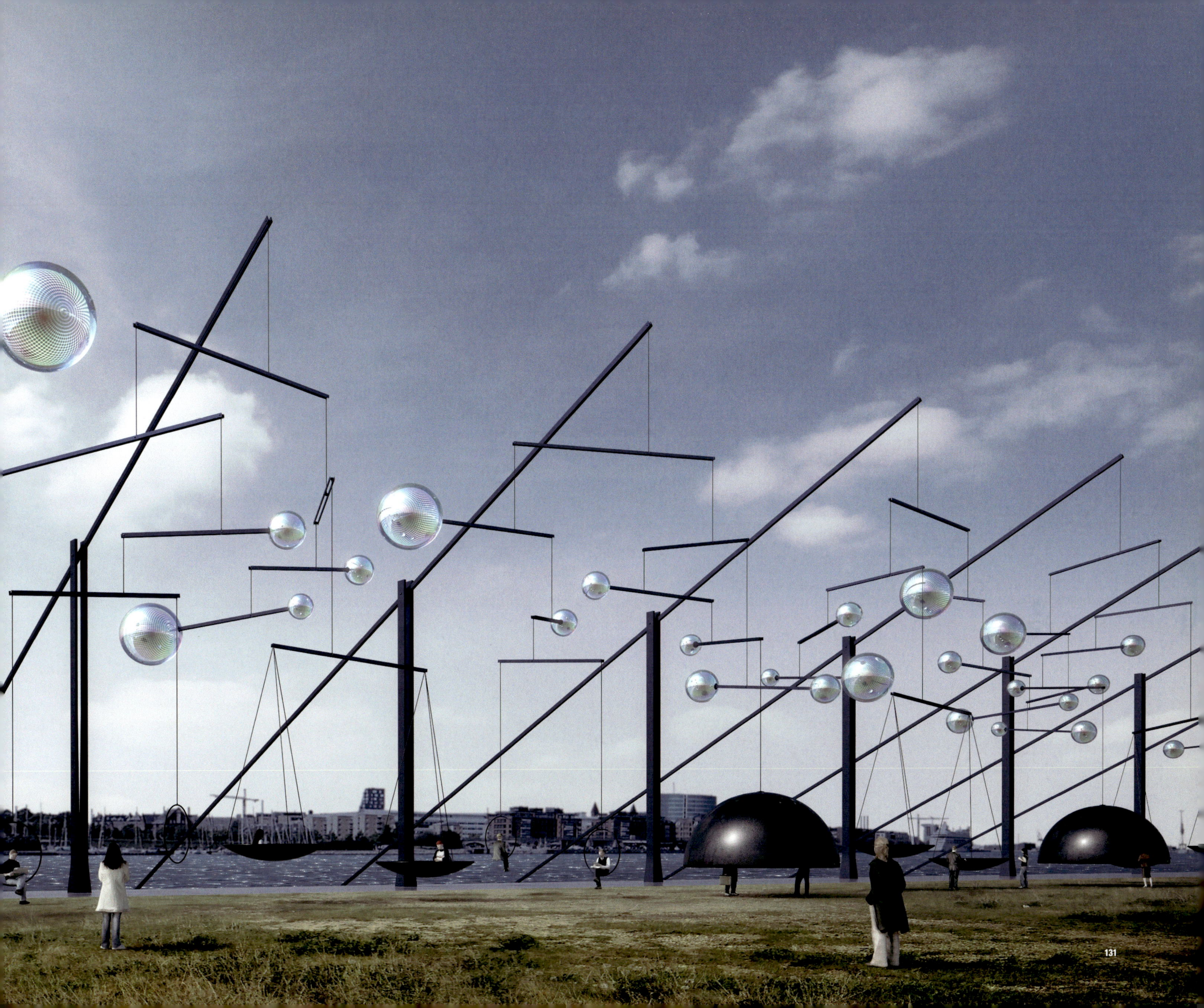

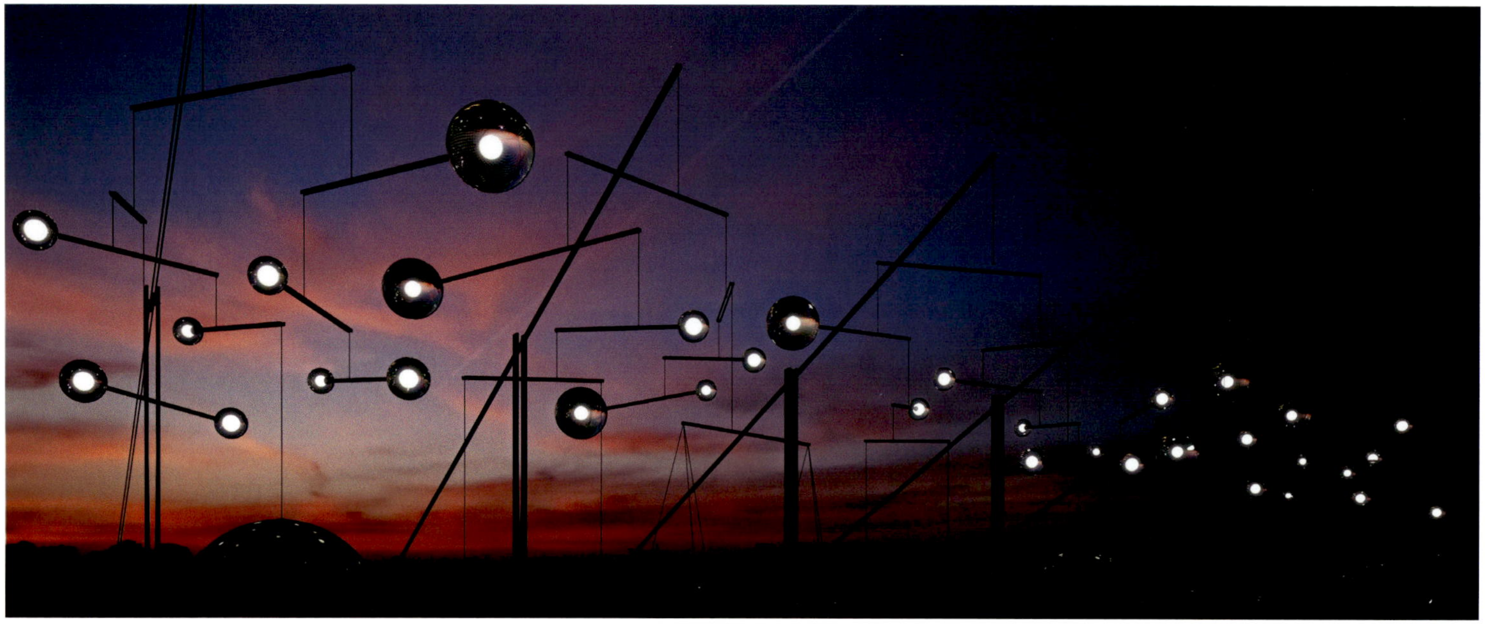

As balance arms sway in conjunction with wave motion, a thin line fluctuates in the air.

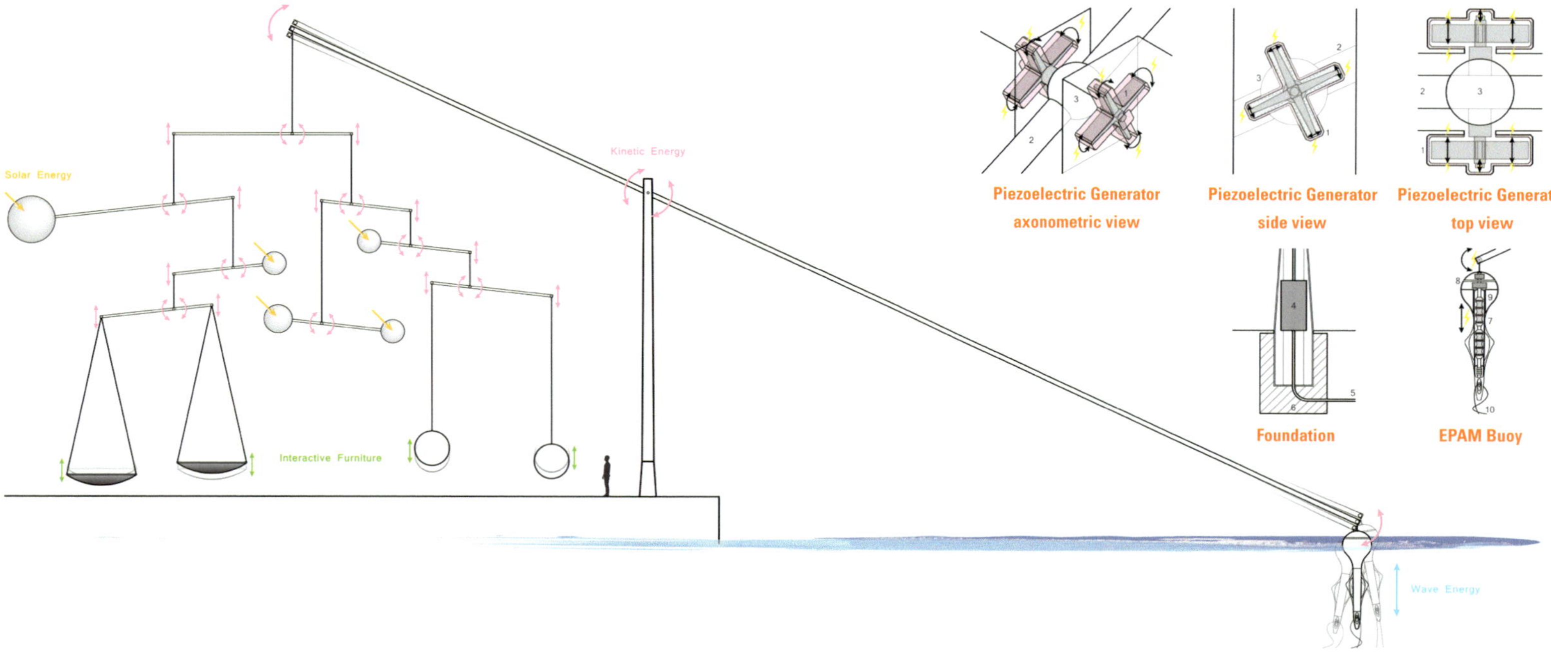

Piezoelectric Generator

1. Piezoelectric film, shock absorber, safety limiter
2. Balance arm
3. Spherical fulcrum

Foundation

4. Battery and transformer
5. Cable to power grid
6. Concrete foundation

EPAM Buoy

7. Electroactive polymer artificial muscle
8. LED round light
9. Buoyancy chamber
10. Cable anchor

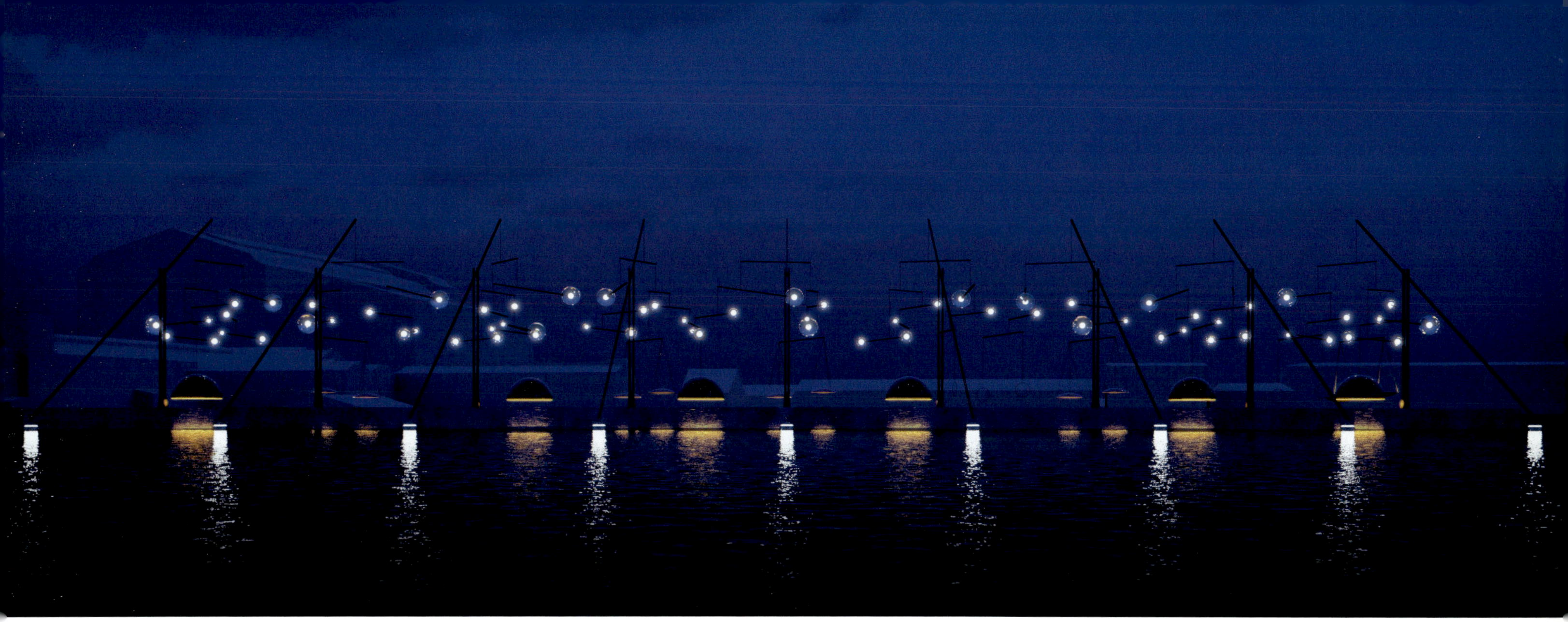

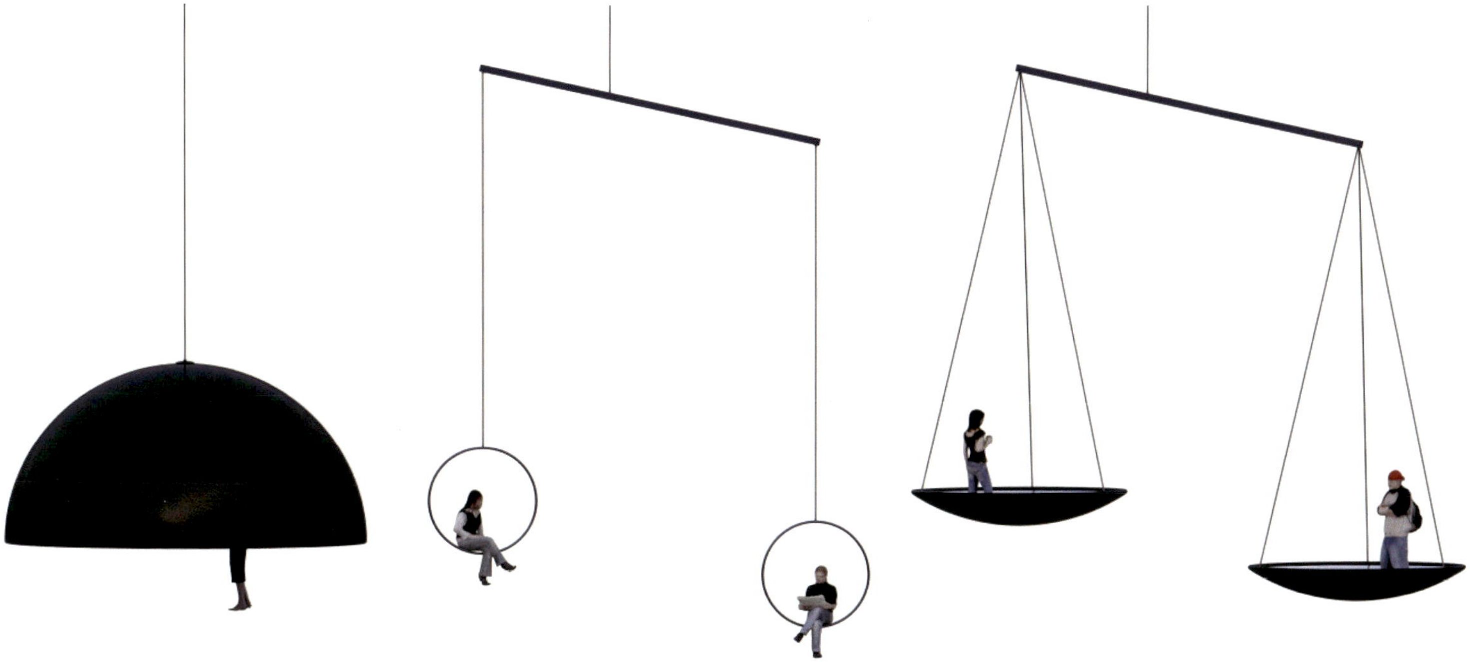

Spherical photovoltaic devices and interactive furniture illuminate the scene at night.

The attached furniture elements allow interaction between visitors and the artwork.

Right: Circular hanging chair

Center: Balance scale bench

Left: Hemispherical suspended shade

Wind Harp:
Wind Energy as Music

TEAM:
Brian Chi Fung Lee, Ken Ka Chun Lee, Terry Chun Yin Chan

TEAM LOCATION:
Copenhagen, Denmark

ENERGY TECHNOLOGIES:
aeroelastic flutter (Windbelt™)

ANNUAL CAPACITY:
135 MWh

Wind Harp is situated as a monument on a portion of the design site in order to stimulate events while allowing the rest of the site to produce natural growth. By intervening at the far end of the harbor side, the generator collaborates with the surrounding buildings to form a frame, enclosing the void to create the center of activity.

A stationary crane structure of 30 meters by 100 meters is erected at the proposed area near the coastline, next to the water taxi station. The structure, a minimal frame of lightweight steel, is a reference to the gantry cranes that once operated in the B&W shipyard during the former life of the site.

On both sides of this crane, a total of 224 "wind strings" are attached to the structure at one end and to the ground at the other. As these "strings" are placed in tension, they generate both electricity and sound through oscillation by wind or by direct human interaction. The subsequent energy is then harvested by small devices attached to the opposite ends of the string. The turbine noises typically associated with conventional wind turbines are transformed into music.

The strings are spaced out at 80-centimeter intervals, allowing people to walk through and between this instrument. The two sets of strings also create a metaphoric tent that welcomes people to linger inside, either by enjoying the symphony of sounds generated by wind passing through, or to play the instrument directly with their hands.

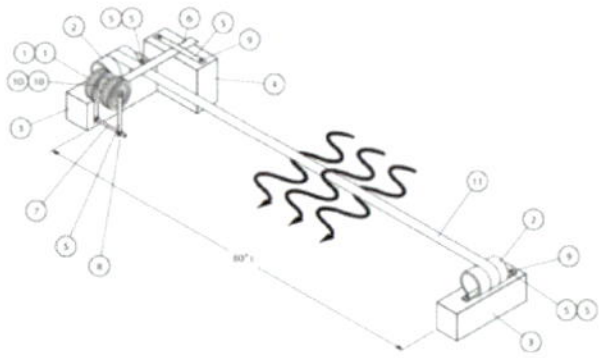

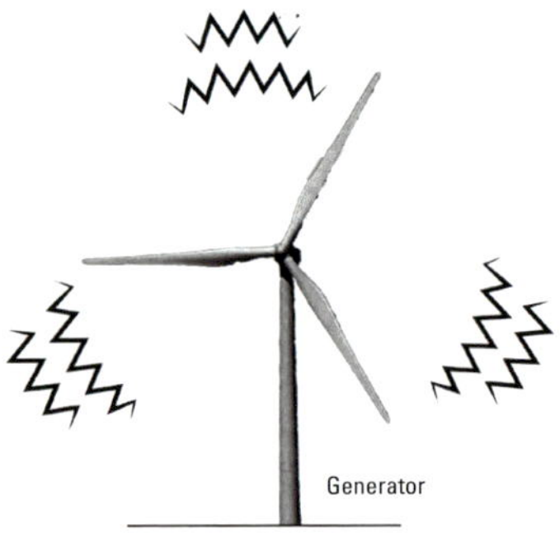

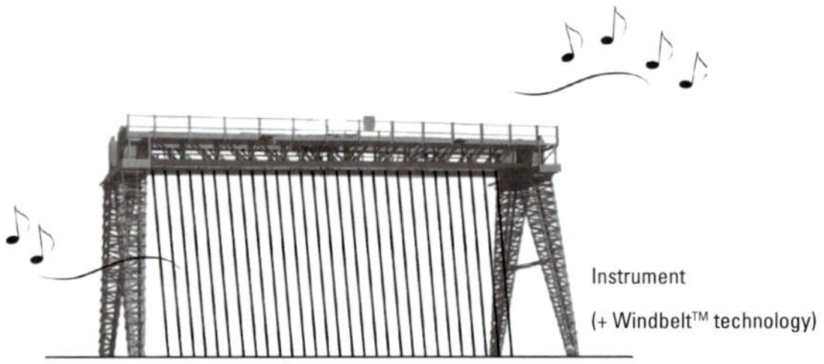

By converting the noise of the conventional wind turbines to musical chords, the generator is transformed into an instrument.

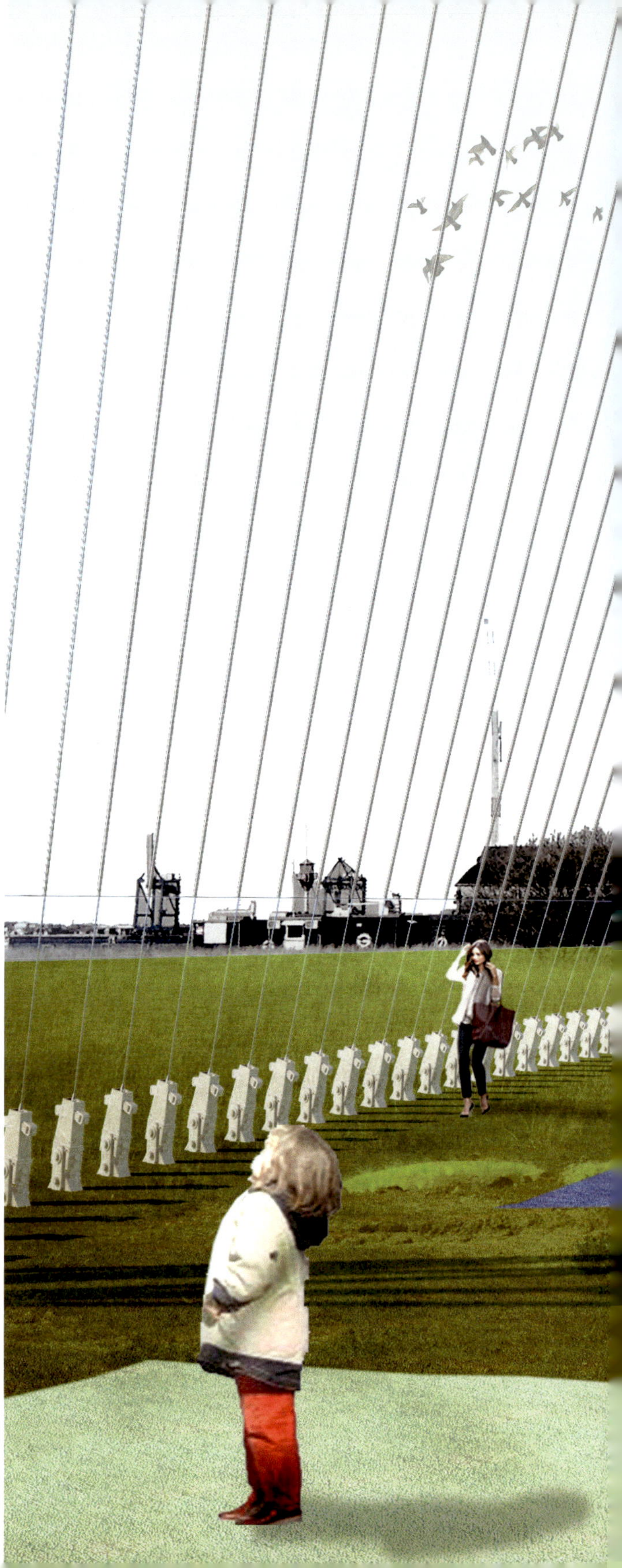

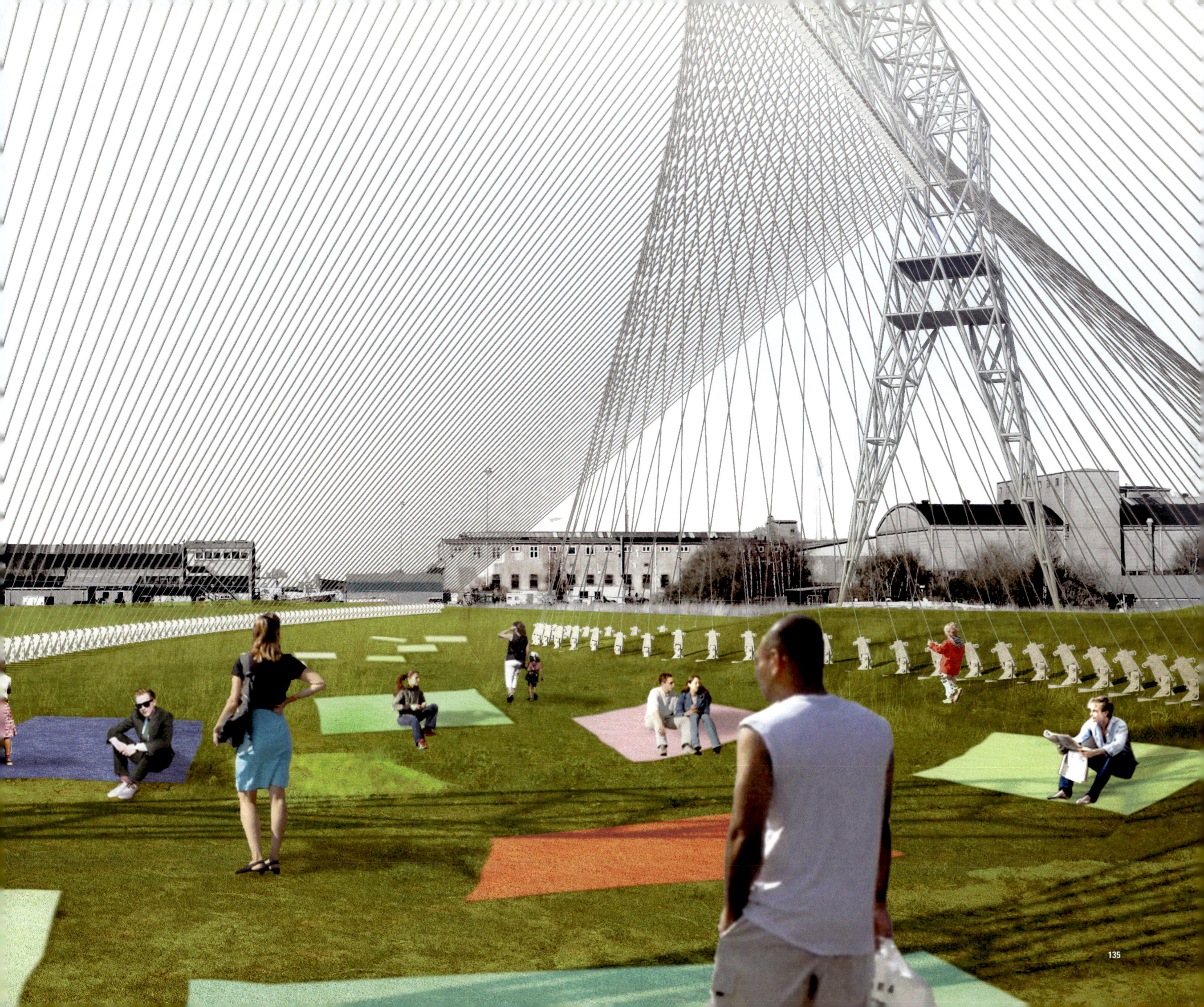

Wind Harp: Wind Energy as Music

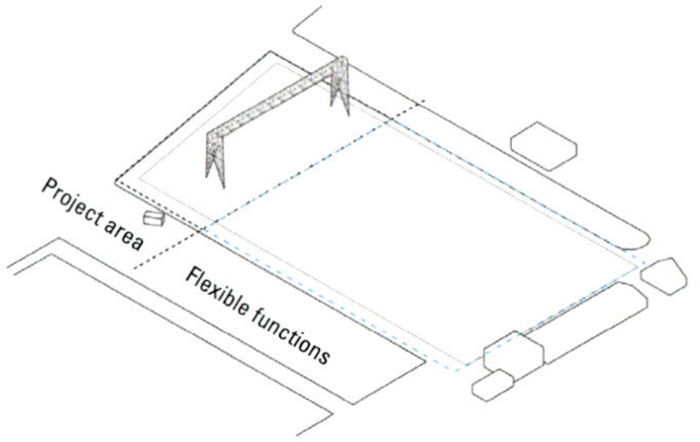

The crane structure placed strategically to frame the plaza space

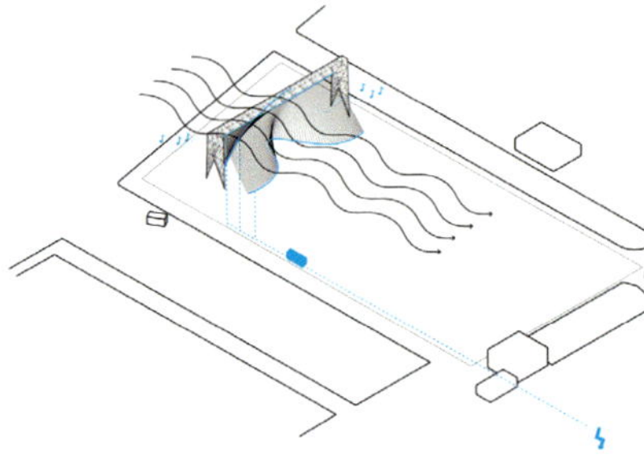

"Wind strings" generate electricity and sound

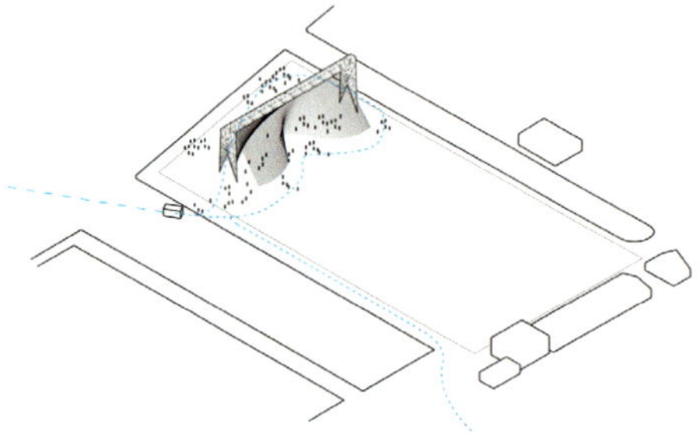

A place for listening and relaxing under the structure

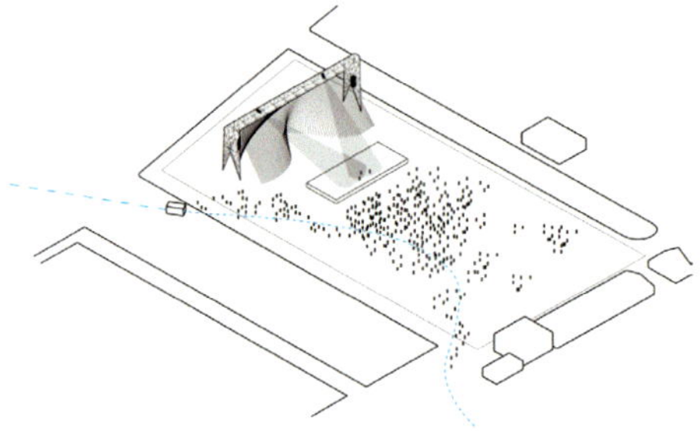

A flexible stage for future events

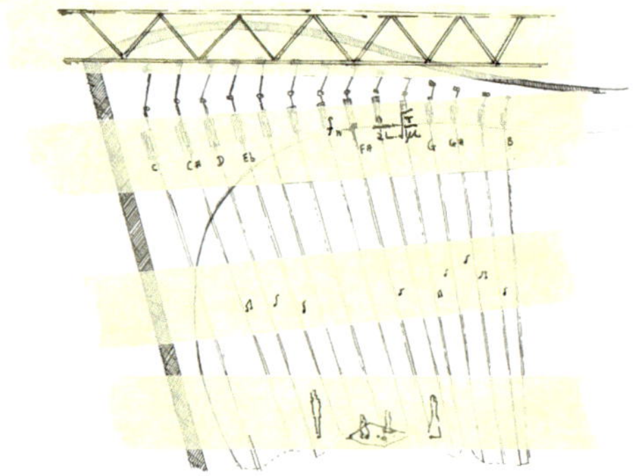

Early sketch on overlay of ideas at different scales based on a harp reimagined as an urban instrument

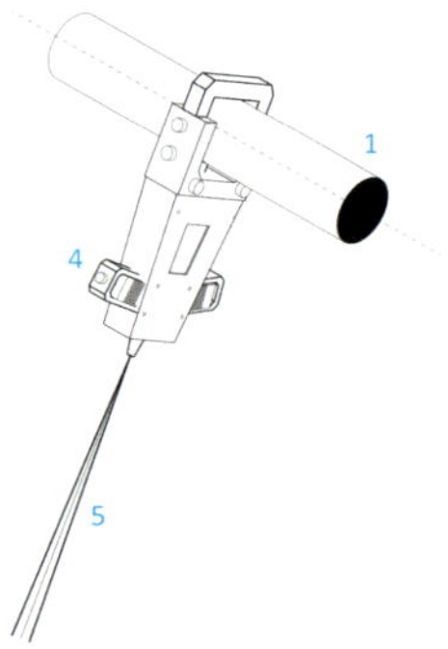

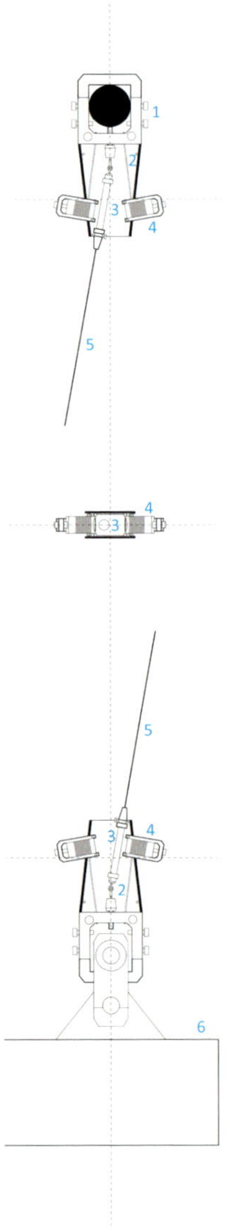

System Details

1. Steel member of the crane
2. Wires connected to synchronizer
3. Diametric bar magnet
4. Coil wound on soft iron core
5. Steel, nylon string in various radii
6. Embedded concrete footing

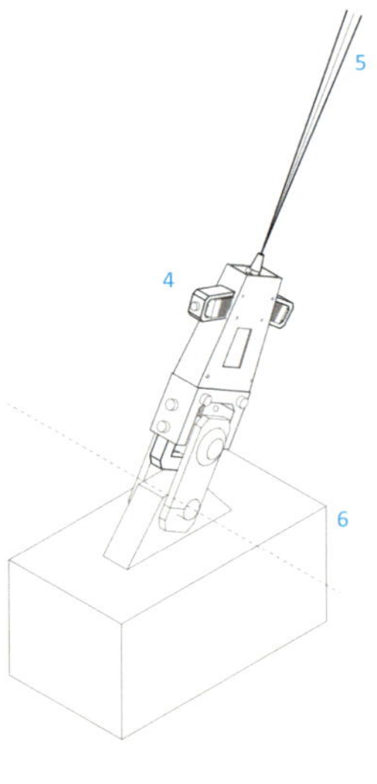

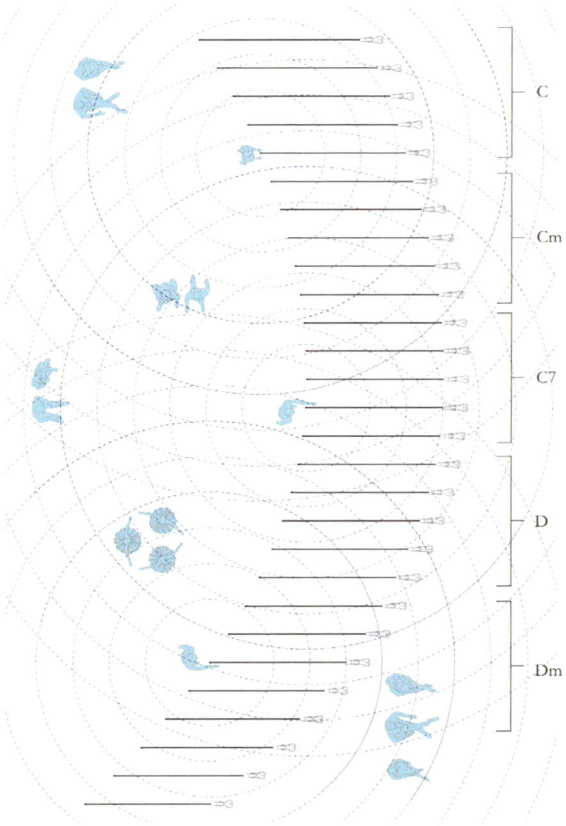

The tension/thickness of the strings is determined by the group of musical notes it plays, resembling chords commonly played on guitars.

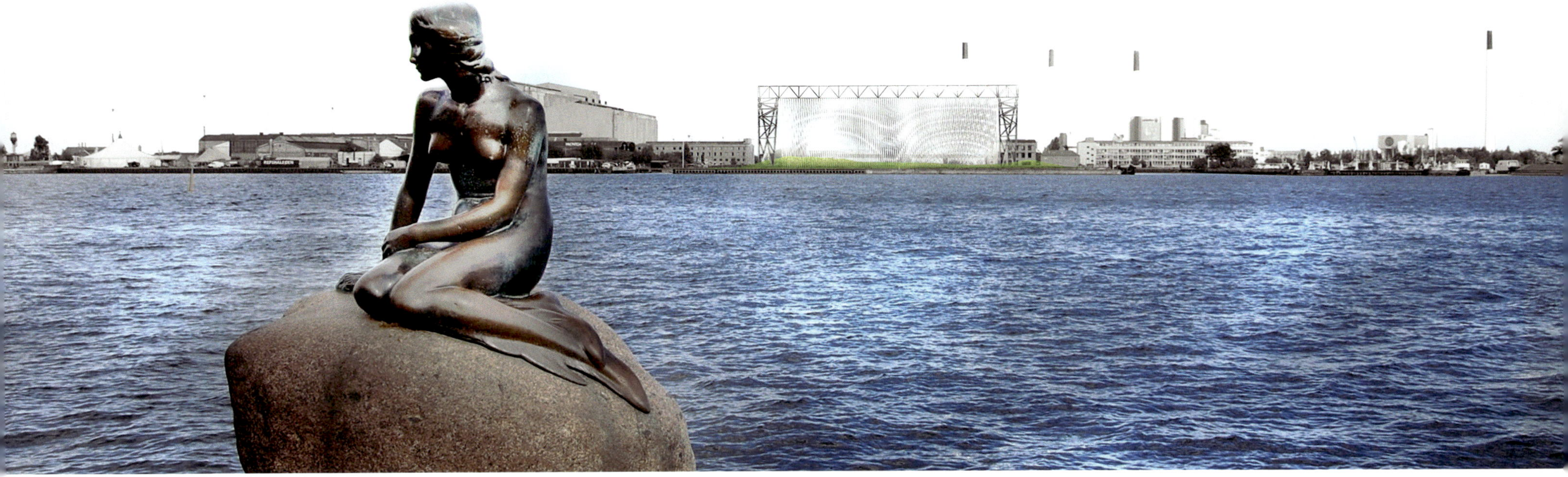

View from across the Copenhagen harbor

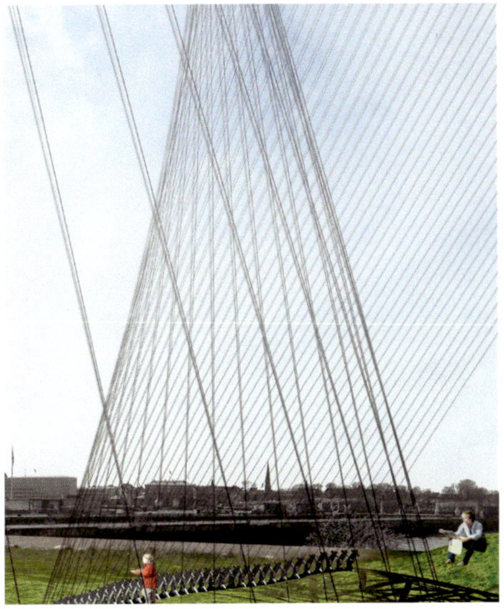

Visitors relax beneath the *Wind Harp*.

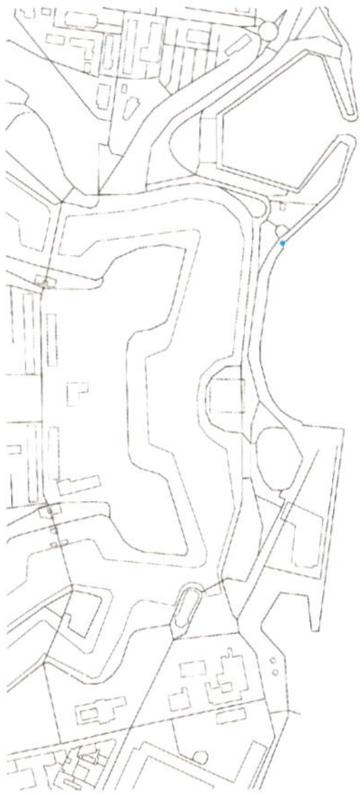

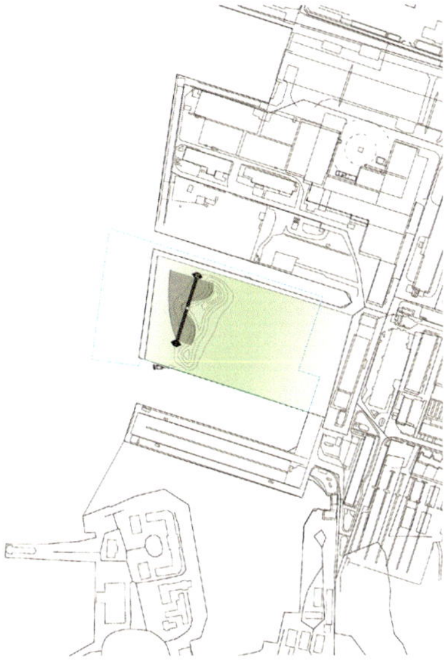

Site plan

Energy Duck

So much more than just a duck! *Energy Duck* is an entertaining, iconic sculpture, a renewable energy generator, a habitable tourist destination, and a celebration of local wildlife.

The common eider duck resides in great numbers in Copenhagen; however, its breeding habitat is at risk from the effects of climate change. *Energy Duck* takes the form of the eider to act both as a solar collector and a buoyant energy storage device.

Solar radiation is converted to electricity using low cost, off-the-shelf PV panels. Some of the solar electricity is stored by virtue of the difference in water levels inside and outside the duck. When stored energy needs to be delivered, the duck is flooded through one or more hydro turbines to generate electricity, which is transmitted to the national grid by the same route as the PV panel-generated electricity. Solar energy is later used to pump the water back out of the duck, and buoyancy brings it to the surface. The floating height of the duck indicates the relative cost of electricity as a function of city-wide use: as demand peaks the duck sinks.

Energy Duck can thus be used to cover short-term fluctuations in available sunlight (modulating its energy generation). When the duck is pumped out, in addition to providing automatic cleaning to the PV panels, all manner of fun and interesting things can be done with the ensuing waterspouts.

Visitors inside the duck will have views upward to reveal the striking pattern of the mesh of PV panels in silhouette, backlit by daylight streaming through the air gaps. Looking downward allows one to see seawater rising and falling within the pressure storage tanks.

TEAM:
Hareth Pochee, Adam Khan, Louis Leger, Patrick Fryer

TEAM LOCATION:
London, UK

ENERGY TECHNOLOGIES:
photovoltaic panels (Panasonic HIT™ or similar), hydraulic turbines (Kaplan, Francis, or similar 100–500 kW capacity)

ANNUAL CAPACITY:
400 MWh

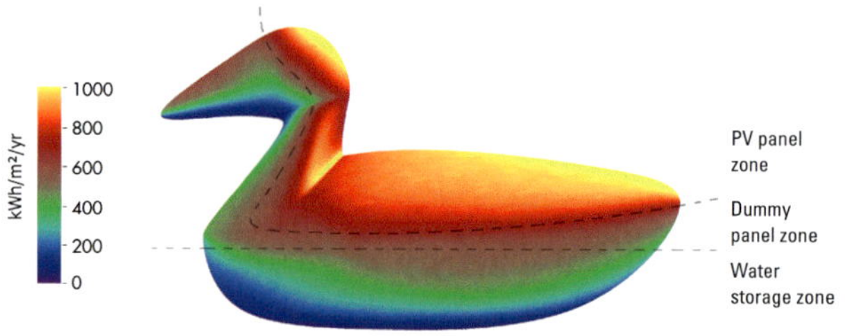

A computer simulation using hourly Copenhagen climate data was used to tune the duck shape to collect the most amount of annual solar energy.

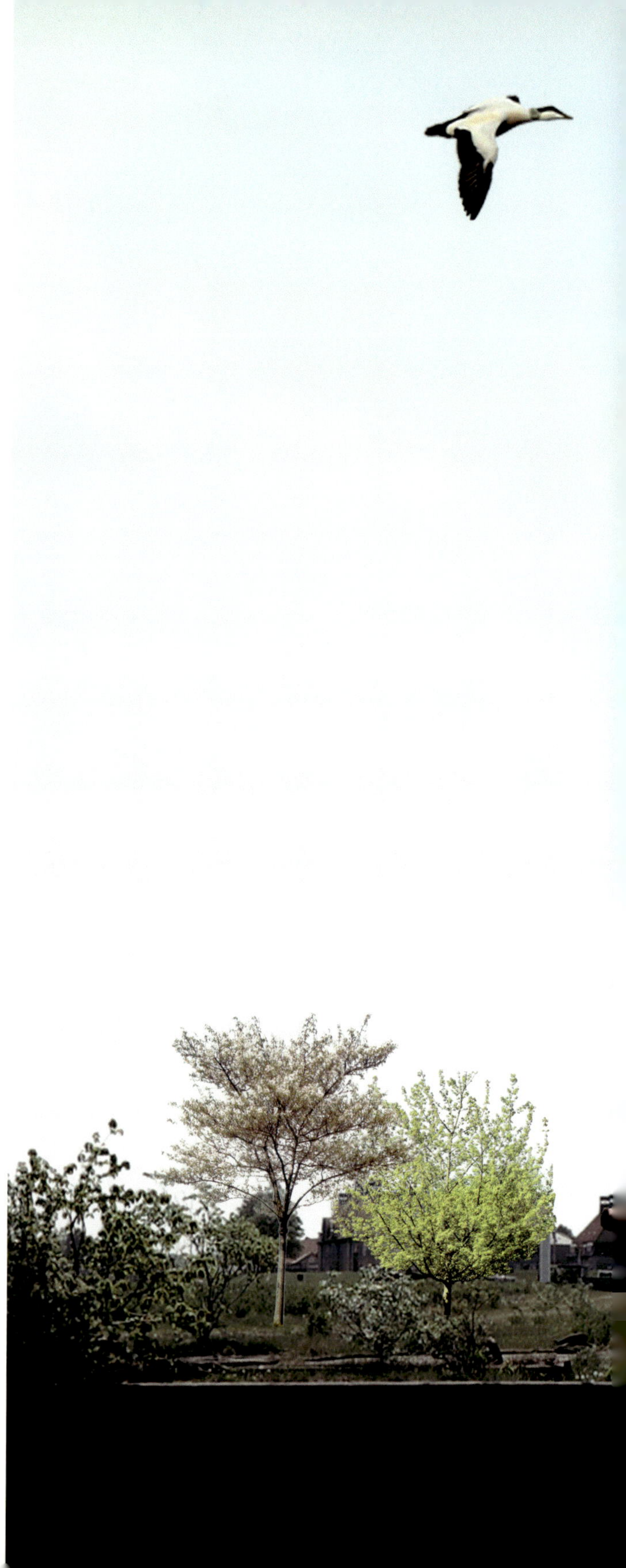

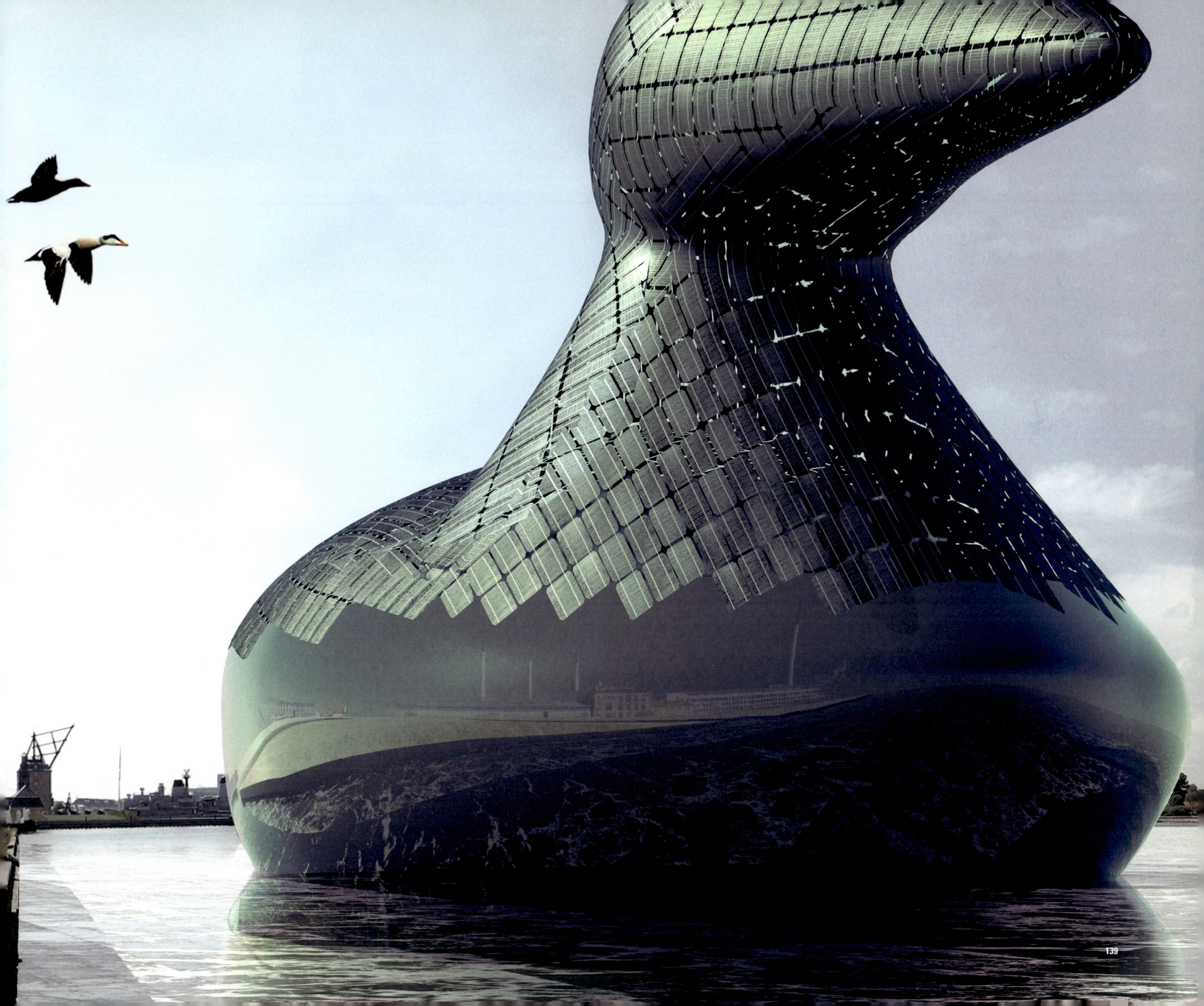

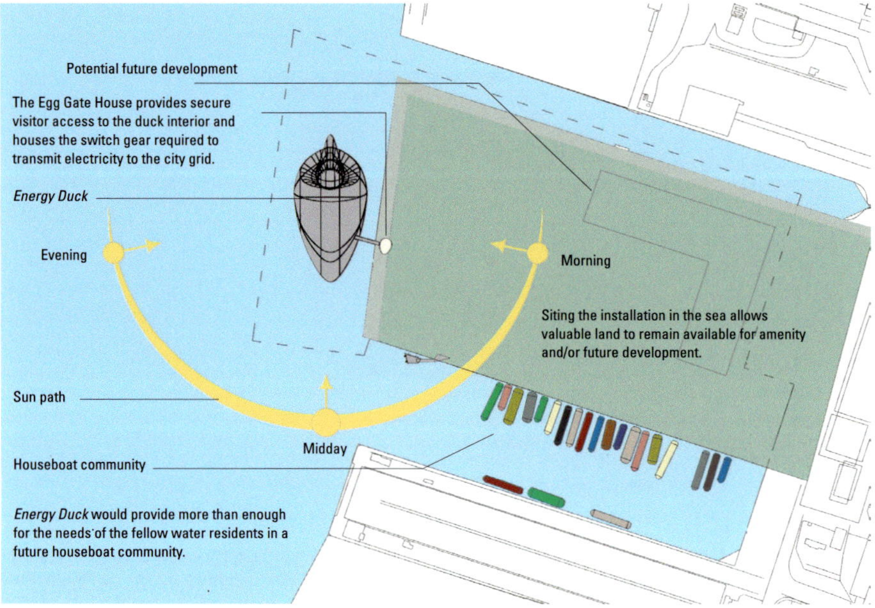

***Energy Duck* has a kinship with the houseboat community.**

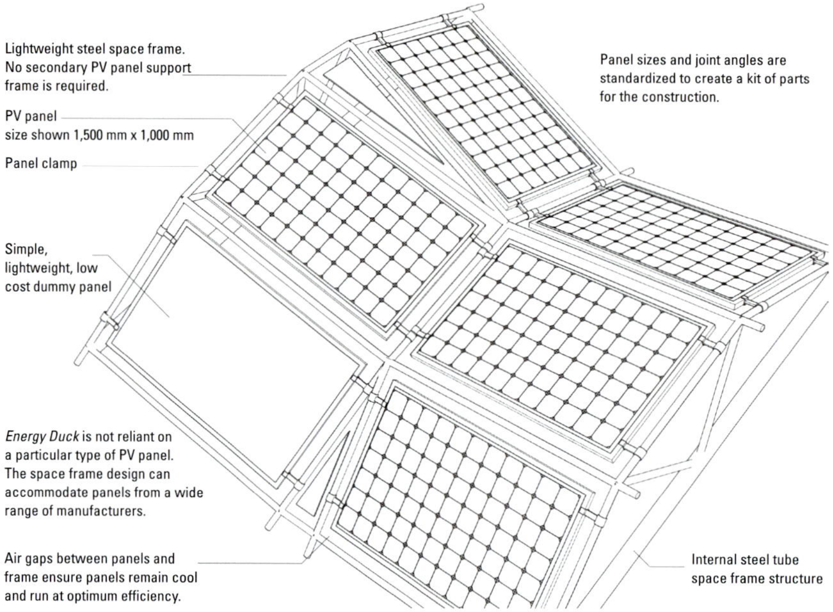

The external space frame uses a combination of quad- and tri-shapes to accommodate the panels and create the duck form.

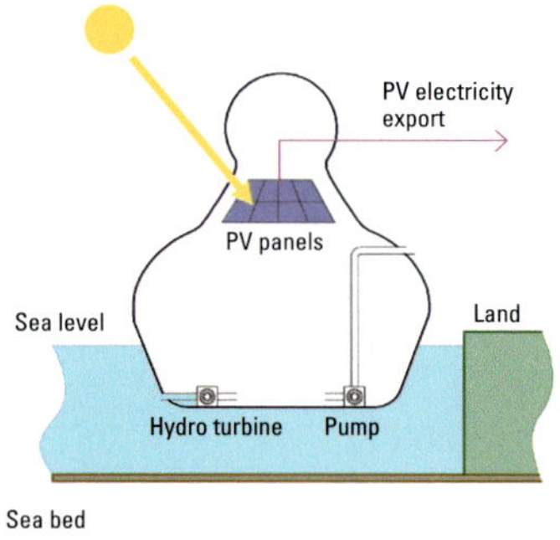

Duck Floating High
Energy storage is full.
All electricity is exported.

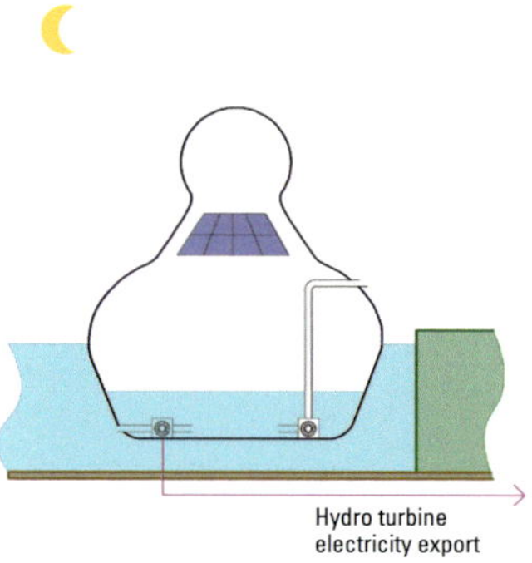

Duck Descending
Hydro turbine runs as storage is flooded. Stored energy is exported.

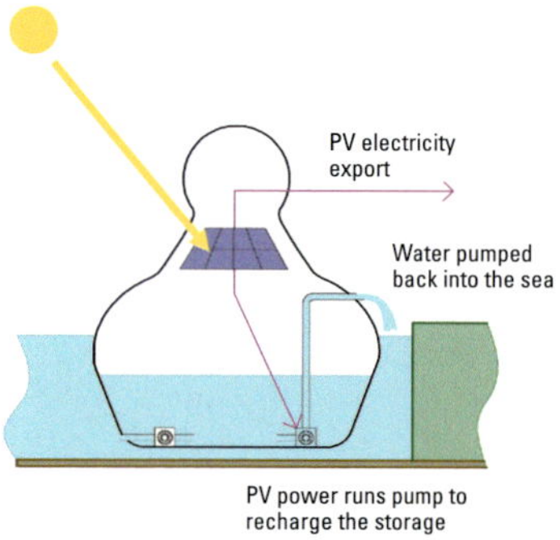

Duck Floating Low
Stored energy has been used.
PV panels power the pump. Duck rises to recharge the storage.

***Energy Duck* can store solar energy by making use of its floating nature. Energy is stored by virtue of the difference in water levels inside and outside the duck.**

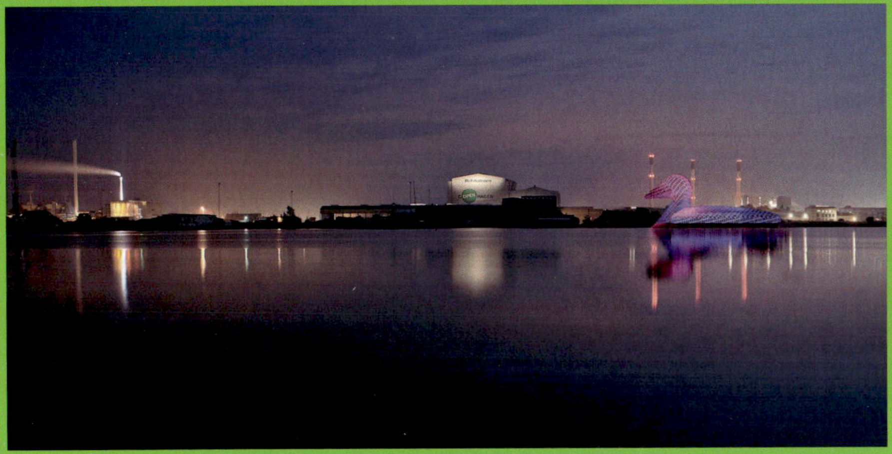

At night *Energy Duck* is lit with very low power color changing LED lamps. The color pattern undulates with a rhythm proportional to the output of the hydro turbines.

Background photograph by Johan Biilmann

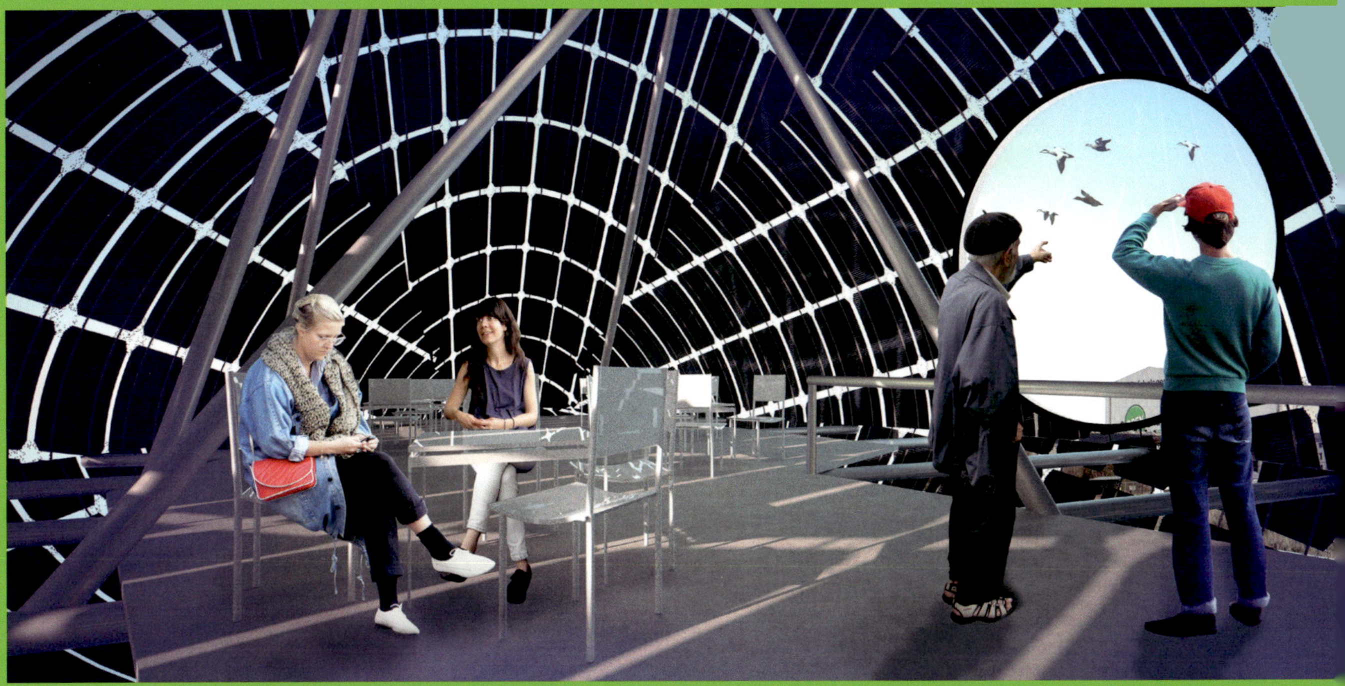

Visitors inside the duck move through an enormous honeycomb mesh of lightweight steel. Views upward reveal the striking pattern of a mesh of PV panels in silhouette.

Oscillating Platforms

Oscillating Platforms invites the surrounding waterway to enter the site, creating a flooded area within the Sønder Hoved Pier into which a series of free-form, floating platforms are anchored. These platforms, reminiscent of inverse ship hulls, have masts and sails integrated into the design, allowing them to catch the prevailing winds coming from the west. In addition, they are meant to be habitable surfaces that visitors can use as viewing platforms out onto the water.

The platforms act as oscillating water columns that convert both wind and tidal energy into electricity. The design utilizes a modified version of an oscillating water column. The device consists of a pressurized air chamber, which is partially submerged in the water. The energy of the waves and the motion of the platform (resulting from the force of the wind harnessed by the sails, as well as the activity and weight of the occupants) forces captured air through a Wells turbine at the top of the water column. As the water level relative to the platform descends, it draws in air from the exposed top of the water column, also powering the turbine, which rotates in one direction regardless of the direction of airflow.

Oscillating Platforms is playful as well as functional, inviting interaction with people of all ages at any time of the year.

TEAM:
Felix Cheong

TEAM LOCATION:
Toronto, Canada

ENERGY TECHNOLOGIES:
wind sails, oscillating water column, Wells turbine

ANNUAL CAPACITY:
34,000 MWh

Sectional diagram illustrating how an oscillating water column converts tidal energy into usable electricity

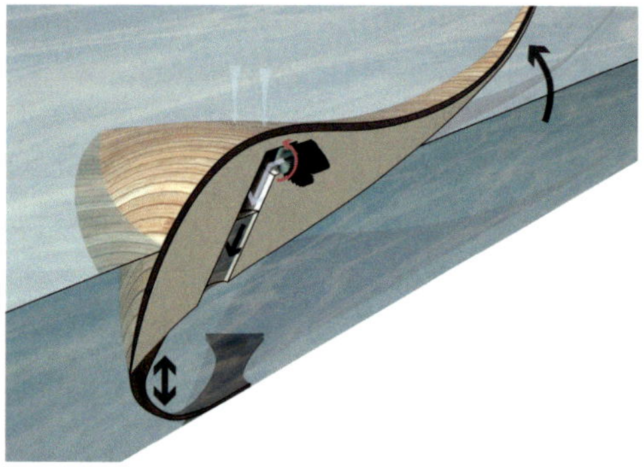

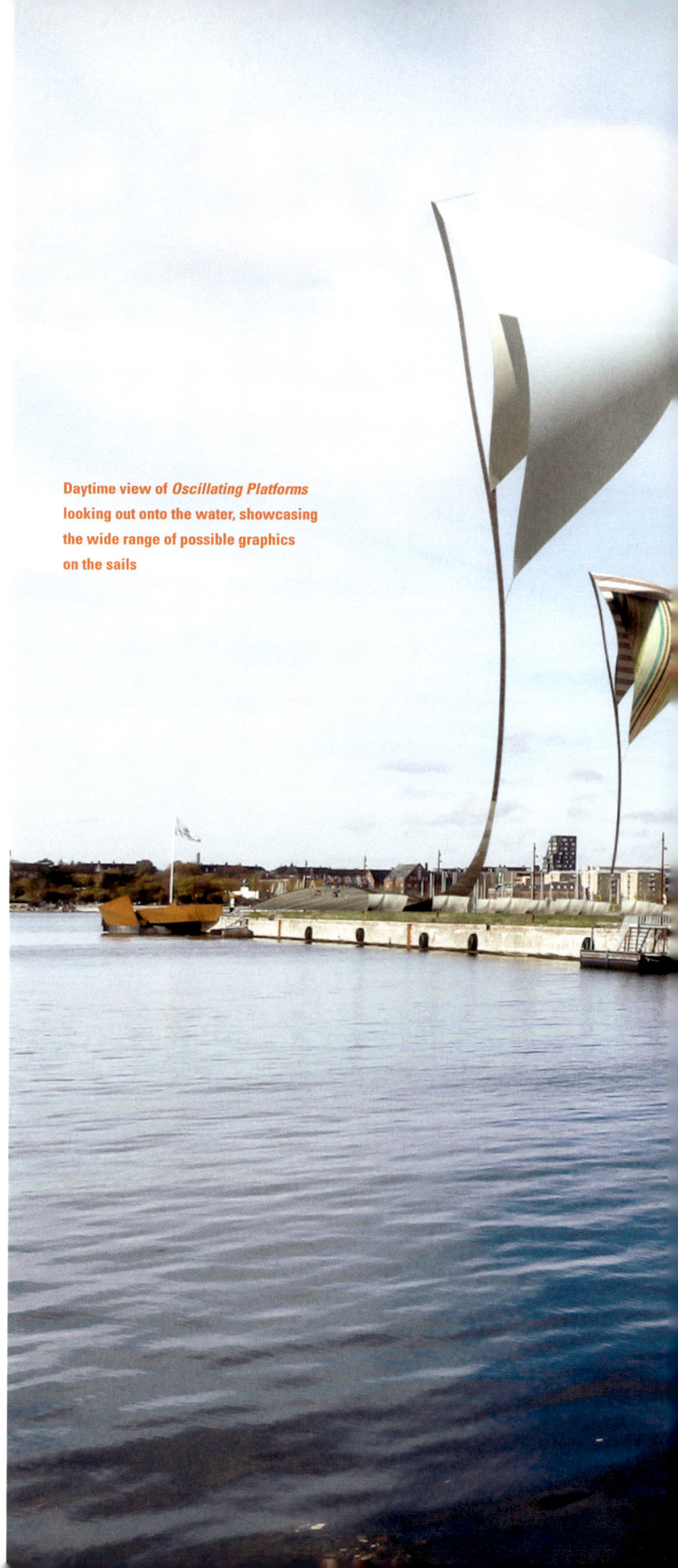

Daytime view of *Oscillating Platforms* looking out onto the water, showcasing the wide range of possible graphics on the sails

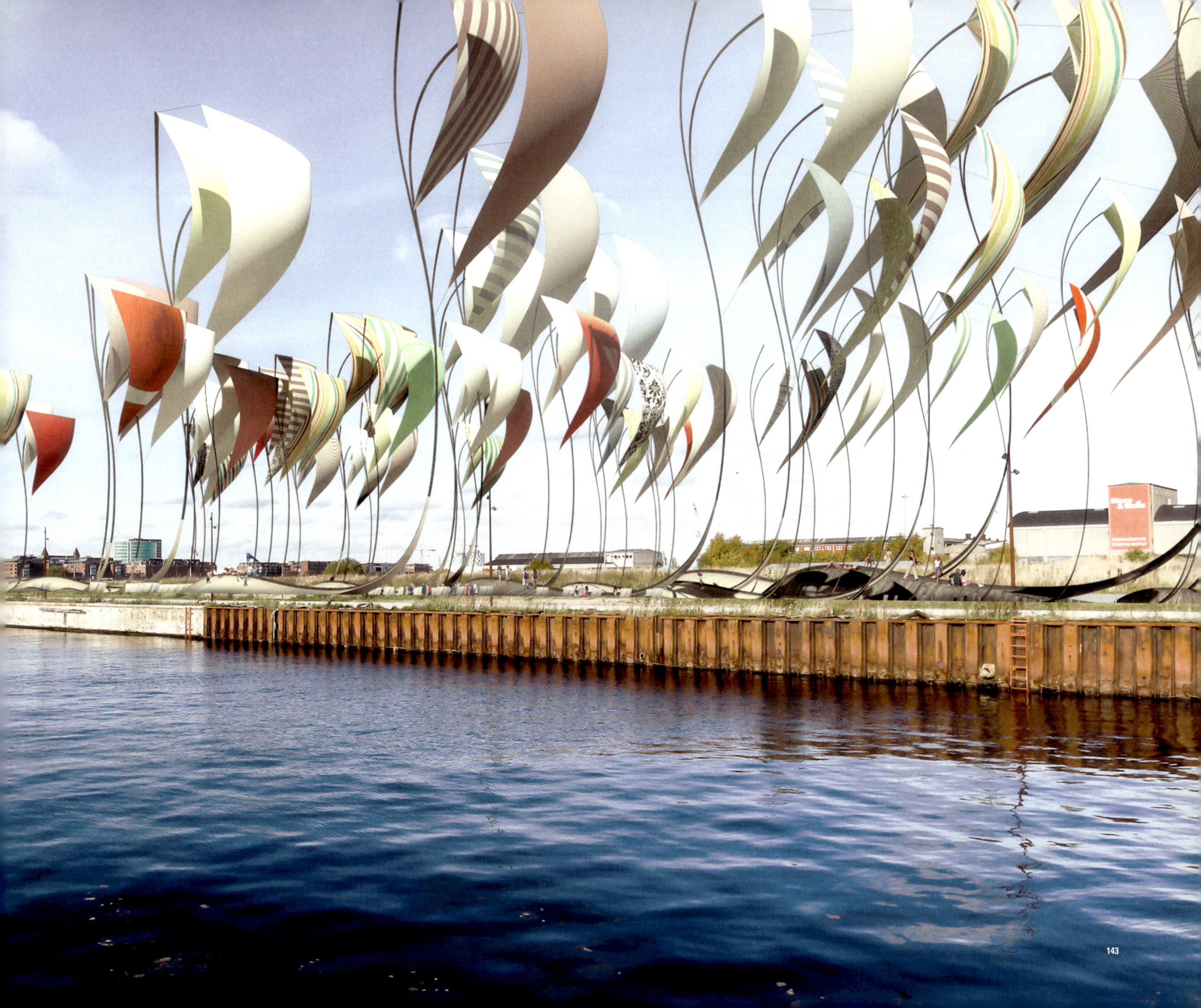

Oscillating Platforms

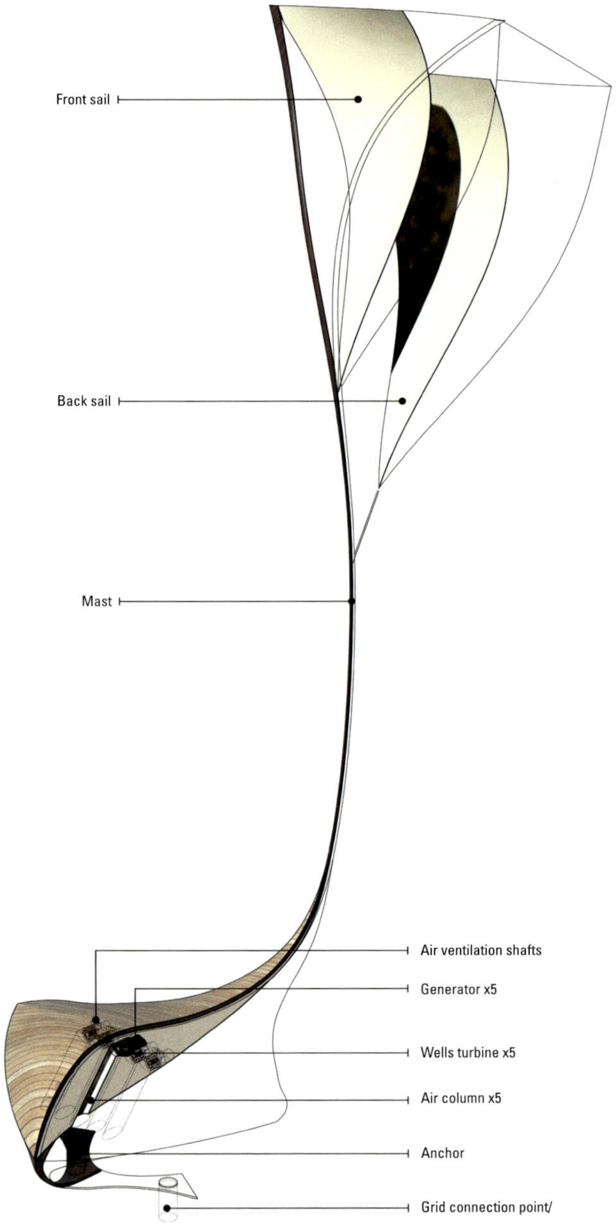

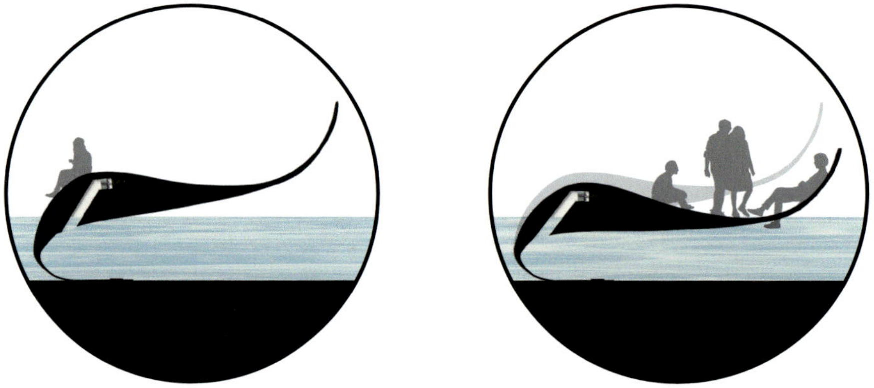

Diagram illustrating live load conversion to sustainable energy

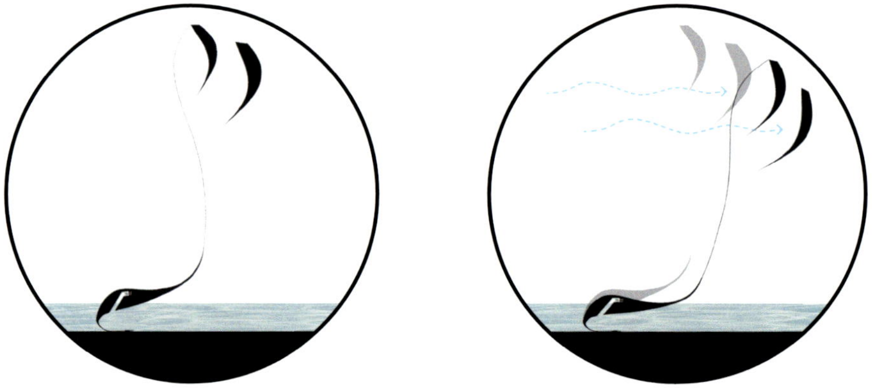

Diagram illustrating wind load conversion to sustainable energy

Three sizes of platform and sail

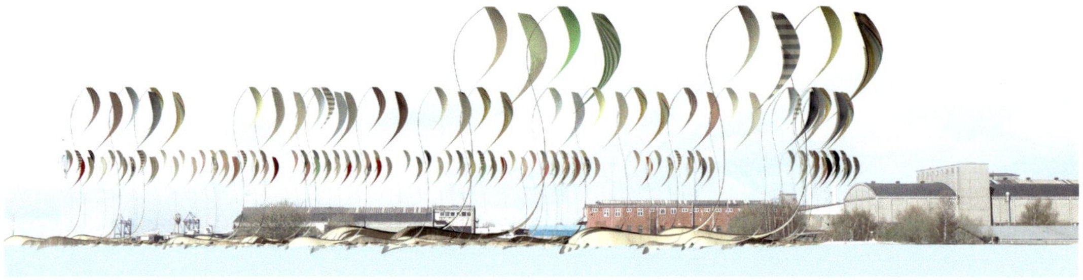

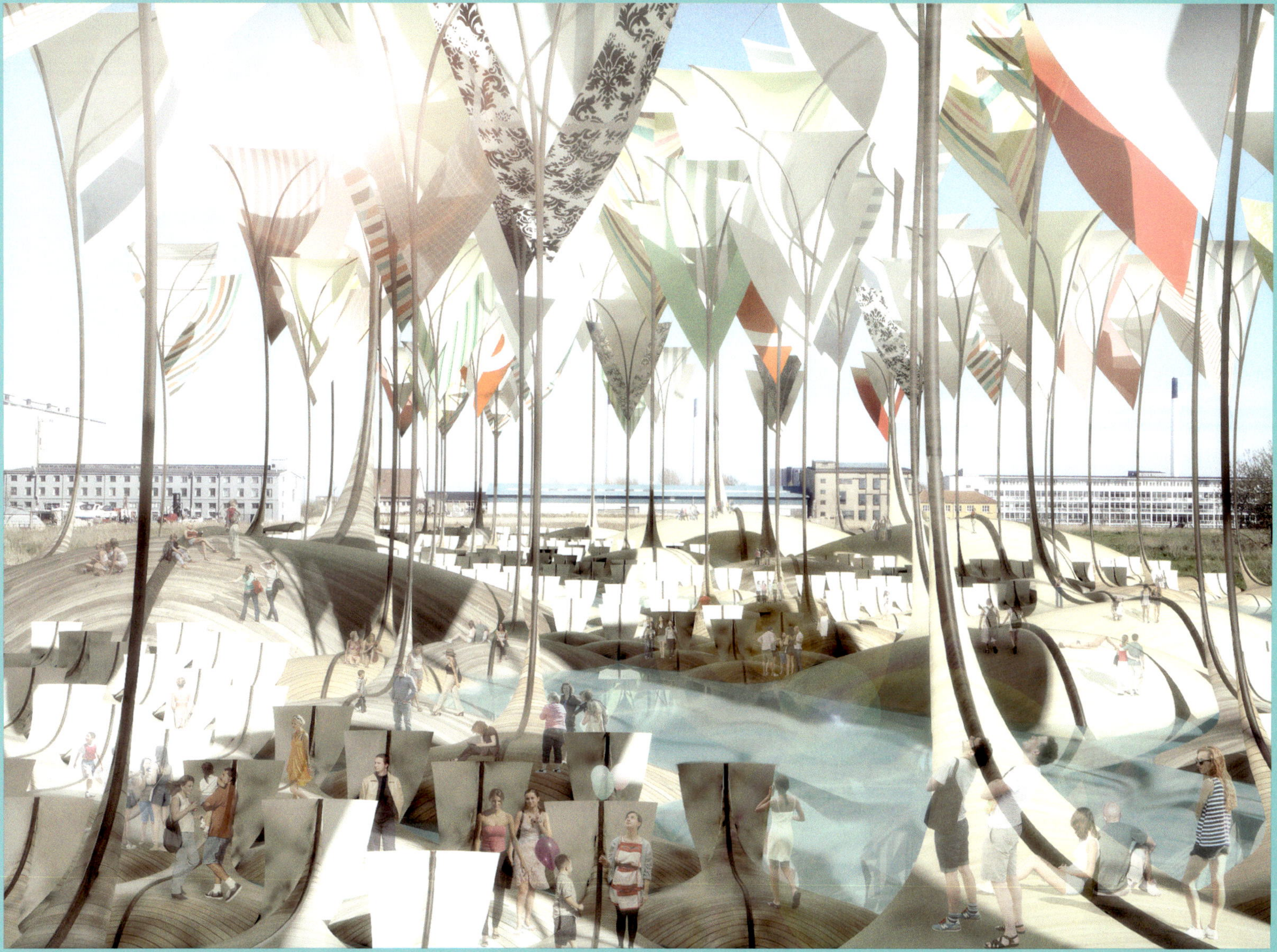

Visualization of activities on the platforms, depicting the playful relationship the installation cultivates between visitors and the site.

Super Cloud

TEAM:
Lucas Jarry, Rita Serra e Silva,
Lucas Guyon, Marianne Ullmann

TEAM LOCATION:
Prague, Czech Republic

ENERGY TECHNOLOGIES:
piezoelectric discs

ANNUAL CAPACITY:
2,487 MWh

Super Cloud mimics the movement of wind patterns. As each tube floats between ground and sky, vibrating in the wind, it acts as a dynamic element within the landscape. Art, energy, and the land act together in a cohesive system, delivering clean, renewable energy.

The *Cloud* embraces, enhances, and interacts with its environment, amplifying its changing states. Shadows, vibrations, and reflections create a unique aesthetic experience through time, weather, and season.

Super Cloud is made up of pliable carbon fiber tubes. A steel space truss structure (270 meters x 60 meters) lifts from the ground and anchors approximately 14,500 tubes (ranging from 2 meters to 10 meters). Half of these tubes are fastened to the top of the structure while the other half hang from the bottom. The varied sizes of circular, sectioned tubes capture the ever-changing direction of the wind.

Located on a site that is largely exposed to wind patterns, the basis of energy production comes from converting kinetic energy into electrical energy. Piezoelectric generators at the base of each carbon element harvest the kinetic energy of the wind and transform it into electrical energy.

From afar, an enormous oscillating mass appears to rise from the land and drift over the sea, revealing an intriguing complexity of reflections and fluctuating light levels. Shivering softly with the wind, it clearly establishes its main purpose: permanent and continuous movement.

Closer to the Island, the structure's enormous dimensions start to unfold and create a distinct image. What originally looked like a mass is now recognized as individual tubes that continuously resonate with the wind.

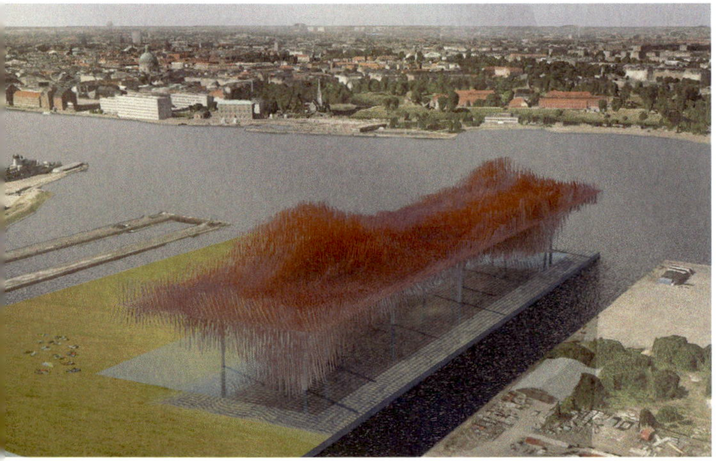

View from the top of the large B&W ship hall

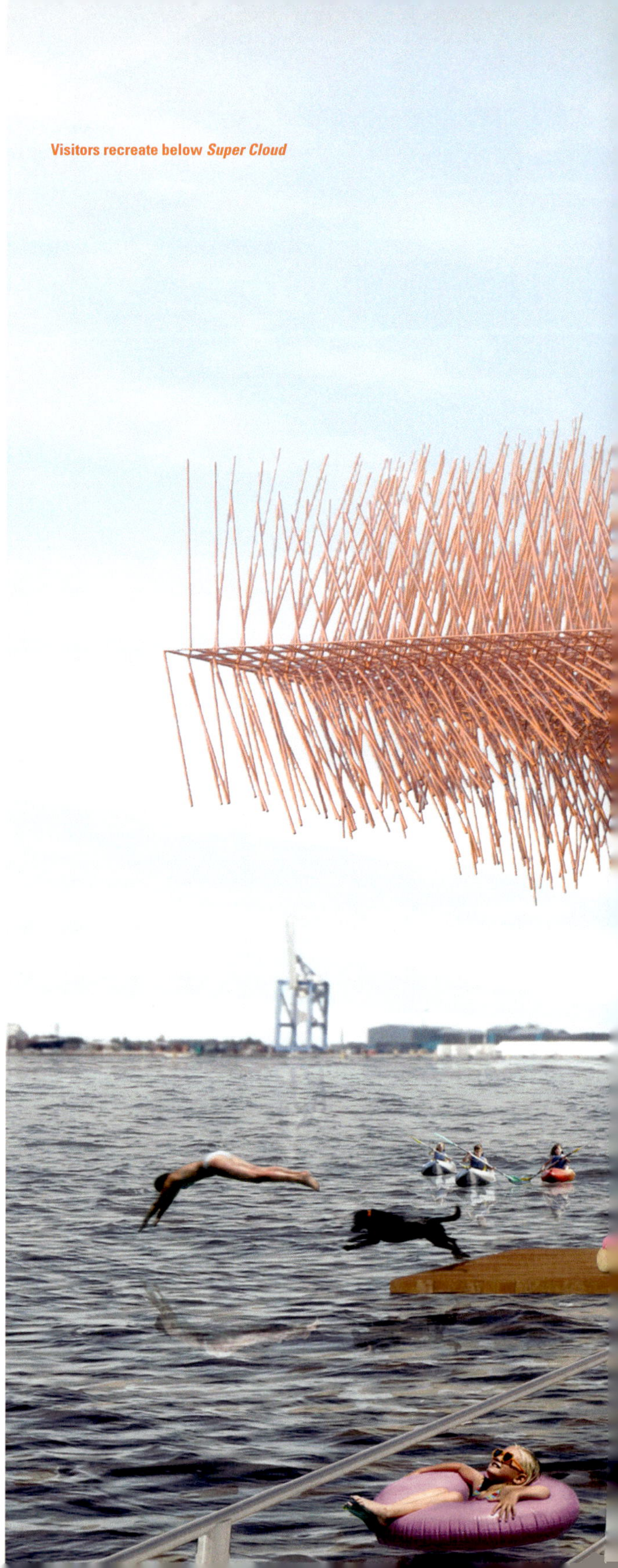

Visitors recreate below *Super Cloud*

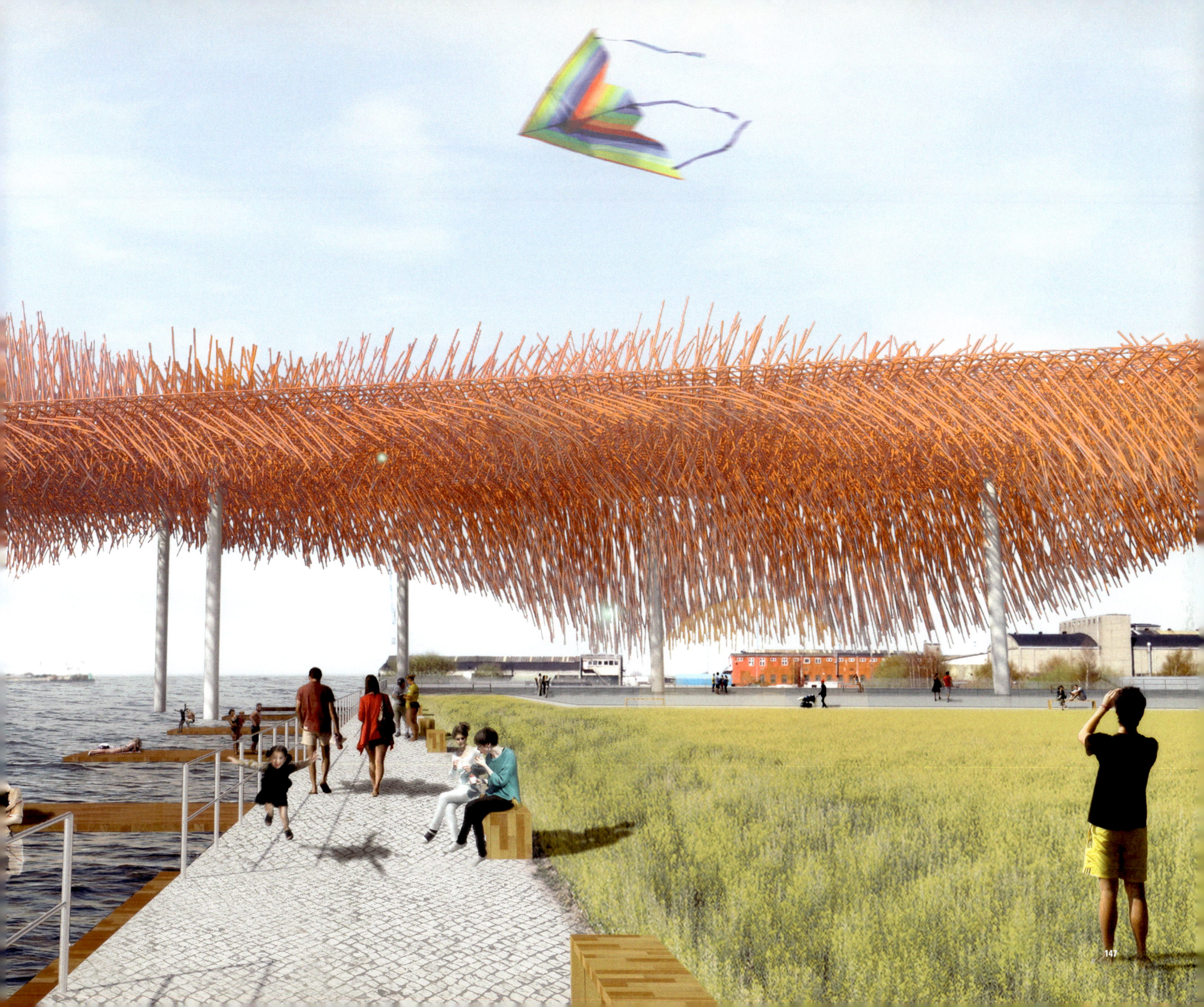

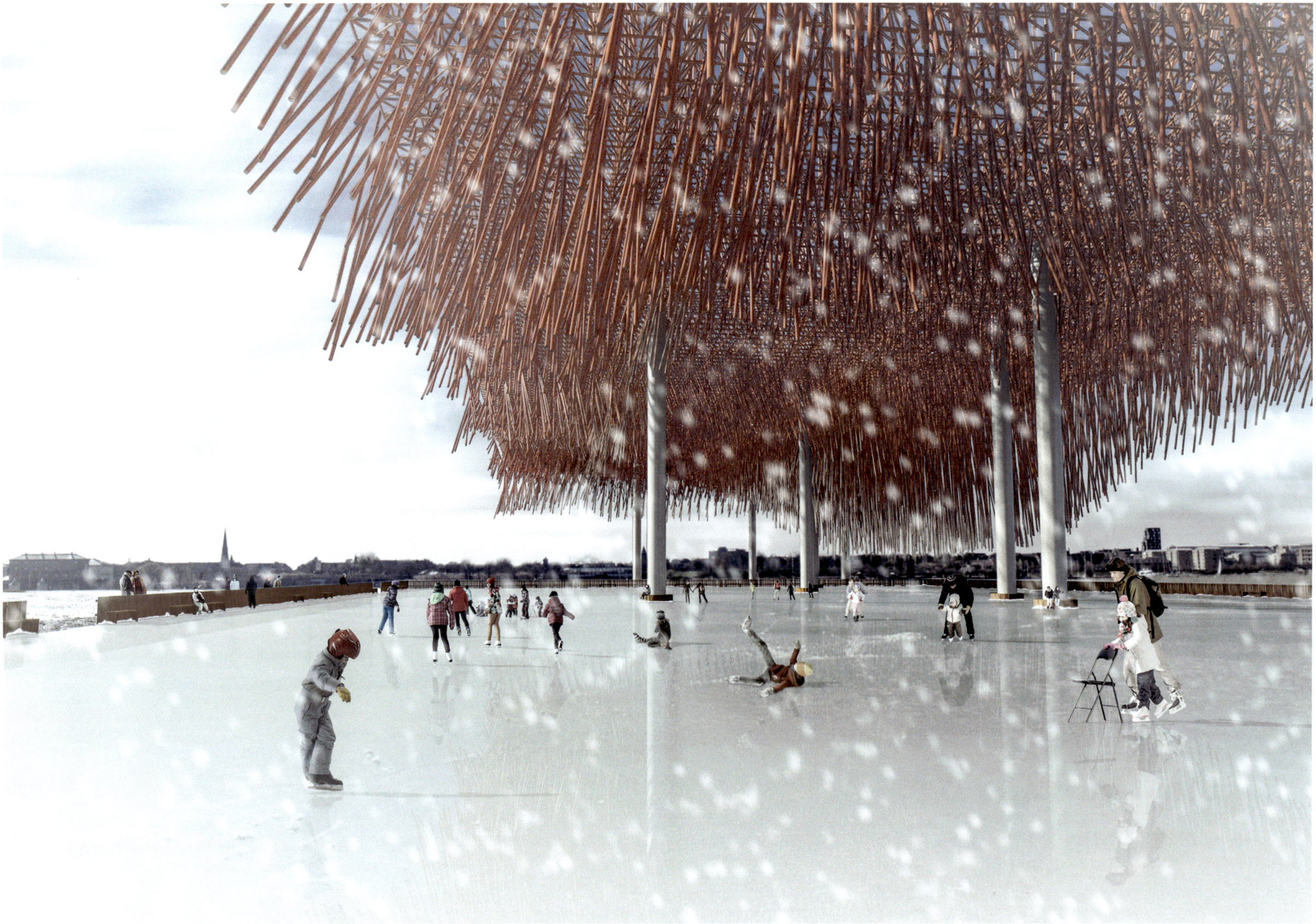

Winter recreation

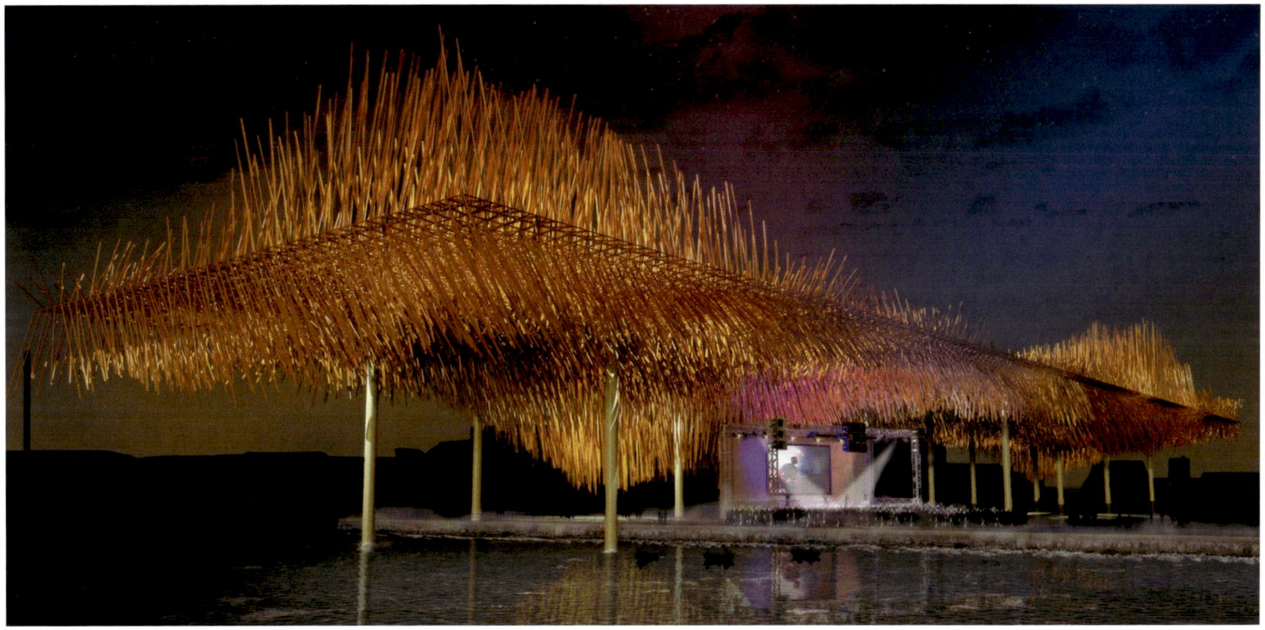

View of a nighttime event

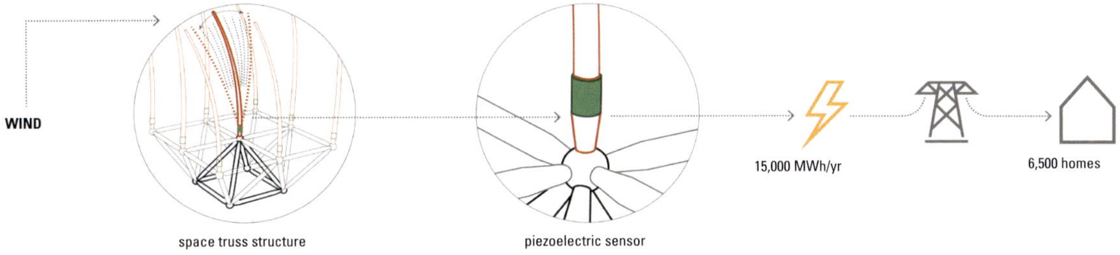

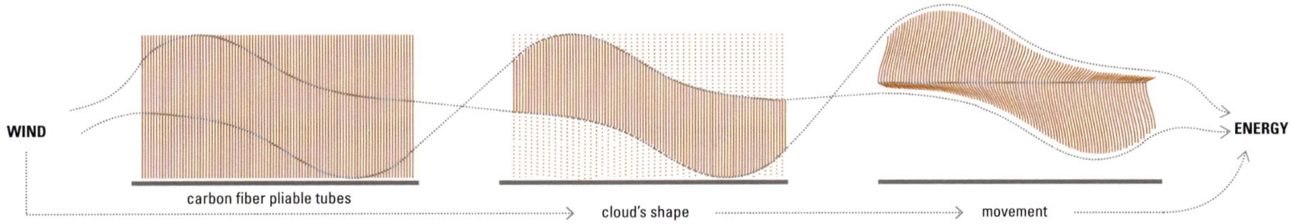

Algaescape

TEAM:
Tobias Anderson, Adam Pajonk

TEAM LOCATION:
Münster, Germany

ENERGY TECHNOLOGIES:
microalgae bio-reactors (algae biogas production)

ANNUAL CAPACITY:
436 MWh

Algae tubes, which are typically used in algae reactors, are replaced by flexible plastic sleeves and woven into a textile for *Algaescape*. During the weaving process, different layers can be applied to the algae mesh along with a supply and control system, which is necessary for algae growth. This textile can then delineate and connect the spaces below and above it. Variations in transparency change the appearance of the mesh with the increase and decrease in the algae population. With all stages in the process open to view, *Algaescape* becomes a place to learn about the science of biomass production.

The installation provides a closed and efficient resource loop between the city, its infrastructure, and energy production. *Algaescape* takes carbon dioxide and nutrient-rich wastewater and produces algae biomass and oxygen through photosynthesis. It works like a filter for the city, cleaning the air and water. The algae can be processed for food production, cosmetics, and for energy production in the form of either liquid fuel and/or gas.

The *Algaescape* mesh is designed to be inhabited by visitors who can enjoy a variety of activities, such as playing sports, viewing the cityscape, or walking off its edge to go for a swim. Illuminated by energy efficient LED lights at night, *Algaescape* will be visible from across the harbor.

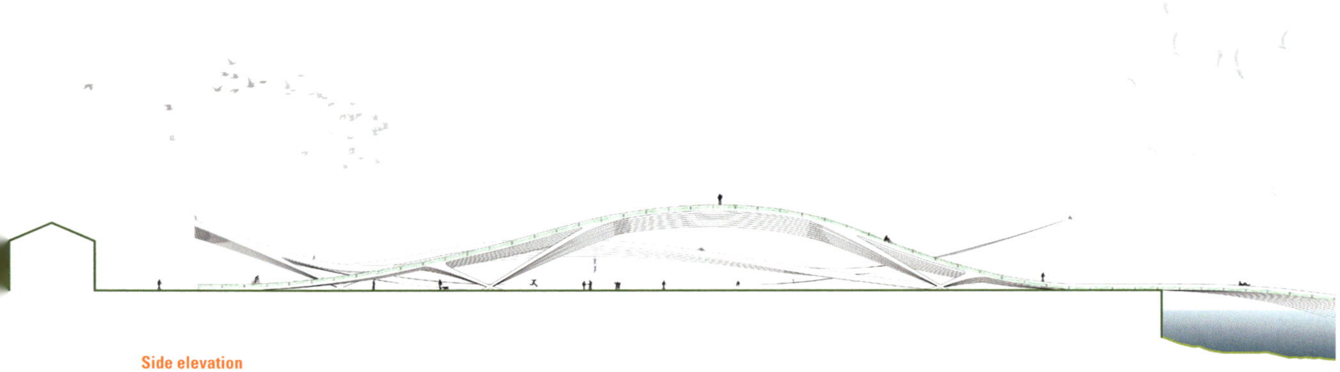

Side elevation

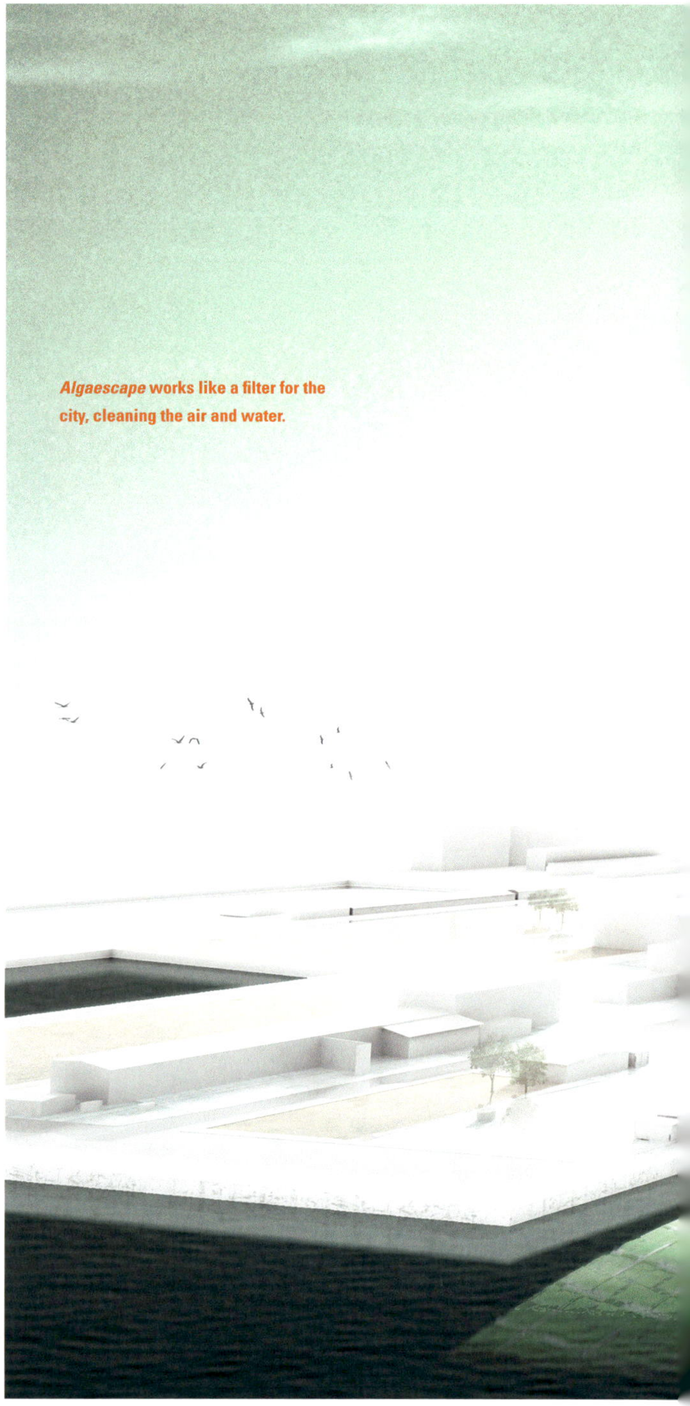

Algaescape works like a filter for the city, cleaning the air and water.

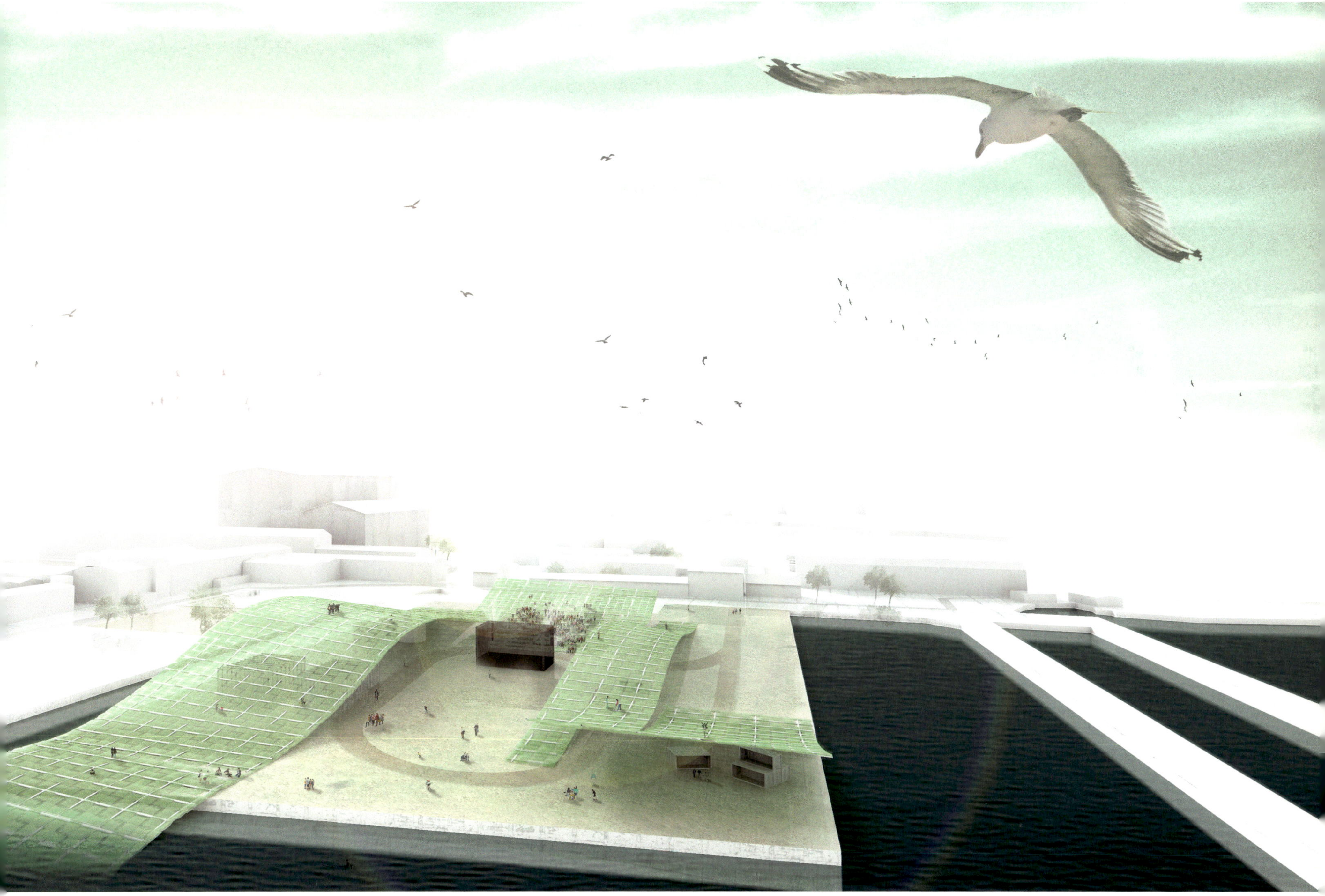

Algaescape

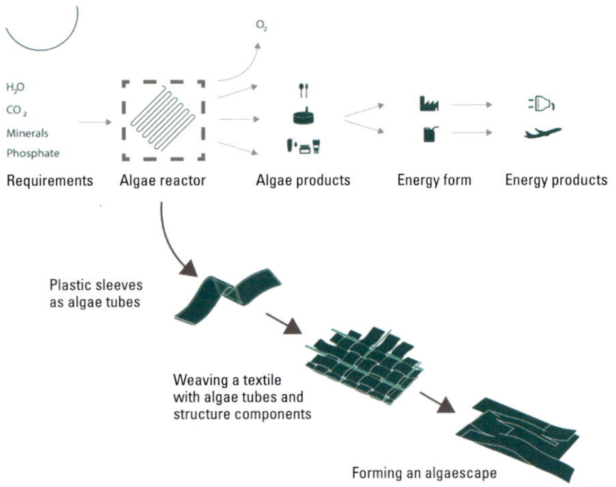

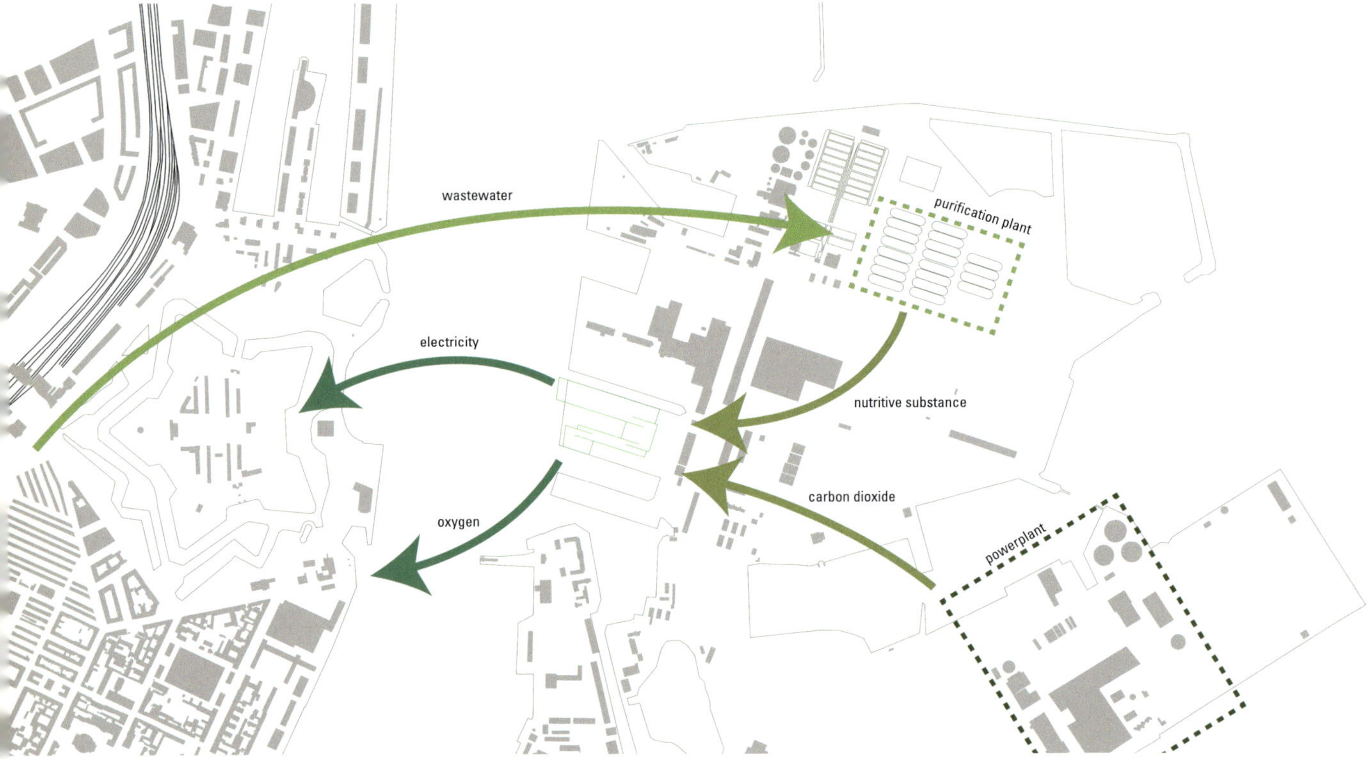

Diagram of the complete cycle of resources

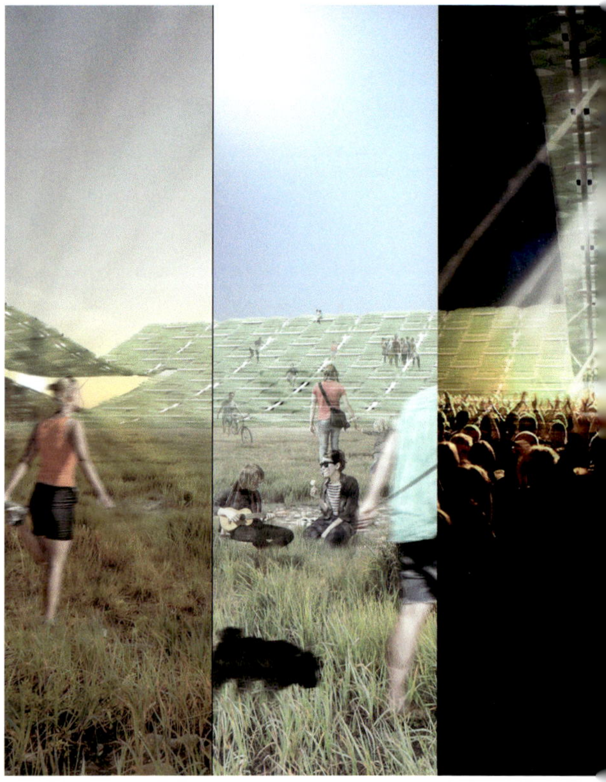

The installation has different uses throughout the day and night.

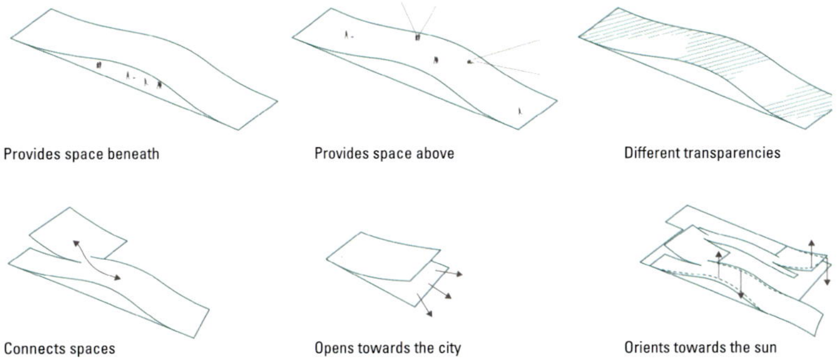

Form derivation

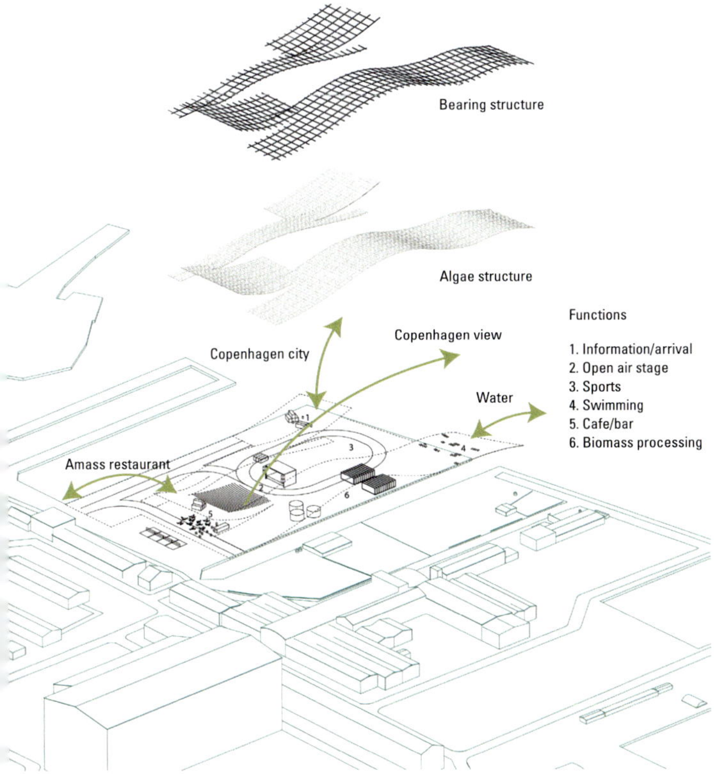

Site diagram

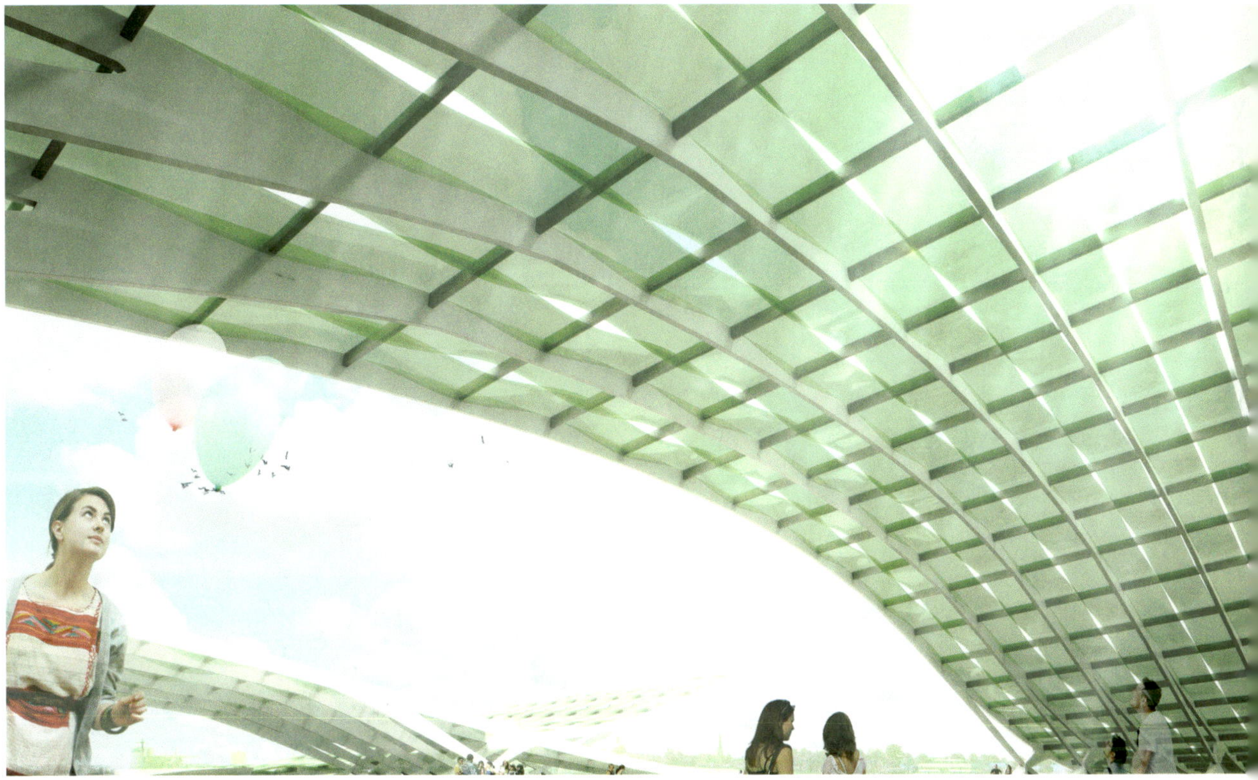

Visitors underneath the mesh

Sail

TEAM:
Andrew Jepson-Sullivan,
Grayson Morris

TEAM LOCATION:
Eugene (OR), USA

ENERGY TECHNOLOGIES:
aeroelastic flutter (Windbelt™)

ANNUAL CAPACITY:
603 MWh

The word "sail" is defined as an object extended on a mast in such a way as to transmit the force of the wind into power. Used as a verb, "sail" suggests a voyage, as in moving toward a renewable energy future.

Sail projects prominently from the west corner of the former shipyard into the prevailing wind, providing an elegant symbol of clean energy for the City of Copenhagen. The form recalls the shipbuilding past of the harbor, transforming from the rigid cup of a ship's hull to the graceful, unfurled surface of an open sail. A curled base results in limited ground-level impact, while the open face at the crest maximizes surface area for wind collection through the use of Windbelt™ technology.

In this design, the site at Refshaleøen is left almost entirely undisturbed, allowing for large gatherings and events to occur, with *Sail* acting as a visual backdrop. Visitors to the site are invited to "enter the wind" at the apex of the tower, becoming part of the energy infrastructure that powers the city.

As night falls, *Sail* produces a bright, fluctuating spectacle. Each Windbelt™ membrane on *Sail* is coated with a thin OLED layer (organic light-emitting diodes that emit light in response to an electric current) that glows when the membrane oscillates in response to the wind. OLEDs can be printed onto flexible and transparent substrates, making the membrane of the Windbelt™ an ideal surface. The OLEDs fluctuate in brightness depending on how much power is being generated by each individual membrane, resulting in a shimmering light display that reflects how much power is being produced by the entire structure in real-time.

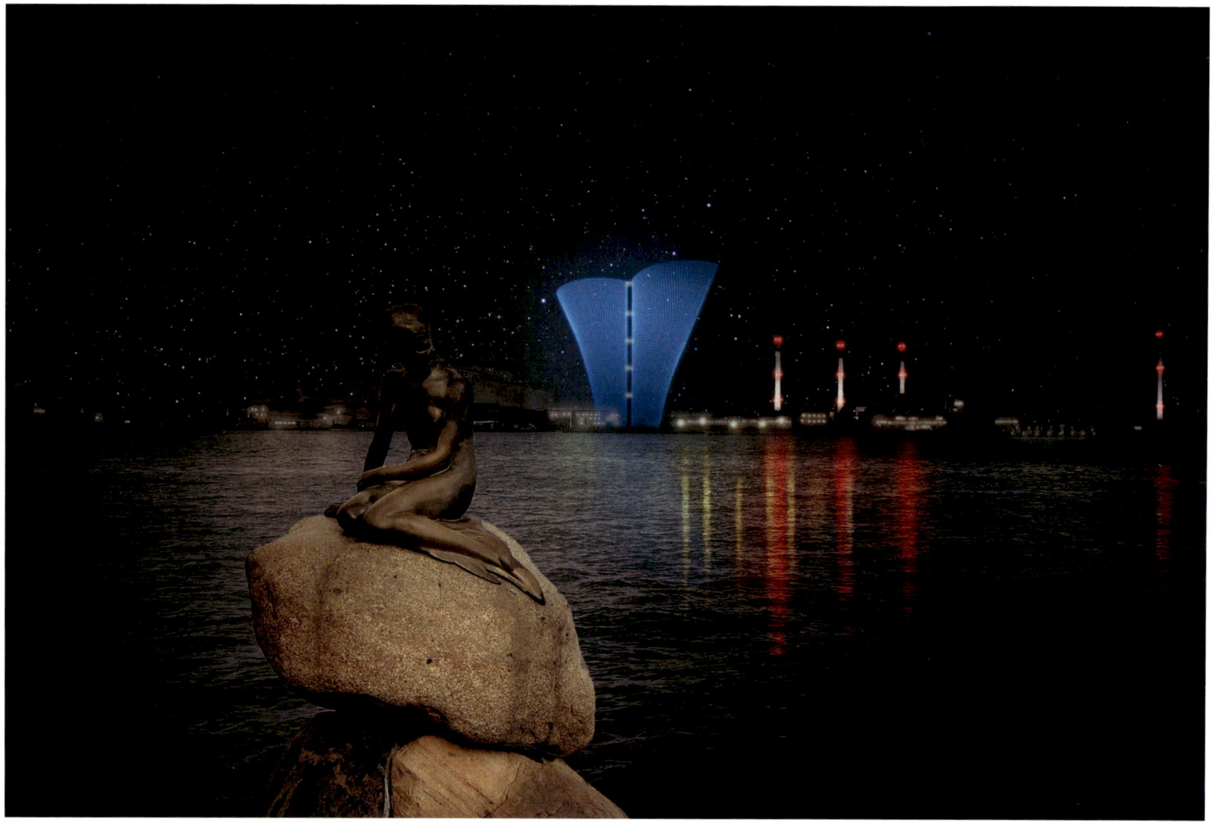

Nighttime view from across the harbor

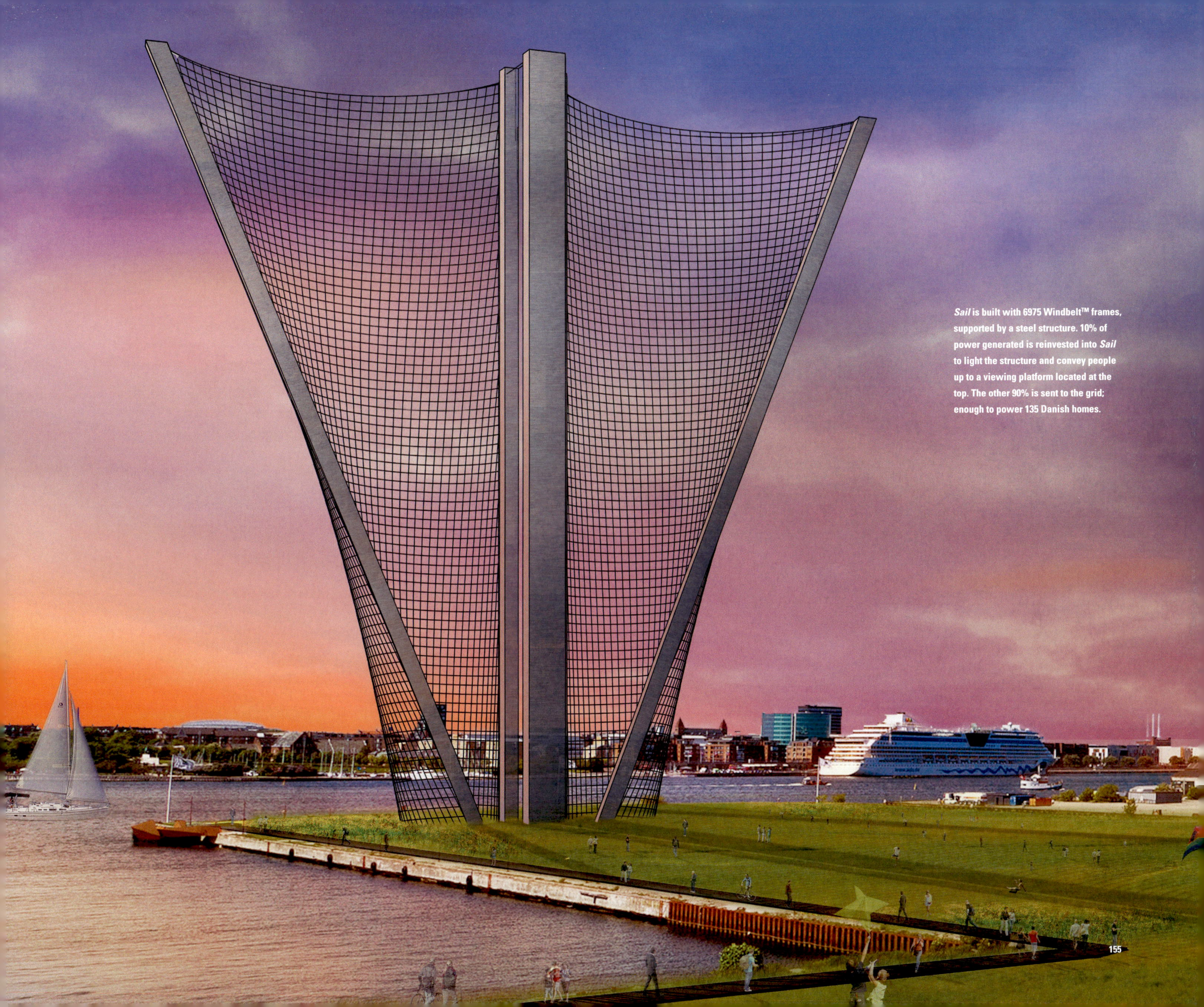

Sail is built with 6975 Windbelt™ frames, supported by a steel structure. 10% of power generated is reinvested into *Sail* to light the structure and convey people up to a viewing platform located at the top. The other 90% is sent to the grid; enough to power 135 Danish homes.

Sail

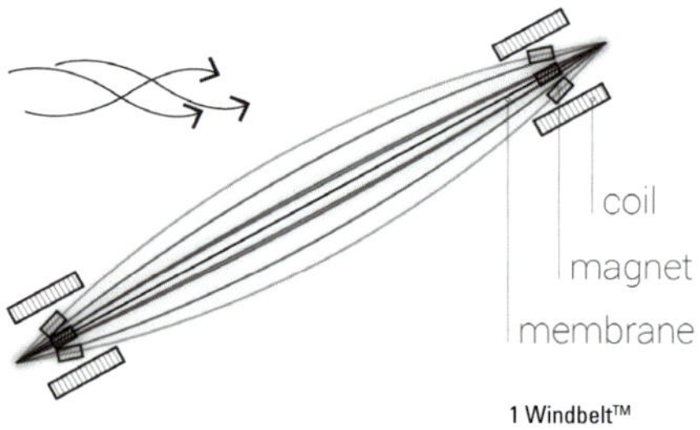

1 Windbelt™
0.36 kWh monthly
4.32 kWh annually

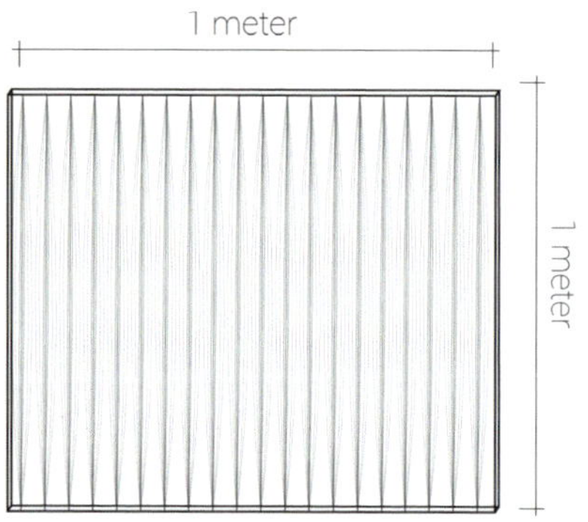

1 Windbelt™ Frame
7.20 kWh monthly
86.40 kWh annually

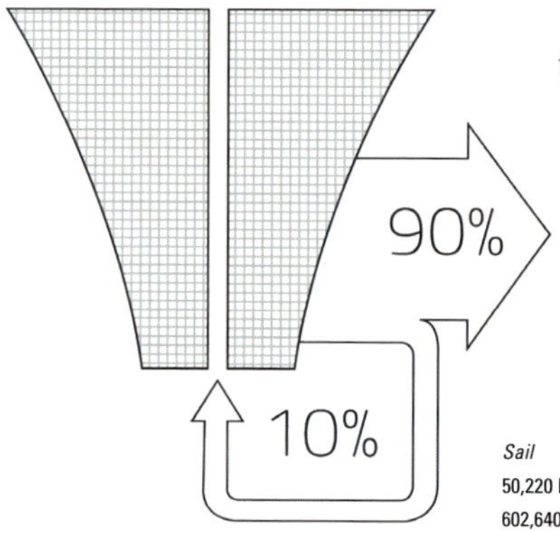

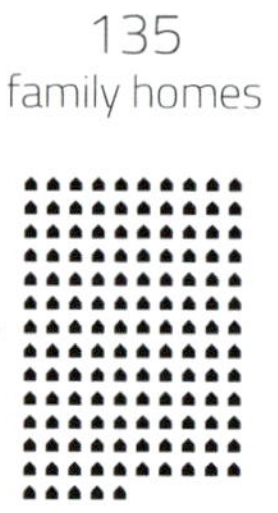

Sail
50,220 kWh monthly
602,640 kWh annually

Windbelts™ use a thin tensioned membrane that catches the wind and vibrates though a phenomenon called aeroelastic flutter. Magnets on the end of the membrane oscillate between copper coils, generating an electric current. *Sail* uses an array of 1x1 meter aluminum frames (each frame housing 20 Windbelts™) to generate energy from the wind at Refshaleøen.

Each Windbelt™ membrane is coated with a thin OLED layer that glows when the membrane oscillates in response to the wind. These glowing Windbelts™ can be programmed to playfully celebrate significant cultural occasions, civic holidays, and important events, allowing *Sail* to become an icon for the City of Copenhagen.

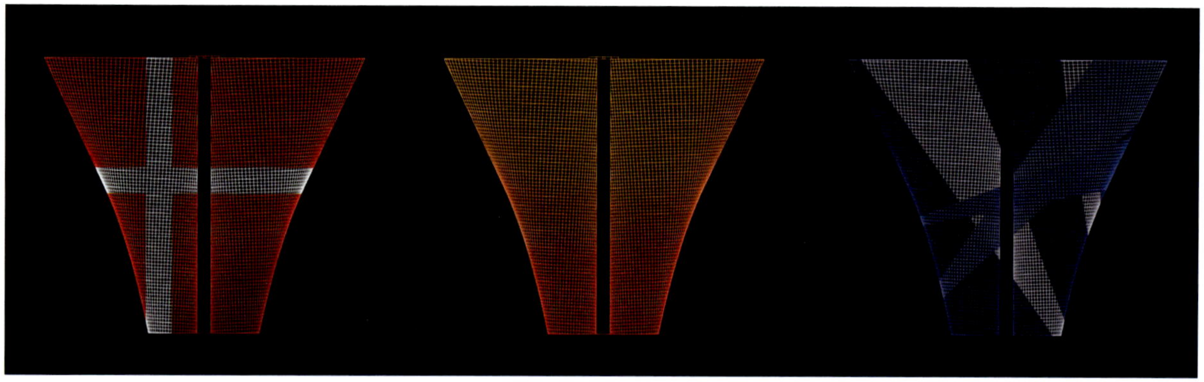

From left to right: Valdemar's Day, St. John's Eve, music event

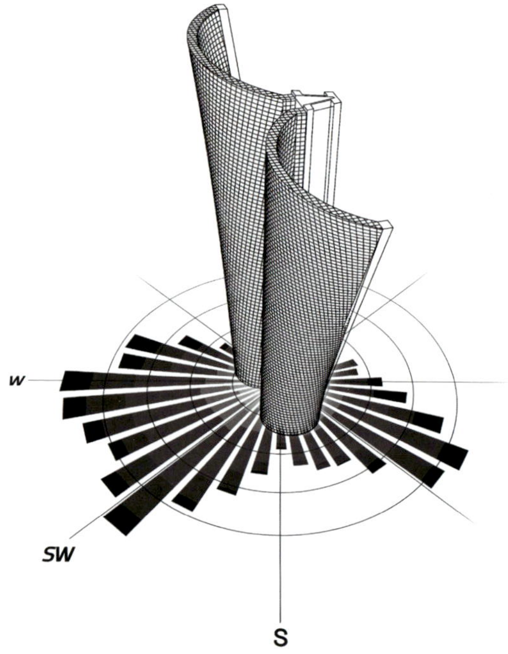

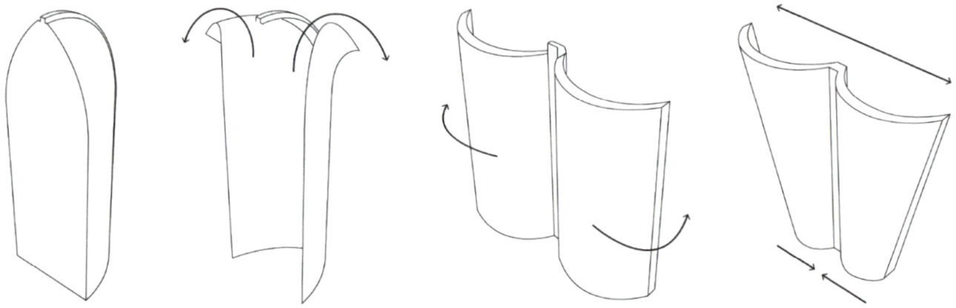

Sail's form is an allusion to the shipbuilding history of Refshaleøen. Starting with the shape of a ship's hull, the structure peels back at the bow and curls open to resemble an unfurled sail. This form also limits impact on the site by providing a small footprint, while at the same time increasing the surface area of the structure where gusting winds are most prevalent, maximizing power generation capabilities.

Sail is oriented toward a steady and predominately westerly wind. Wind speed and direction make Refshaleøen an ideal place for an artful and iconic wind power generator.

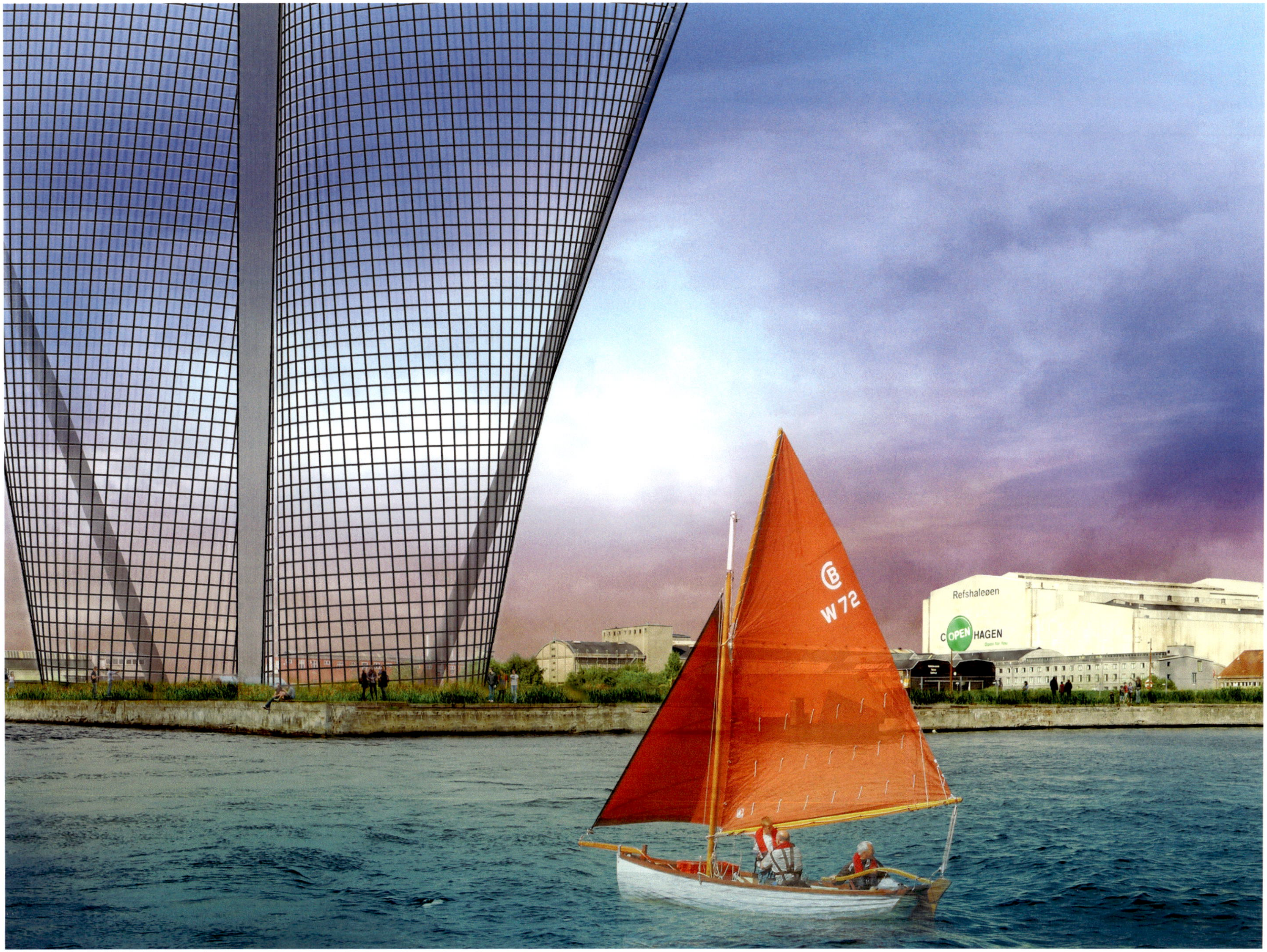
W 72
Refshaleøen
COPENHAGEN

Stone Wave

As land in Copenhagen was reclaimed from the sea, the sea now becomes the land. *Stone Wave* stands at a meeting point between the energy of water and human energy. It allows the water to move the ground, and it allows visitors to walk on water… almost.

The installation is composed of 5,742 paving units, suspended in a water channel carved into the land from the sea. The basaltic stone is extracted locally to reduce transport waste.

The energies resulting from the meeting of human and wave are converted into electricity by harnessing the forces of movement. The weight of visitors walking on the paving units that float on the water causes the units to slowly sink. By this action, a small turbine at the base of each unit is driven to create electricity. Water movement from the wake of a passing boat or from the subtle action of the tides and winds act on each unit to drive the turbines in the opposite direction, while the stone pavers rise back up to a resting position, buoyed by watertight boxes fixed under the stone blocks.

The technologies used are inspired by mechanical spinning tops and by R. Buckminster Fuller's Synergetics concept, an empirical study of systems in transformation.

Stone Wave creates a balance between two attributes—a calm and diffuse, yet constant energy.

TEAM:
Nicolas Violette,
Christofer Cornevaux

TEAM LOCATION:
Paris, France

ENERGY TECHNOLOGIES:
hydraulic cylinder microturbine generators

ANNUAL CAPACITY:
60 MWh

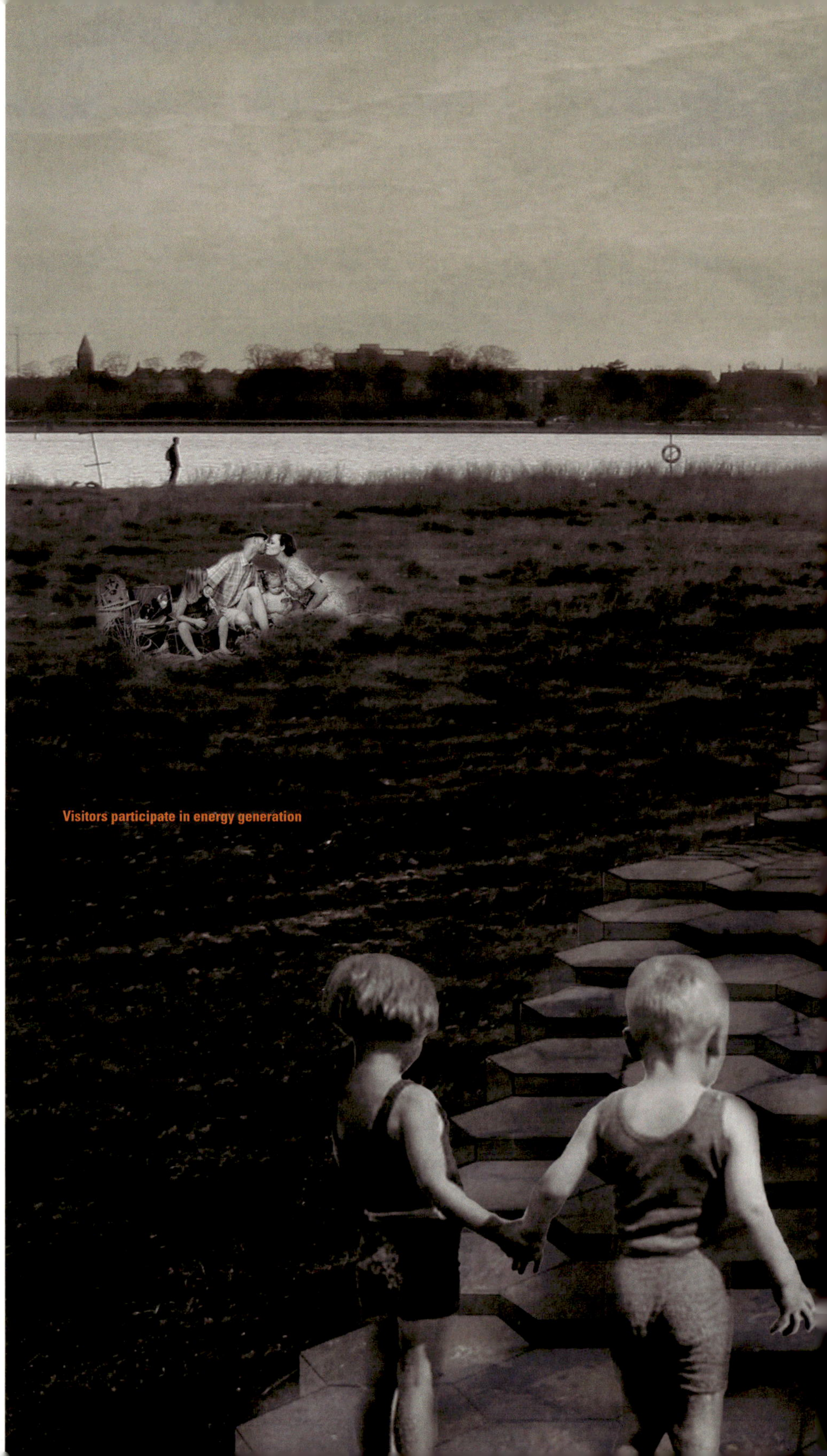

Visitors participate in energy generation

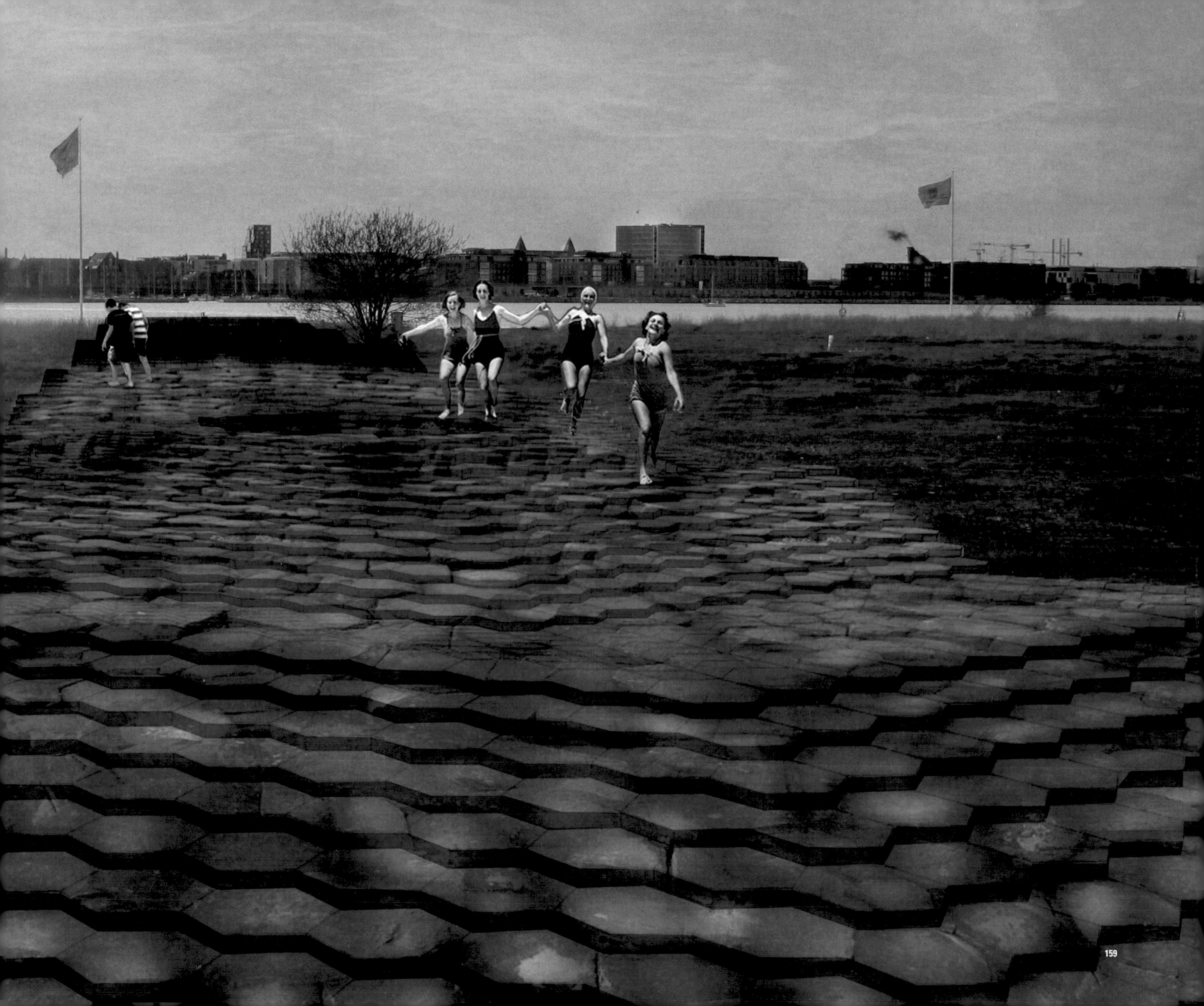

Stone Wave

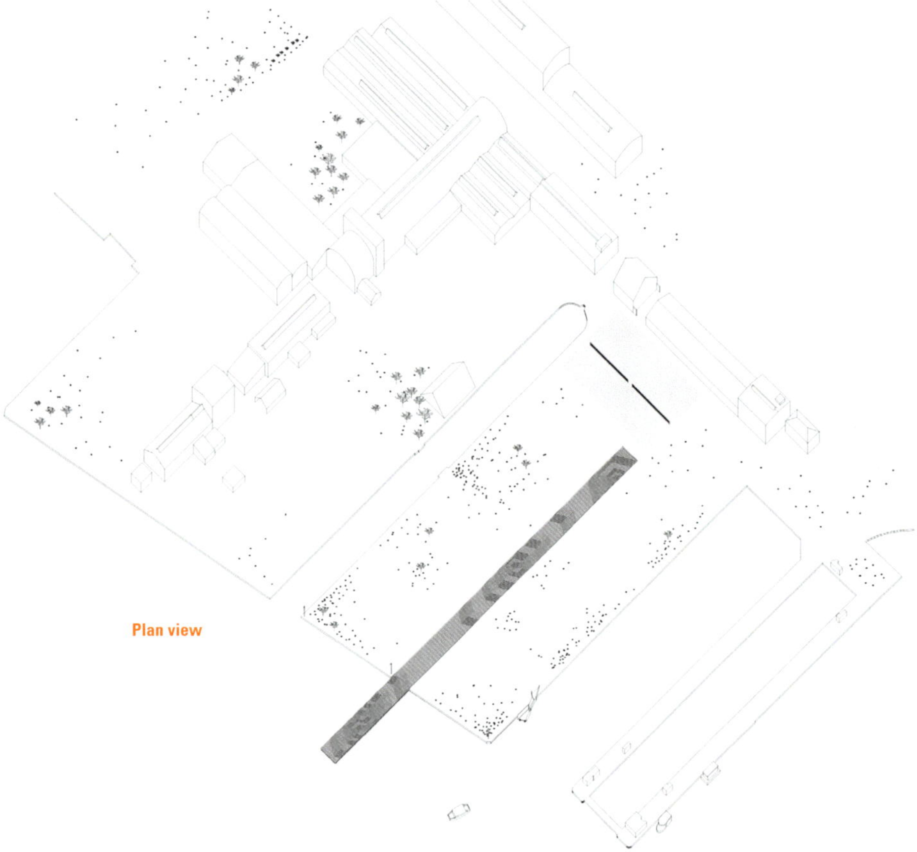

Plan view

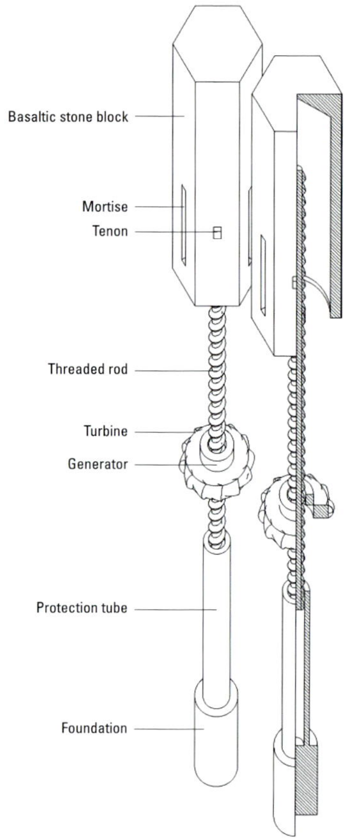

Detail of generator assembly

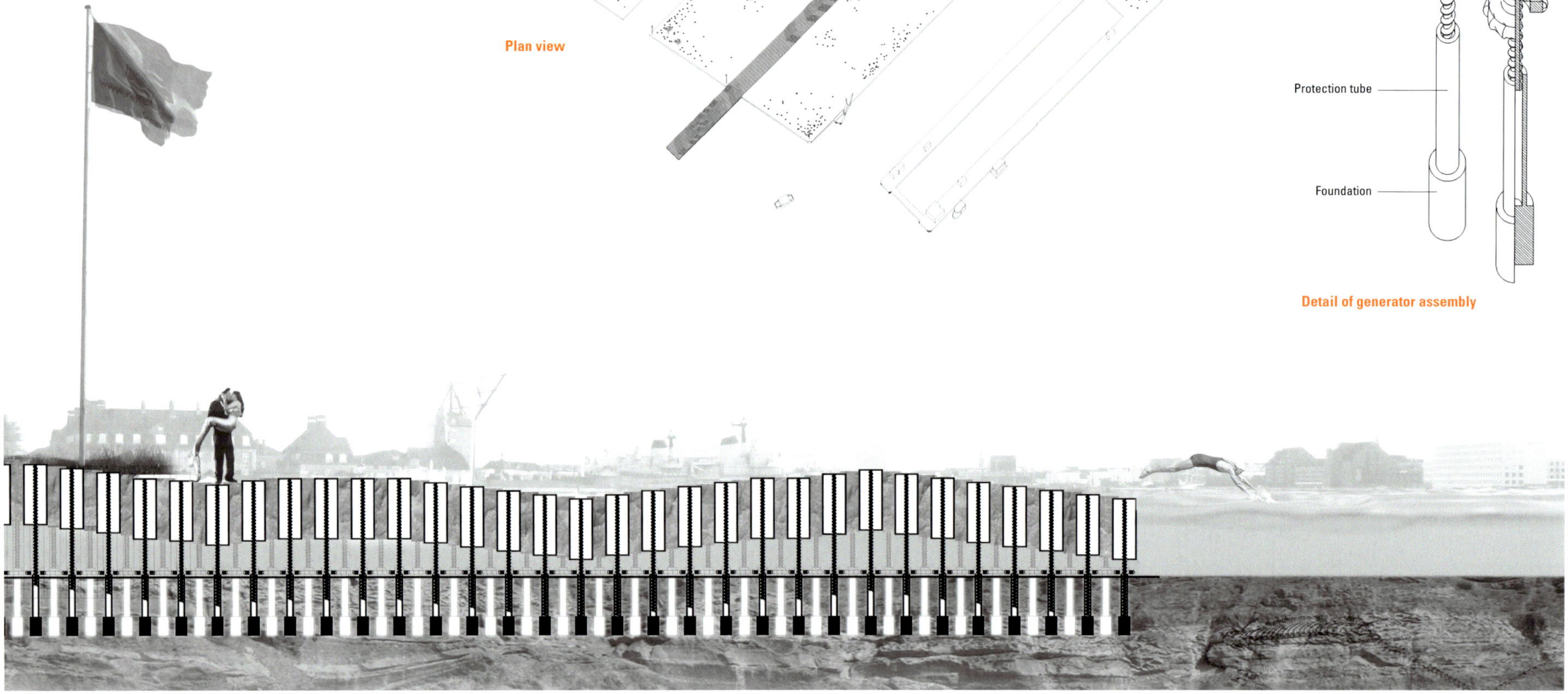

The weight of visitors helps to run the turbines.

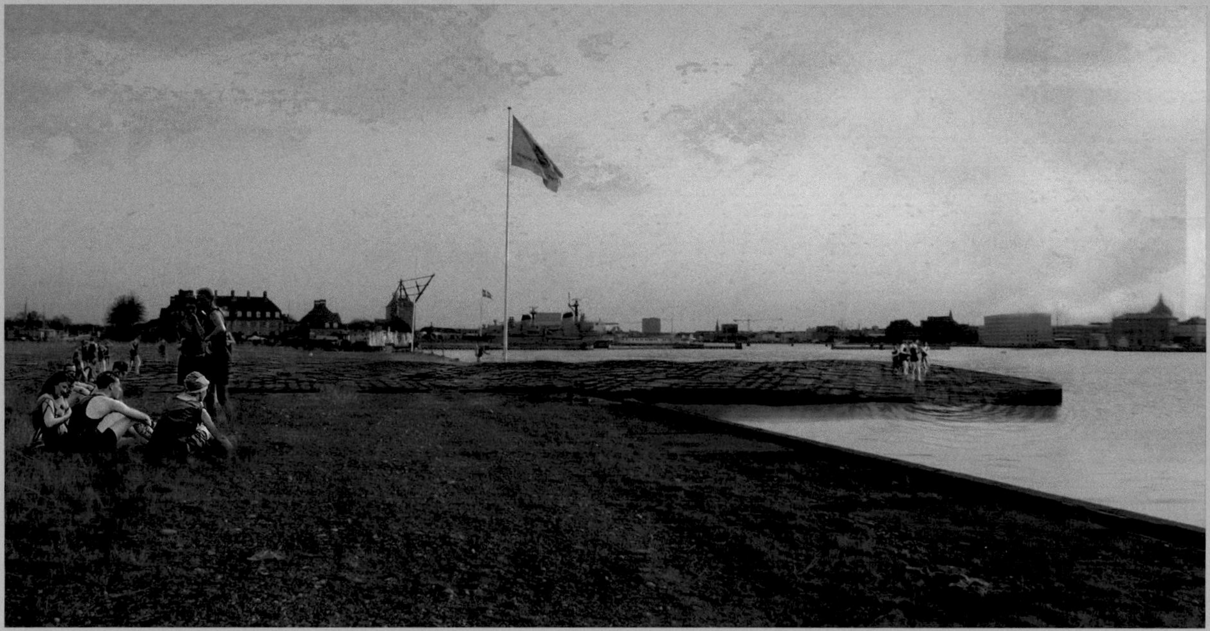

The *Stone Wave* reaching out into
the harbor

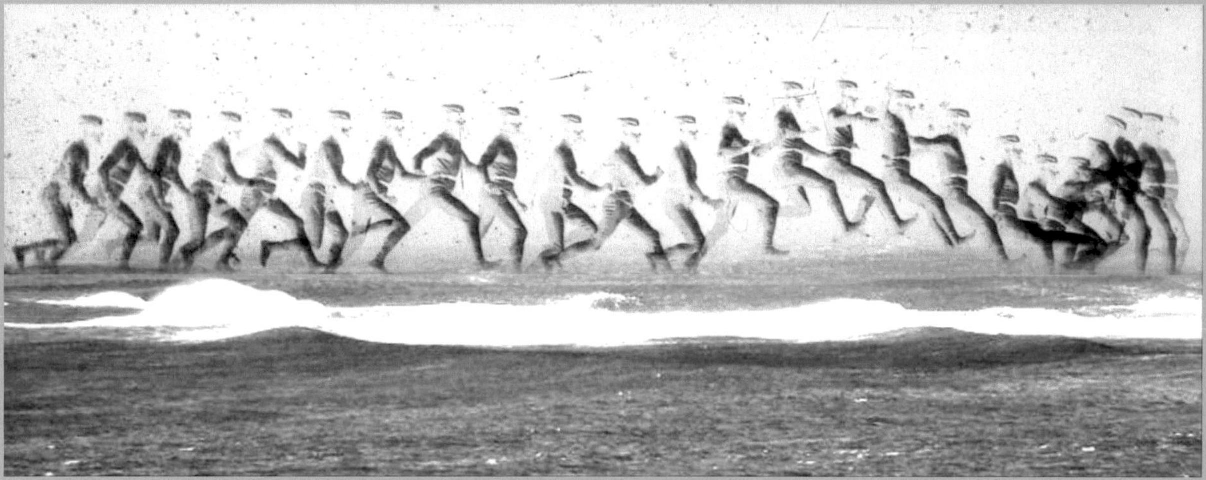

The kinetic energy of running visitors
is transformed into electricity.

SELECTED ENTRIES

The Hybrid Garden:
The Story of the Chimera

Postulate

The By-Cycle

Factoryscape

Emotional Fog

Ripple

WindWaker

Refkløver

Sails: An Homage to Seafaring

Regatta Fields

Frame

Cloud

100*100*100 Wind Tower

The Kyst

Floating Forest:
Between the Trees

Golden Roots

Kinergic

Light Guard //
Know Your Flood Risk

A Place for a Chocolate
Ice Cream

Lightfoam

Celebrating the Sun

Tree of Knowledge

Windbrator

Wind Plaza

Rør

Solar Lake

Energy Hive

Blowing Horn

#HelioTweet

Metamorphosis

Kinectart Generator

Palace of the Prince

Tornado Tower

The Mermaid's Forest

Sight

Raised Farm

Silva

Driven by the Wind: Towards a
Green Horizon

Pixelated Wind

“The Land Art Generator Initiative provides new and exciting proposals for approaching the green transition. We saw in the previous competitions in Dubai in 2010 and New York City in 2012, where creative forces of art, architecture, and engineering together brought forth innovative ideas, concepts, and solutions that can produce green energy while being integrated to beautify the local environment.”

—Martin Lidegaard

Former Danish Minister of Climate, Energy, and Building

The Hybrid Garden:

The Story of the Chimera

TEAM:
Alexandra Arènes, Axelle Grégoire, Sophie Fernandez, Paul Riley

TEAM LOCATION:
Paris, France

ENERGY TECHNOLOGIES:
photosynthetic hydrogen production, algae (Cyanobacteria) biofuel

ANNUAL CAPACITY:
384 MWh

Algae are at the heart of *The Hybrid Garden*, covering a large canopy and forming the installation's centerpiece. Visitors will find themselves under an ocean of algae, submerged in an unrecognizable environment. Above, light is filtered through the algae and its water. The canopy is the sky and the sky is a moving, living sea.

The algae's ecological universe unfolds in the installation's garden, a closed and ordered entity similar to the Greek cosmos. It is constituted by the ecosystem's varied parts and therefore embraces the entropy, the chaos that such a space can possess.

The manufacturing of energy is connected to a broader ecosystem. The algae are contained in tubes within which shrimps and other tiny organisms clean the water. The tubes resemble a greenhouse, reheating the air below and creating a micro-climate between the canopy and the ground. When the algae are exposed to the sunlight, they release hydrogen as a byproduct of photosynthesis. The hydrogen can be used to generate electricity in fuel cells, or it can be used directly.

The microalgae's different elements are separated by gravity, which also results in liquid biofuel and solid biomass. The biomass created by the algae will be reused in the installation's garden, the purpose being to avoid the use of pesticides and chemical fertilizers.

Rainwater is collected and recycled as a mist, producing a humid atmosphere for the plants. The plants also purify the soil, absorbing pollutants like metals and oil residue.

On the surface: A green layer of algae tubes is a sea seemingly taken over by machines. The canopy resembles the sea, mimicking its movements rhythmically and refracting any natural light—a window to the sky.

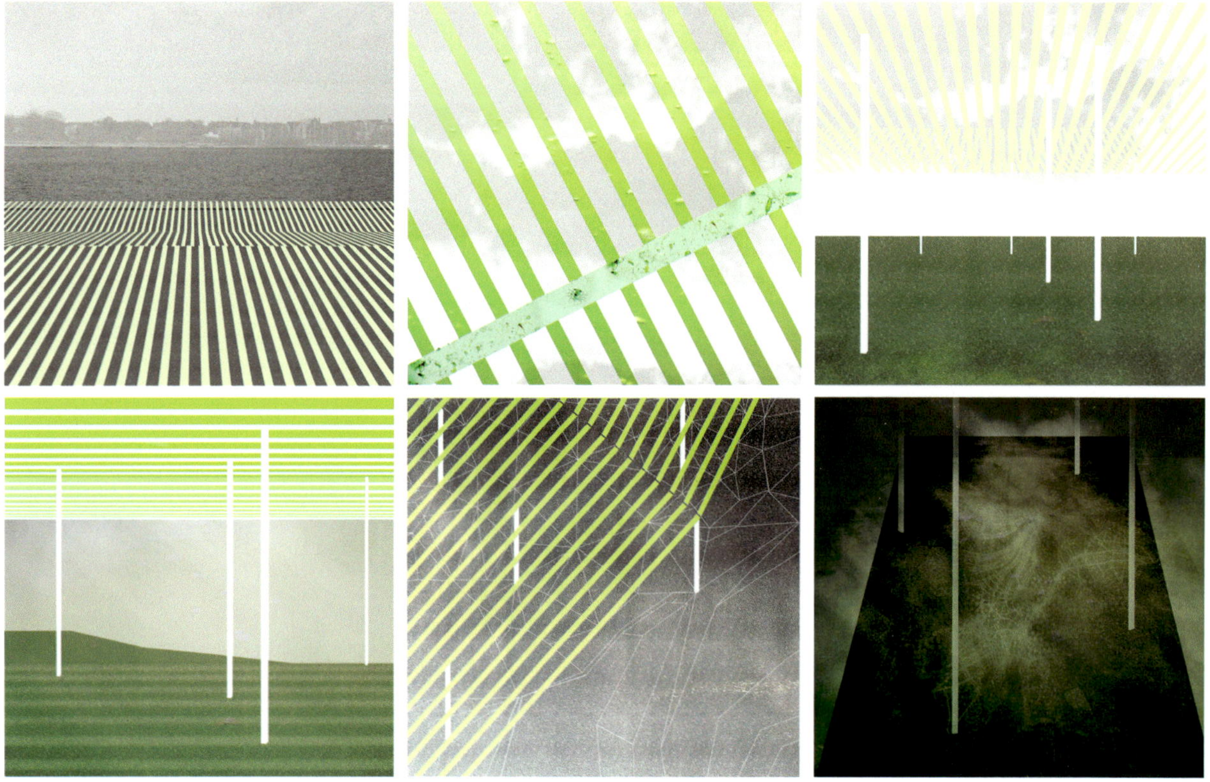

Under the surface: Entering the installation recreates the feeling of diving into the sea and of being underwater. Three different gardens form the route through the installation. The narrative of each garden leads the visitor to search for traces of the imaginary Chimera.

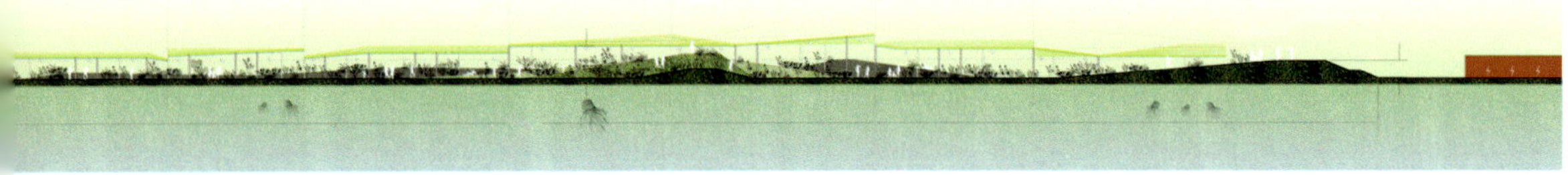

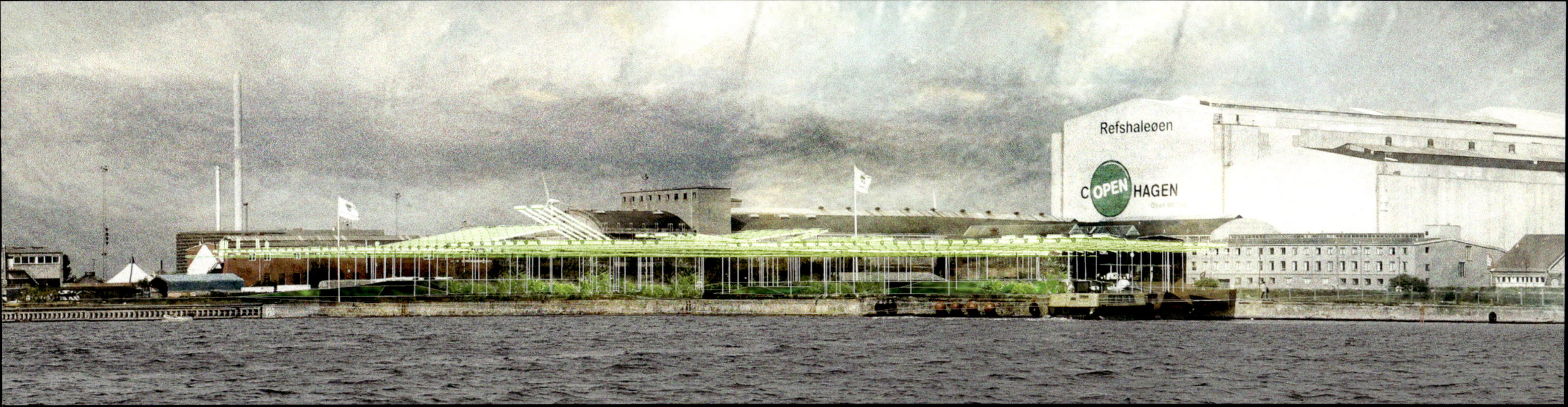

View of the installation from across
the harbor

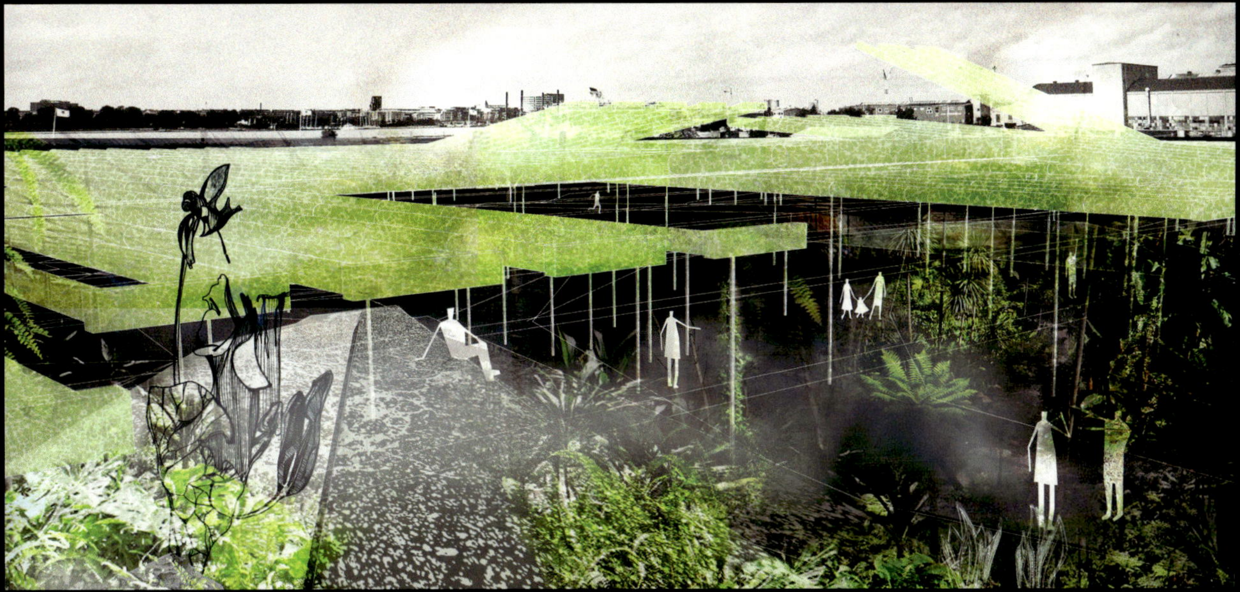

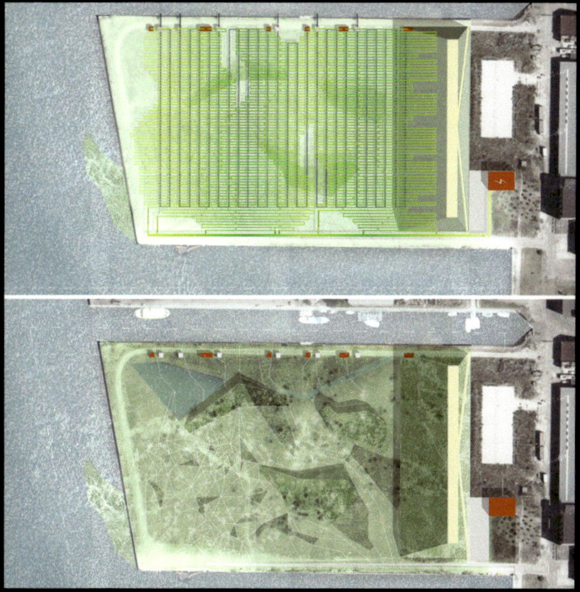

Visitors interact under an ocean of

Site plan both above and below the

Postulate

TEAM:
Luka Stojanovic, Djordje Subaric,
Milica Didic, Marko Babic

TEAM LOCATION:
Nis, Serbia

ENERGY TECHNOLOGIES:
photovoltaic panels

ANNUAL CAPACITY:
5,000 MWh

Postulate represents the time distance between the Big Bang and the present, and the stages that have marked the path of the universe by reflecting the constellation of Orion onto the earth at Refshaleøen.

The artwork is also a constructed transposition of the sub-linguistic systems that undergird our universal grammar—the fundamental building blocks of our communication. The dodecahedron is the starting point geometry onto which are mapped the twelve stages of the life of the universe and twelve universal primordial representations of language and meaning.

These symbols modify the shape of the dodecahedron and come together to create the entire sculptural form. Photovoltaic panels create the skin of the structure and generate energy—a dialogue between the scale of the universe and the subatomic scale of the electron, which forms our thoughts and our language expression while also powering our economies.

Visitors to the installation will be taken on a journey of discovery through time and space as they progress through the twelve-stage walking path of cosmos-city-human thought. Infinite space meets infinite complexity.

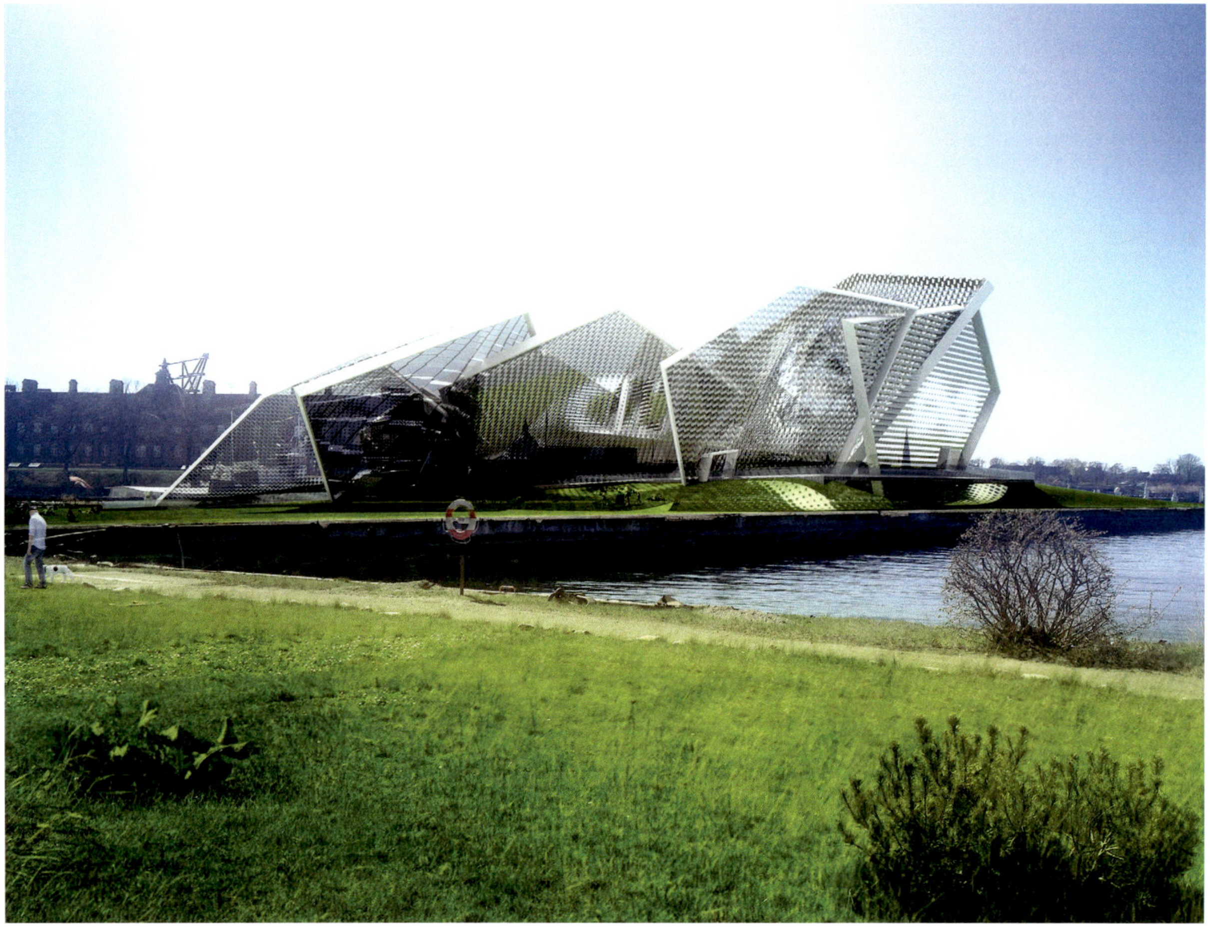

The dodecahedron is one of the most suggestive symbolic representations of the cosmos. The shape is decomposed/deconstructed and left suggestively for visitors to find a new harmonic order.

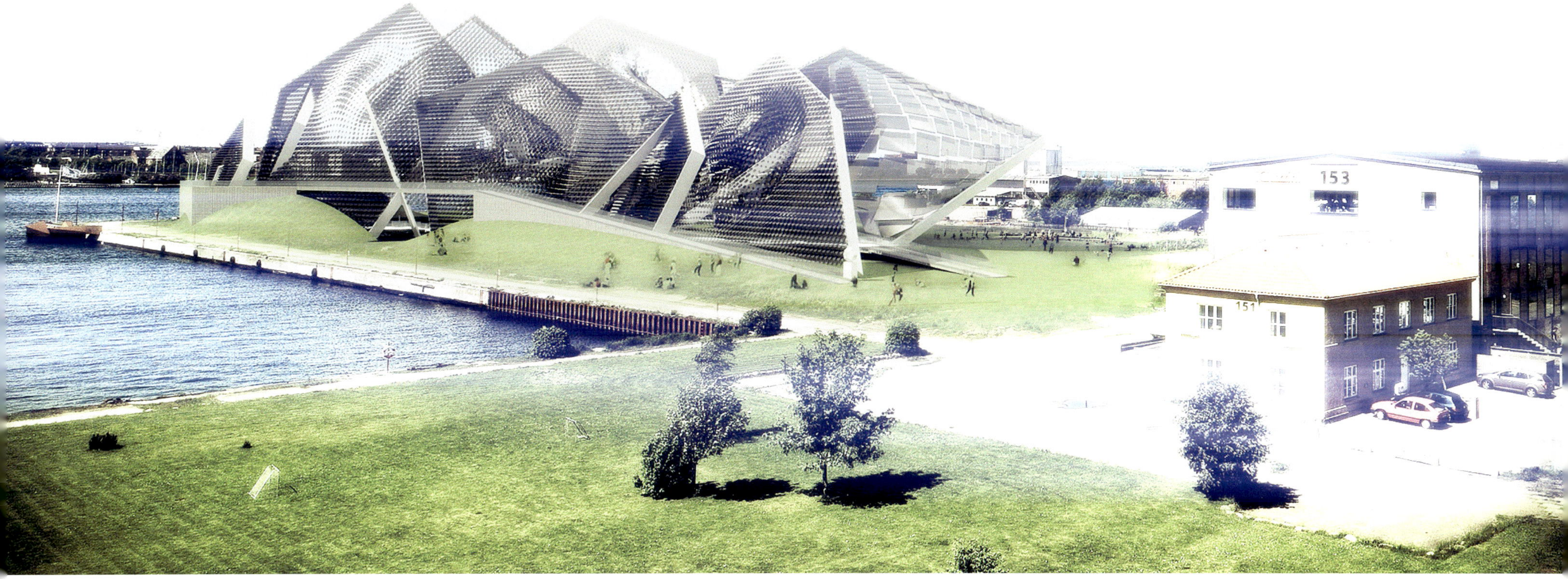

The installation is a reflection of the constellation of Orion. In this way, one city salutes all others through something pure and authentic that exists everywhere.

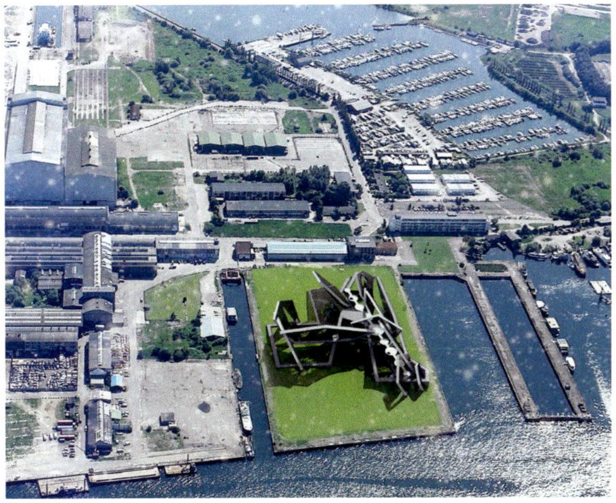

The installation is also a transposition of language in constructed form.

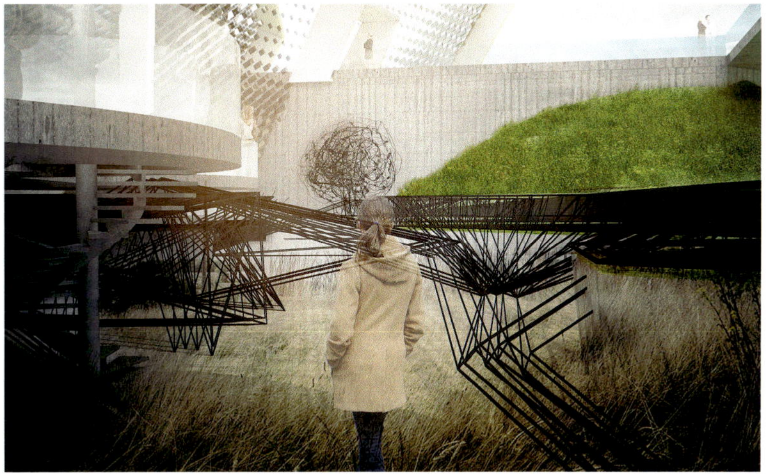

Visitors can pass through the twelve universal primordial representations of our collective unconscious and the grand archetypal implications, which influence language and imbue it with ethical and aesthetic values.

The By-Cycle

A well-known and widely discussed feature of Copenhagen's sustainable development is its bicycle infrastructure. Over one third of people living in Copenhagen commute to work or school by bicycle, and bicycle ownership in the city is upwards of 90%.

Bicycles produce no pollutants when they are being peddled around the city, but what becomes of the bicycle when it has passed its usable lifespan? And what becomes of the abandoned bicycles that amount to approximately 13,000 each year?

What if old and abandoned bicycles were reintroduced back into the sustainable ecosystem of Copenhagen? Although they may no longer be suitable to be used on the road again, their basic framework and structure allows them to be recycled for another purpose. Instead of disposing of them as waste, these bicycles could be transformed into a meaningful part of Copenhagen's ambitious plan to be carbon neutral by 2025.

The By-Cycle transforms old bicycles into a wind farm, harnessing the natural resources of the city, while making use of what would otherwise be wasted resources. The bicycle is deconstructed into its basic elements—its frame and its rims—and the individual parts are reassembled into a tree-like structure. Each rim is then fitted with polycarbonate fins to catch the wind, thus allowing the disused bicycle to be readapted as a windmill. Waste becomes useful again.

TEAM:
Kenneth Ip, Joey Yim

TEAM LOCATION:
Hong Kong, China

ENERGY TECHNOLOGIES:
wind microturbines from repurposed bicycle rims

ANNUAL CAPACITY:
1,000 MWh

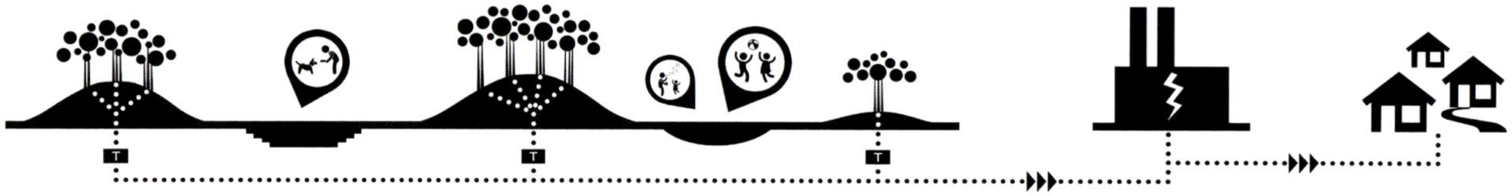

Each disused bicycle wheel is reassembled and fitted with polycarbonate fins to catch the wind. The output shaft is connected to a generator that converts the rotational movement into electricity.

The electricity flows down electrical cables to the transformer, which increases the voltage of the electrical power to the distribution voltage.

The distribution-voltage power flows through underground lines to a collection point where the power will be combined with other power trees.

The electricity is transmitted to the sub-station and goes into the local grid to serve the community.

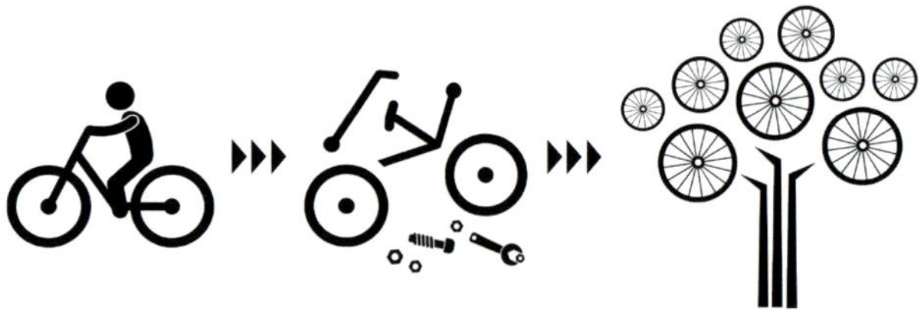

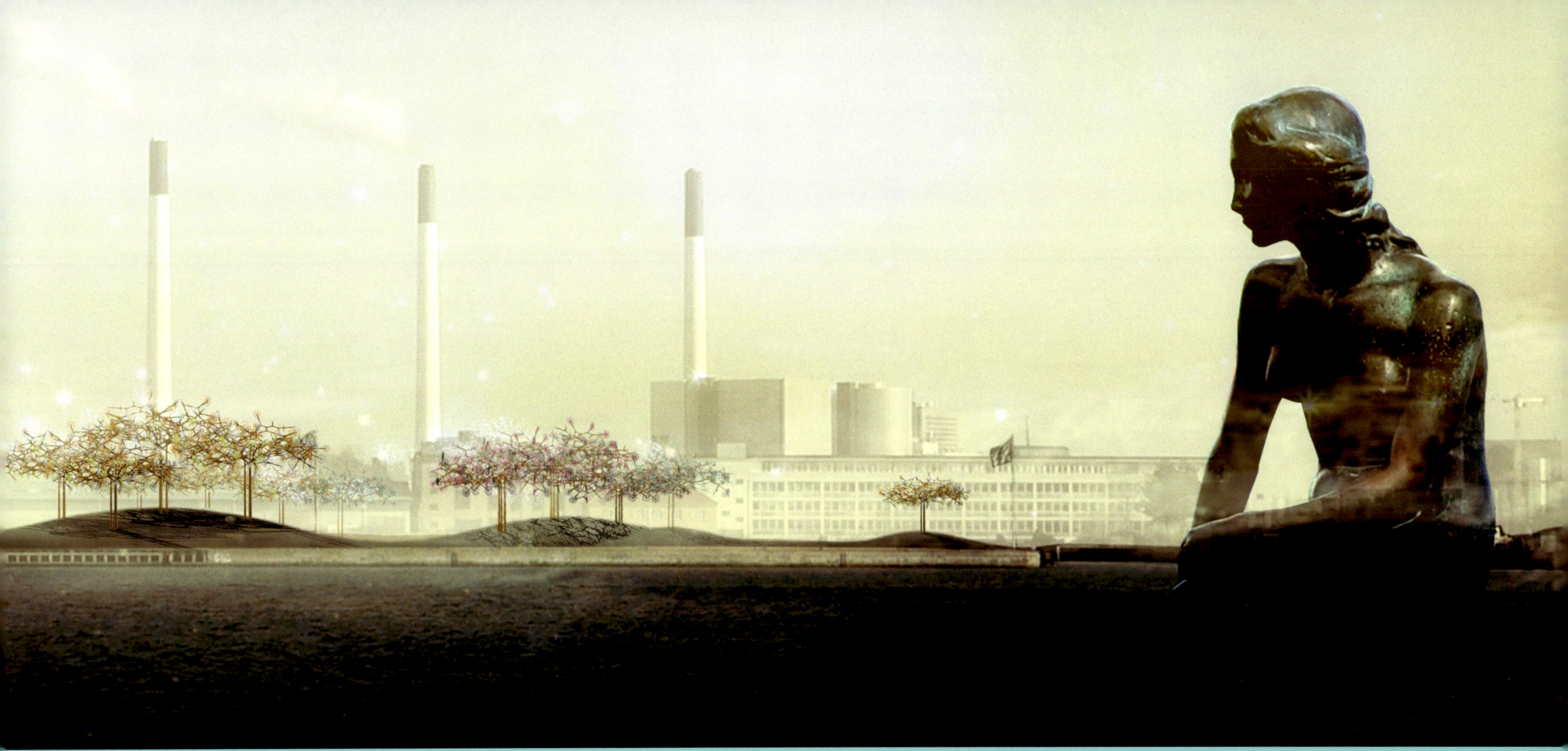

View from across the Copenhagen harbor

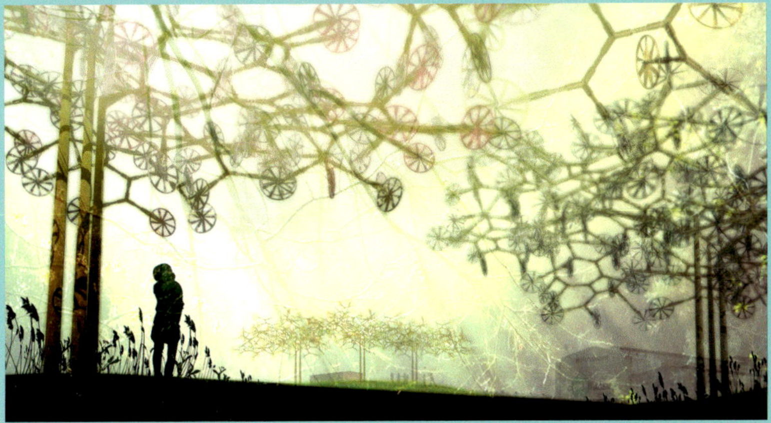

Light filters through the *By-Cycle* branches.

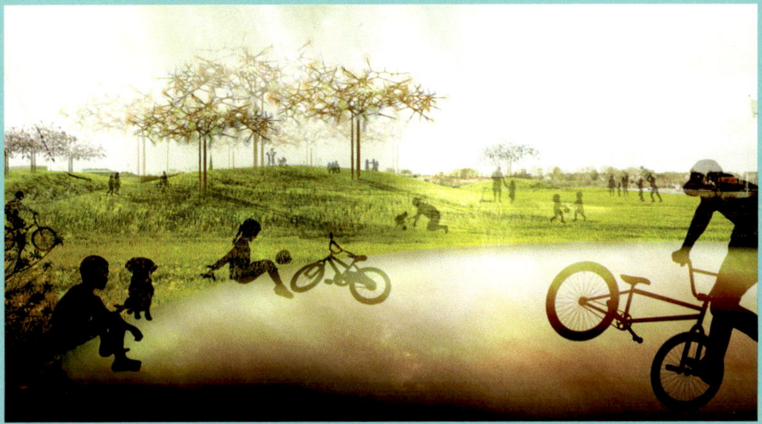

Someday the visitor's bicycles could become a part of an expanded energy installation.

Factoryscape

Factoryscape removes existing solid land, creating space for algae photobioreactors—an urban-agriculture area that uses the sun, CO_2, neighboring waste, and the forces of the sea to create biofuel from algae.

The algae farm is interlocked with a complex engine comprised of time-tested technologies that dewater, compress, transesterificate, and burn the algae, then produce steam out of its energy, and ultimately convert the steam into electricity. All these machines stand together on the site as a central *objet d'art*, reminding us of the accumulation of human inventions that are needed to create cycles of self-sufficiency. The complexity and the materiality of the installation create an image that we are all familiar with and of which we may be wrongly suspicious. Love it or hate it, the machine is the child of our collective, accumulative endeavor.

Contrary to our industrial past, this complex machine is not designed as an object of pure utility. Rather, it is recontextualized and dwarfed by a gigantic red surface that encapsulates it—creating a dense, cave-like environment. Within this grotto, the sounds from the machine are echoed, its connectivity is accentuated, and its aesthetics are exploited.

Visitors to this environment encounter the technologies in an entirely new manner that will cause them to reflect upon the relationship between man and machine.

TEAM:
Yasin Toparlar, Onur Can Tepe, Huseyin Penbeoglu

TEAM LOCATION:
Eindhoven, The Netherlands

ENERGY TECHNOLOGIES:
algae photobioreactors, boiler/steam turbine

ANNUAL CAPACITY:
1,400 MWh

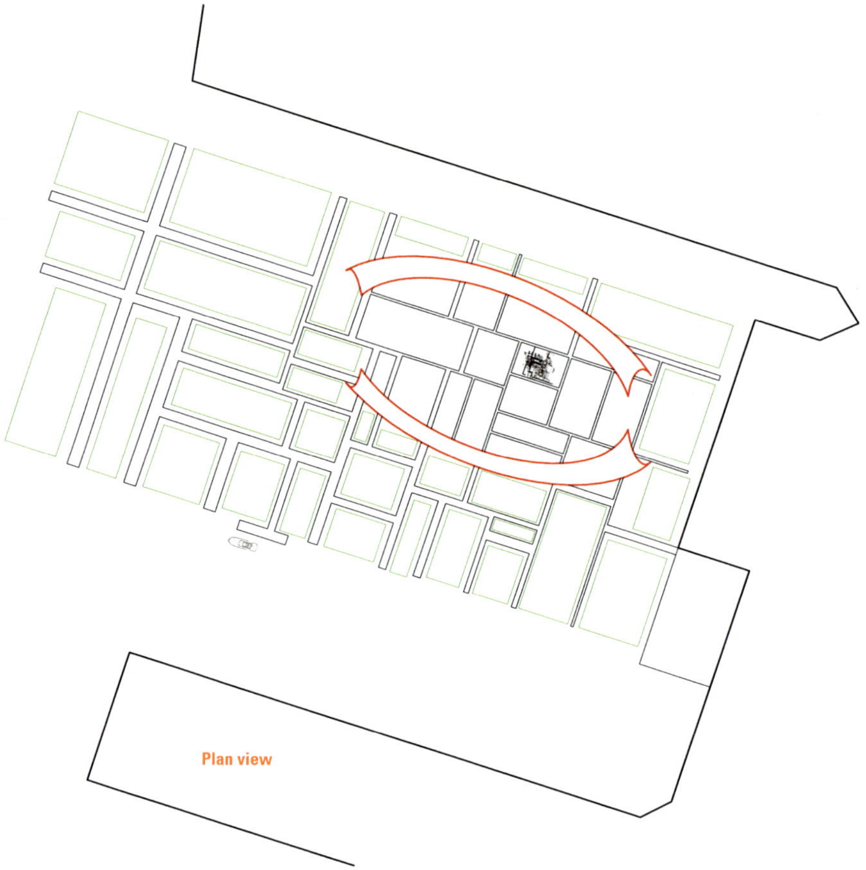

Plan view

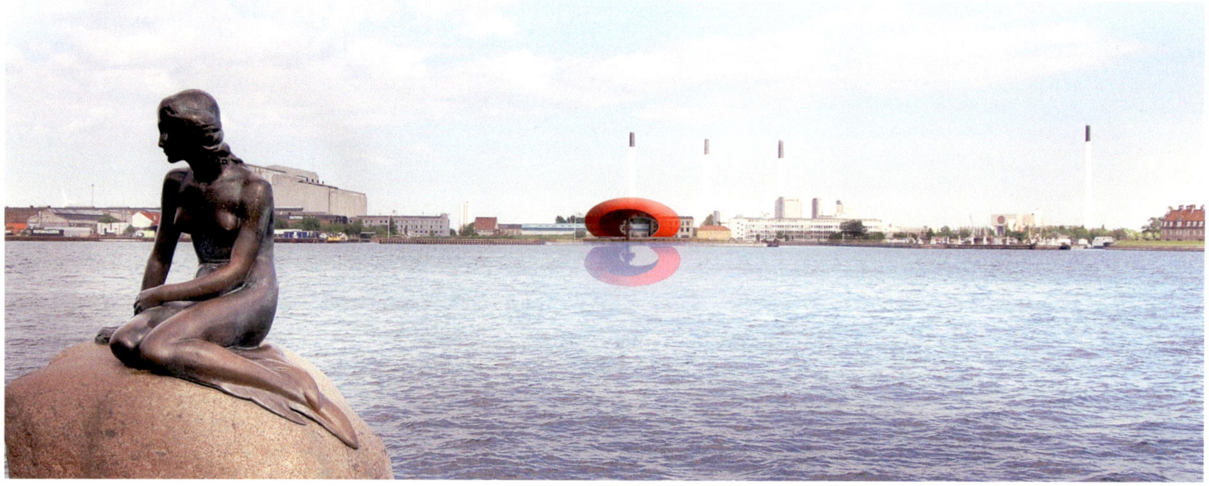

View of *Factoryscape* from across the Copenhagen harbor

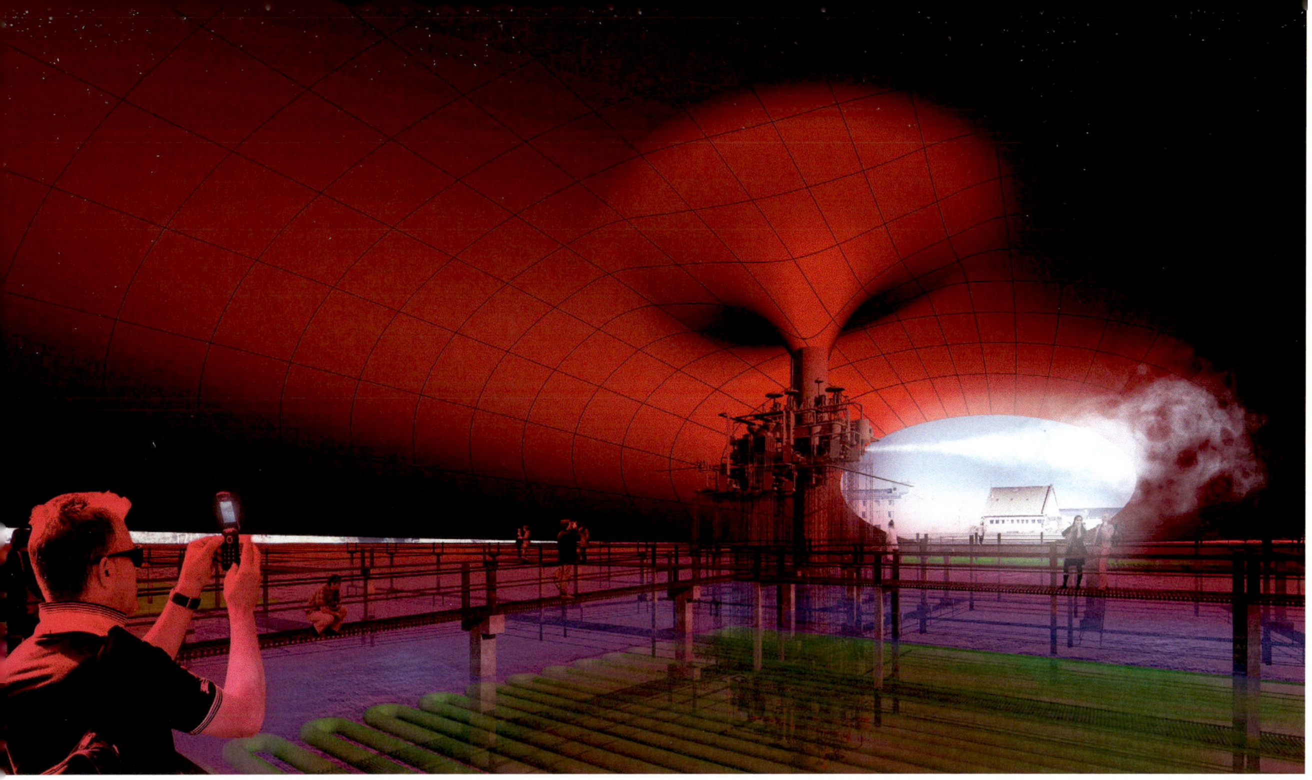

Inside the machine. The 21st century factory is a phenomenological space where humans and technology critically encounter each other.

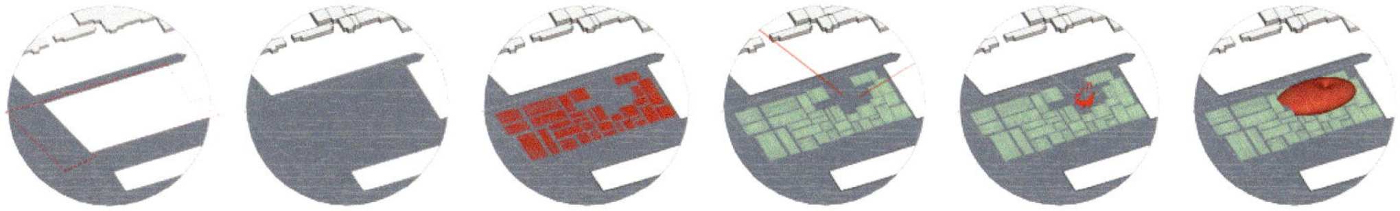

How to harness the greatest amount of natural energies on a small site?

Industrial land takes a step back for the needs of the 21st century.

Algae farms fill the site to provide the resources for urban energy generation.

Wastewater from the treatment plants and from the marina is transferred to the site, feeding the algae farms.

An ecology of technologies is built up, combining the infrastructure for energy generation.

A red surface covers the factory. Sounds of the machinery are echoed and connectivity is visualized.

Diagram of the site transformation

Section through the site

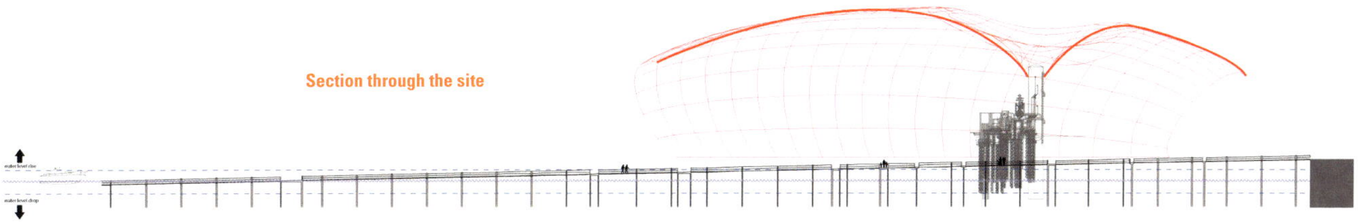

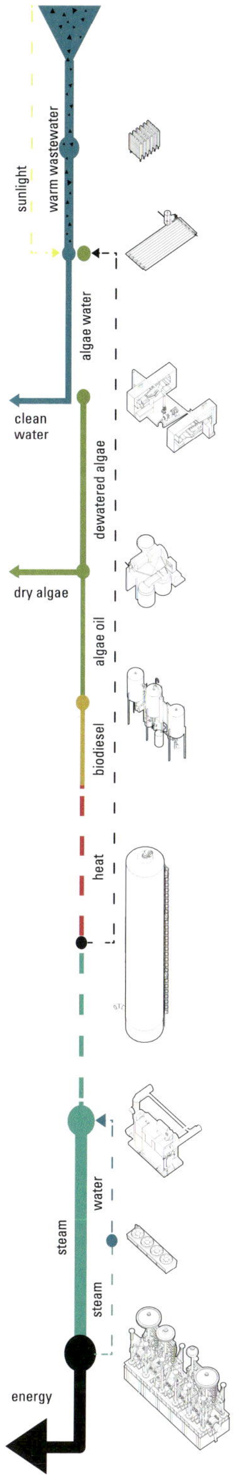

Heat exchanger
Used for efficient transfer of heat from one medium to another.

Photobioreactors
Used for algae cultivation. They float on sea level which helps for mixing, cooling, and achieving maximum sunlight. Tubes contain warm wastewater (rich in nutrients) and CO_2 for algae growth.

Belt filter
Algae slurry is separated into algae and clean water with the help of gravity and the density differences of algae and water. Vacuums can be used to achieve faster results.

Screw press
Uses compressive forces to extract oil from algae and to produce dry algae and algae oil. Dry algae can be used as a fertilizer in the agriculture industry.

Transesterification
A chemical process using an alcohol and an ester to produce another type of alcohol and another type of ester. The algal oil will be mixed with methanol to produce glycerol and biodiesel. The glycerol can be used in pharmaceuticals and food products.

Combustion chamber
The heat produced by combustion of fuel in the combustion chamber will be used by heat exchangers and boilers.

Chimney
Using the stack effect and pressure differences, the warm and polluted air is carried out and processed before release for emission regulations, and the CO_2 will be collected for use in the algae harvesting process.

Boiler
Using the heat generated at the combustion chamber, water is heated to produce steam.

Condenser
Used to convert steam from steam turbines back into water, which is supplied back to the boilers.

Steam Turbine
A device that converts thermal energy from high-pressure steam to mechanical energy. Two steam turbines generate 1.2 MW output of electricity.

Emotional Fog

Emotional Fog is an art and light installation that forms new relational spaces based on the psychology of color as it reacts to relative humidity conditions. A subtle landscape during the day transforms the image of the city into a space of wishes and emotions overnight.

Legend has it that the songs of the Sirens bewitched men of the sea and the City of Copenhagen. The artwork seeks to recreate this emotional fog that can change the course of human action by simulating the effect of natural fog with smoke cannon effects. LED lights are used to replicate Rayleigh scattering and Mei scattering within natural fogs and to provoke different feelings depending on the colors within which viewers of the artwork are immersed.

The structure of *Emotional Fog* is consistent with the idea of fog—it is a lightweight structure made from a double-module tensegrity system based on the work of Kenneth Snelson. The three-meter long anodized aluminum compression rods are fitted with LED lights. Spanning the meeting points between the tensegrity arms is a skin of small propellers that rotate with the wind and produce electricity.

Over the entire site, the tensegrity modules are arranged to form large, hexagonal geometries similar to the pattern in which water molecules bind in a solid state.

TEAM:
Ricardo Morcillo

TEAM LOCATION:
Albacete, Spain

ENERGY TECHNOLOGIES:
wind microturbines

ANNUAL CAPACITY:
100 MWh

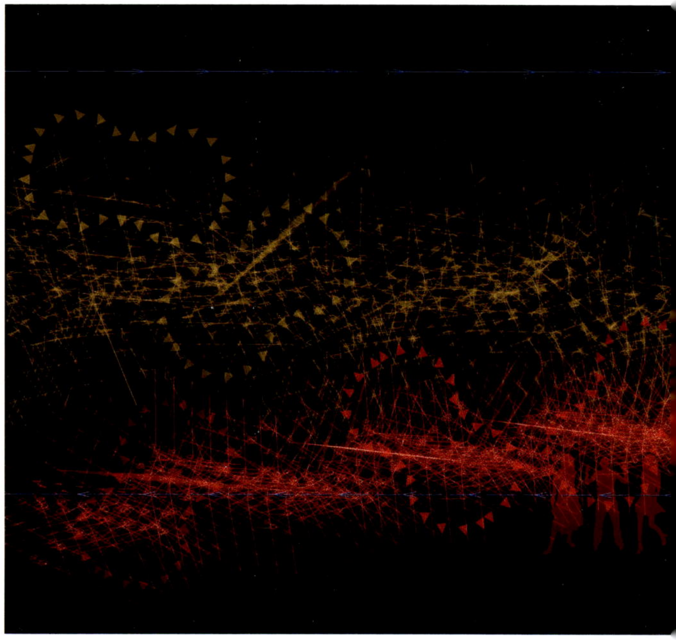

Emotional Fog **is a magical public space where the coexistence of emotions shape the landscape of Copenhagen.**

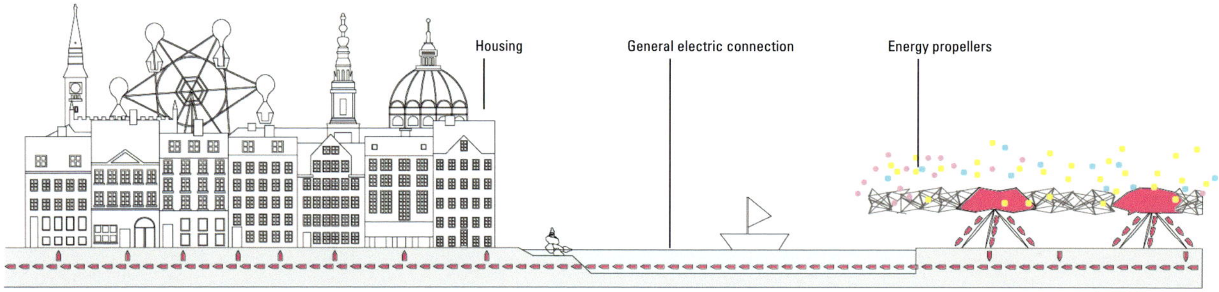

Schematic distribution of energy to the city

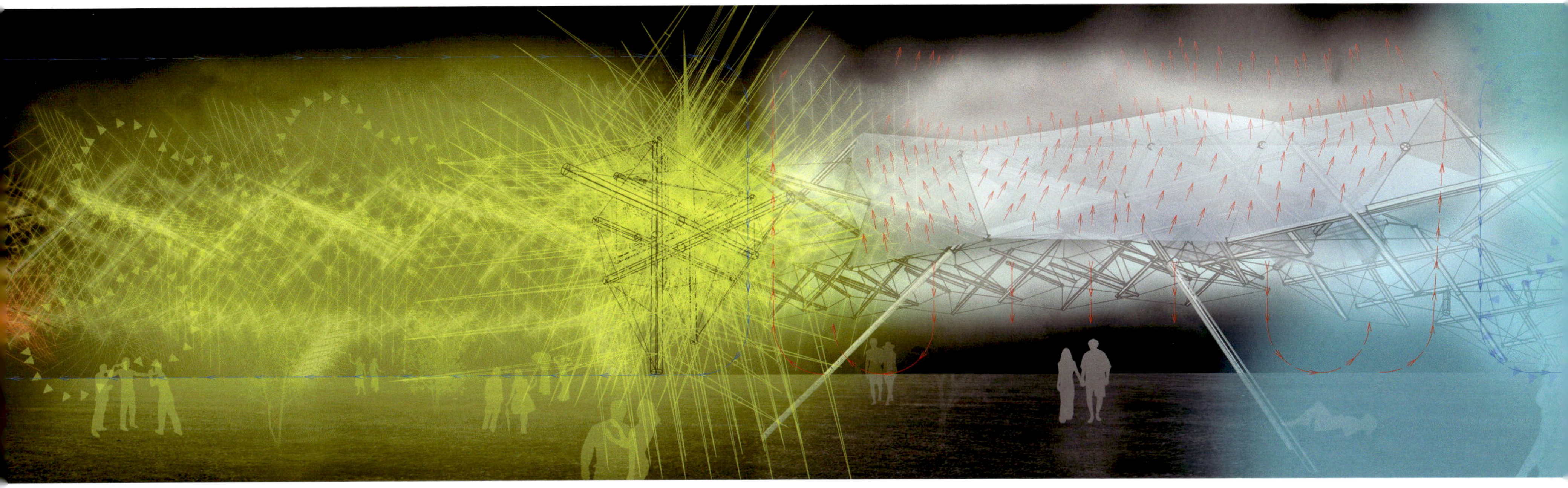

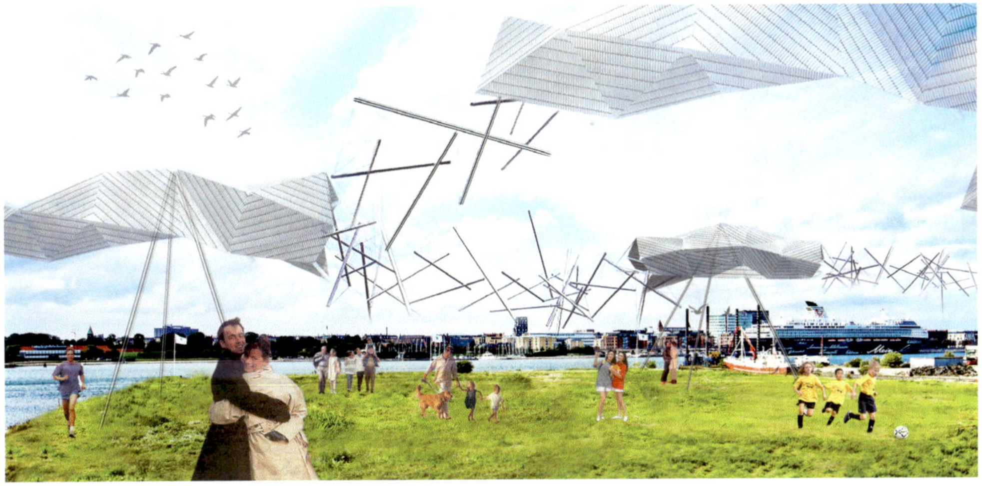

The netting located at the meeting points of the tensegrity arms is formed by small propellers that rotate with the wind and produce electricity that is transmitted by the tension cables up to the light bar assemblies.

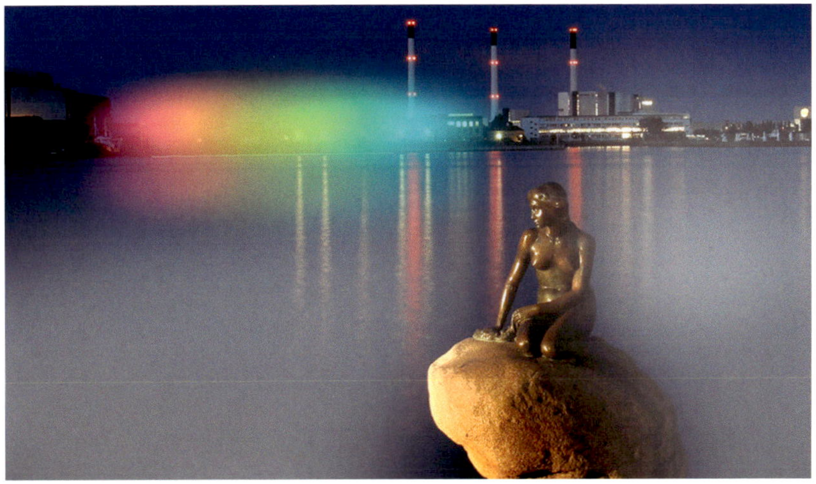

Emotional Fog and the harbor mist

Ripple

TEAM:
Junfeng Wang, Qiyao Li

TEAM LOCATION:
New York City, USA

ENERGY TECHNOLOGIES:
piezoelectric generators

ANNUAL CAPACITY:
5,000 MWh

Bridging from water to sky, *Ripple* settles on a curved surface, creating a harmonious conversation between two parallel worlds. It embodies a rhythmic, unitized shape inspired by the form of water, generating electricity by harvesting kinetic energy.

The light, aluminum panels reflecting the surrounding colors of the sky and water blend into the environment, flickering in the wind or resting in stasis. The process of energy production is displayed as a pleasant experience for visitors, blurring the line between the artificial and the natural.

Ripple employs natural power as well as human engagement as its kinetic power source on site. Movements in the surrounding environment—prevailing wind, birds' intermittent resting, people walking, or activities above the panel—trigger the panel's vibration, which is passed down to piezoelectric generator cells. The spring at the bottom of each rod helps to regulate the vibration cycle and confine the movement of panels, on which visitors can walk.

Specific panel installation and material varies slightly depending on the energy input. Side-wind triggered panels are mainly located at the area close to the water frontage and are placed parallel to the ground plane. They are driven by wind power coming sidewards from almost all directions. Front-wind triggered panels take advantage of prevailing wind from the southwest. Flat panels define the common pedestrian flow path on site, which is distributed around the entrance and connects to the water taxi terminal. Flexible event panels have various degrees of curvature for mixed uses.

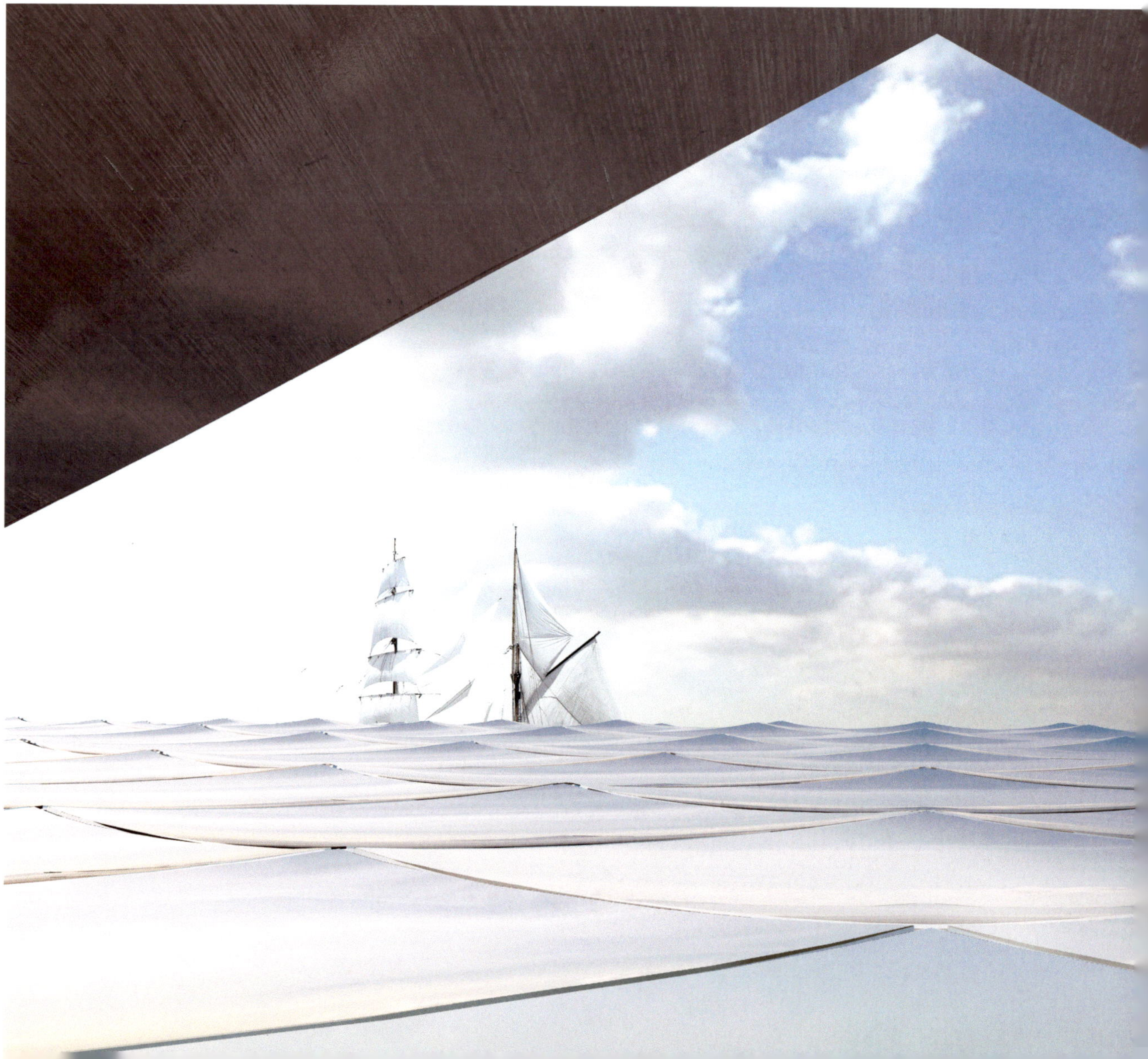

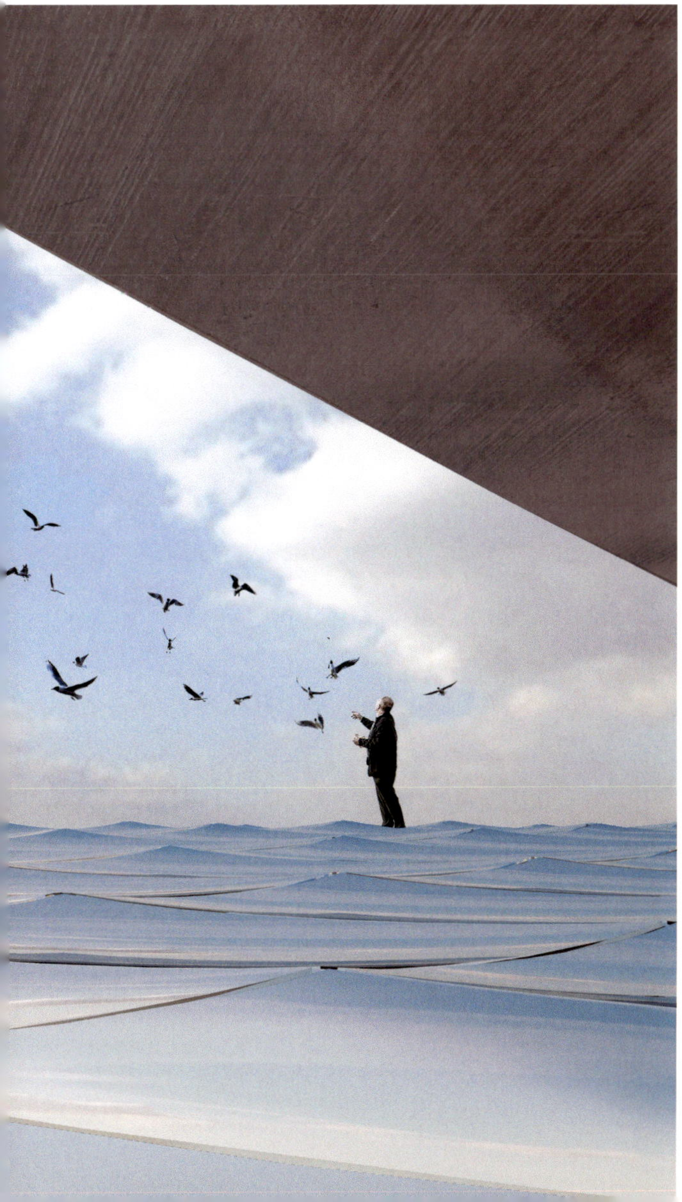

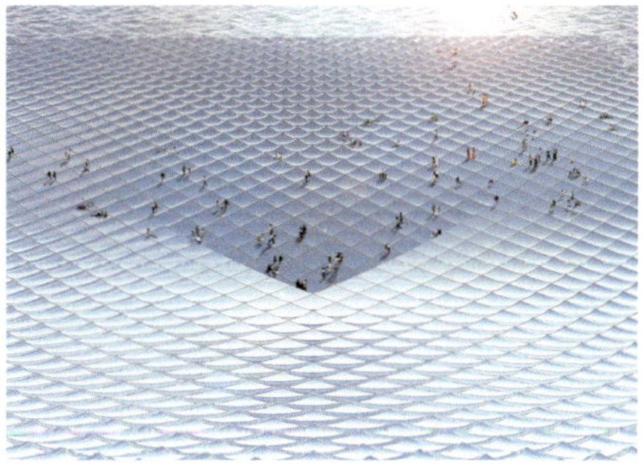

View from the top of the screen wall

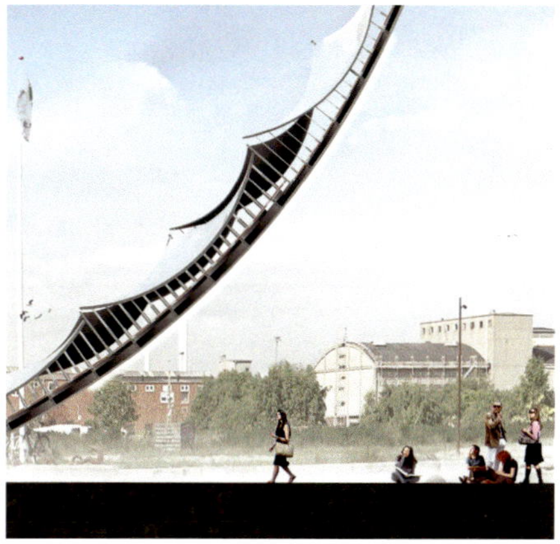

Detail section

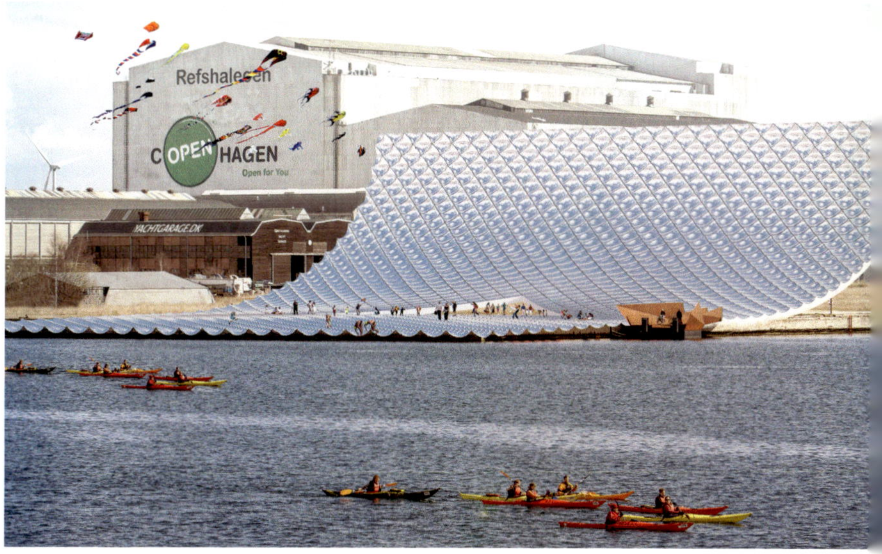

Ripple actively engages in the surrounding urban context. It extends out and into the harbor front and curves upwards into a screen on the urban side, which allows it to be identified from the opposite shoreline.

WindWaker

Emerging from water, *WindWaker* appears within the industrial landscape of Copenhagen's seaport. Masts, like a pillar structure, support 21,860 m^2 of high sails exposed to wind, creating an open and high-tech park, ideal for numerous events.

Nanogenerators (NG) developed over a decade ago by Zhong Lin Wang and his research team (Georgia Institute of Technology) provide the necessary energy output to make the project viable. Because of their nanometric thickness, flexibility, translucent appearance, and size, nanogenerators fit perfectly with the *WindWaker* design.

NGs consist of a free cantilever composed of a five-layer structure: a flexible polymer substrate, ZnO nanowire textured films on its top and bottom surfaces, and electrodes on the surfaces.

NGs are "woven" among a triple layer textile (textile layer + nanogenerator layer + textile layer) and through the entire sail surface to generate energy by wind pressure. Sails can incorporate LED illumination also woven into the textile layers and connected directly to the nanogenerators. During the night, the illumination intensity changes depending on the wind-load pressure. Because the site can be seen throughout the city, color patterns can be programmed to mark holidays and special events.

Maintaining a longitudinal axis free of masts and sails establishes a visual connection between the land art and the sea. It is also a way of channeling wind through the core of the *WindWaker* installation. At the west end of the platform, an elevated pier serves as an observation point, and a variety of programmed spaces are situated along the park: picnic areas near the pier, sculpture gardens, amphitheaters, and a space near the water taxi terminal where visitors can rest before embarking.

TEAM:
Julio Alejandro Romero Alonso, Miguel Ángel López Carro

TEAM LOCATION:
Madrid, Spain

ENERGY TECHNOLOGIES:
nanogenerator (NG) fabric

ANNUAL CAPACITY:
840 MWh

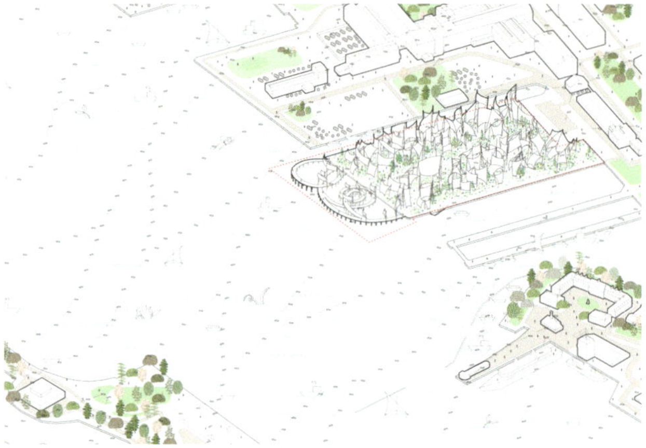

WindWaker in context

Views from across the harbor

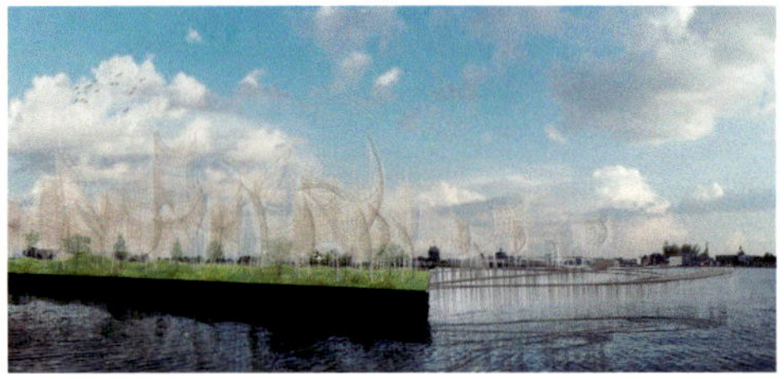

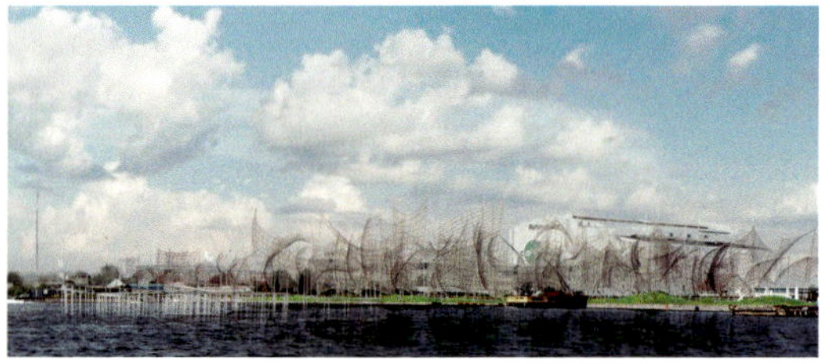

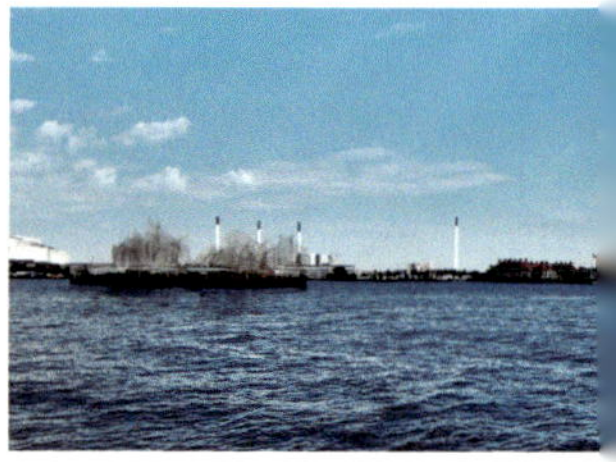

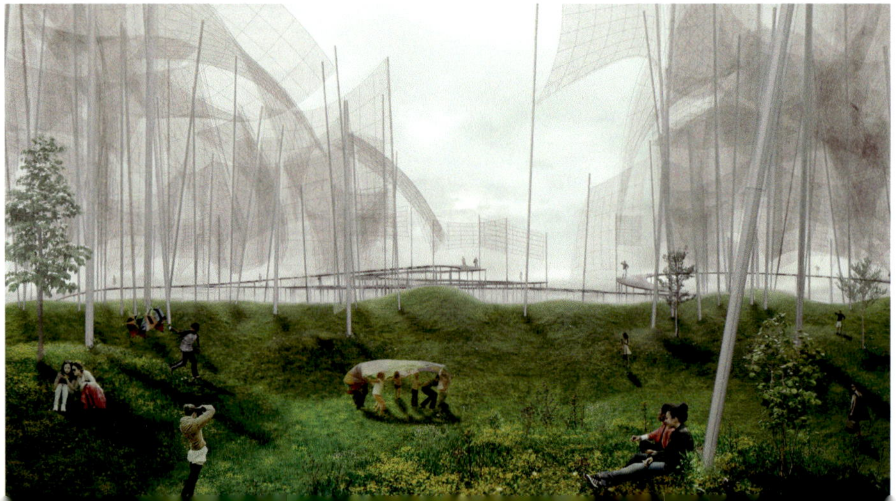

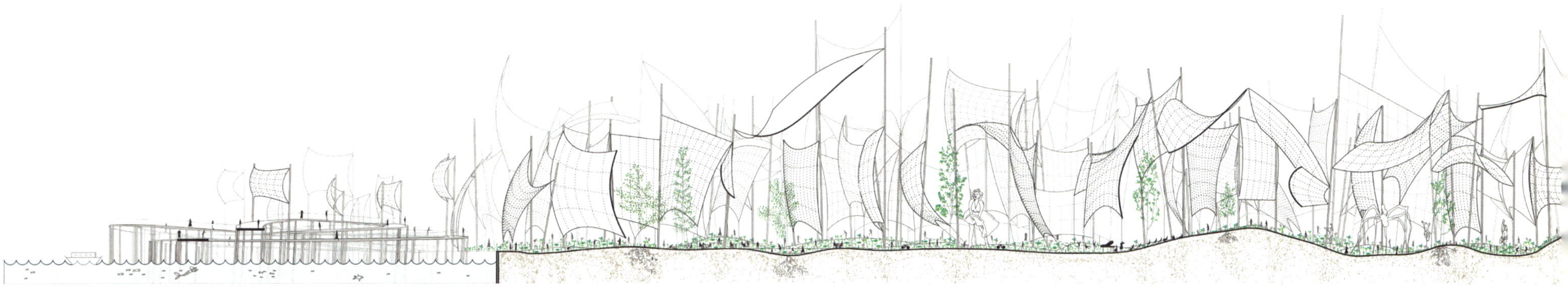

Site section

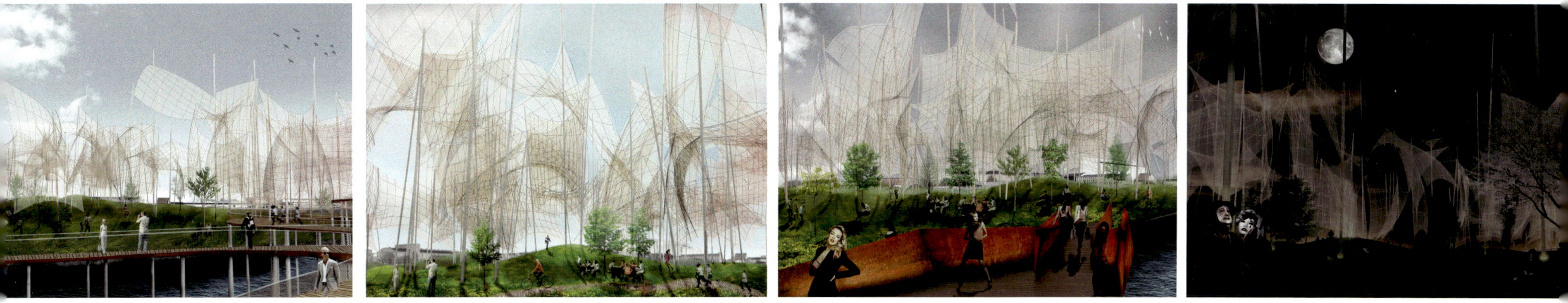

Views from within

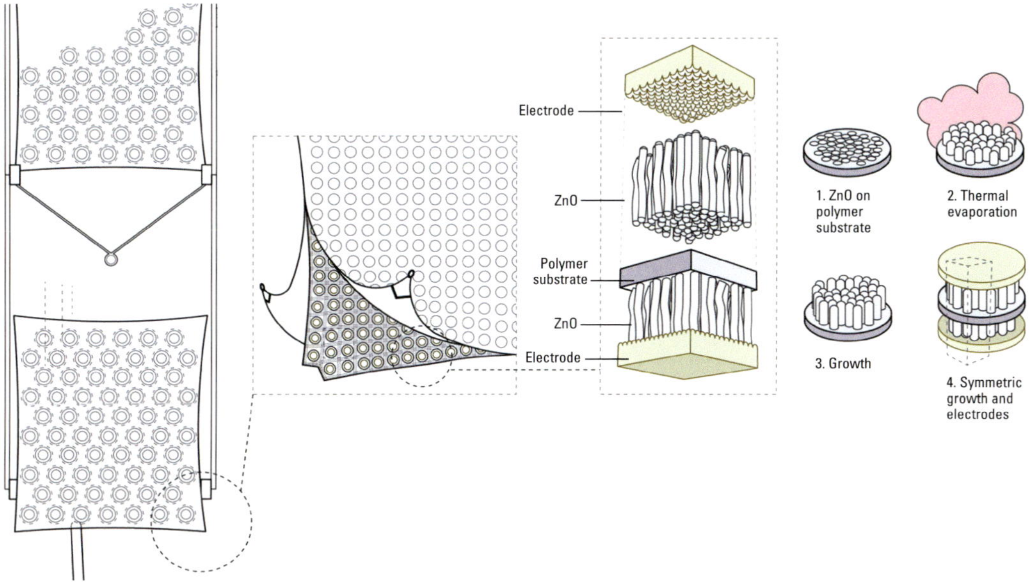

ZnO is a biocompatible and environmentally friendly material. The nanowire films can be grown at very low temperature (<100 °C) on any kind or shape of substrate.

Refkløver

TEAM:
Allison Palenske, Christina Gråberg Røsholt, Akshaya Narsimhan, Javier Vidal Aguilera, Zhao Xie, Diandra Saginatari, Yanli Shen

TEAM LOCATION:
Edinburgh, UK

ENERGY TECHNOLOGIES:
piezoelectric wires, Pavegen™ pavers

ANNUAL CAPACITY:
500 MWh

Geometric forms subtly merge with the organic in *Refkløver*, a contemporary urban adaptation of picturesque meadow drifts that inspires contemplation amidst a new renewable energy power plant for the City of Copenhagen. Hearing the sound of the wind through the grasses and perennials, visitors will feel a part of the energy generation and can participate in it through kinetic interactions.

Taken from microscopic studies of the Danish national flower, the rødkløver (red clover, *Trifolium pratense*), the form of the energy-generating structures mimics the veins found on the leaves and bracts of the plant. These veins act as highways for photosynthetic processes, transporting energy and sugars to the rest of the plant. The design of *Refkløver* acts in a similar way. Using piezoelectric wires to represent the veins of the plant, wind movement will generate energy through these delicate fibers and will be transported to the city grid.

Each energy-producing module consists of a cluster of four recycled stainless steel structures at various heights, with piezoelectric wires spanning across each surface. These wires transform their own movement, caused by wind currents, into electrical energy. The specific piezoelectric technology used is nanoparticle-coated yarns patented by Perera & Mauretti in 2009. These fibers are embedded within 1-millimeter braided stainless steel threads.

The 300 small clusters, 168 medium clusters, and 260 large clusters of triangles found on the site are oriented primarily to take advantage of the prevailing wind direction and also intermittent variations in wind patterns. Additionally, 1,557 Pavegen™ kinetic flooring tiles are interspersed throughout a paved path and convert the weight of passing visitors into electricity.

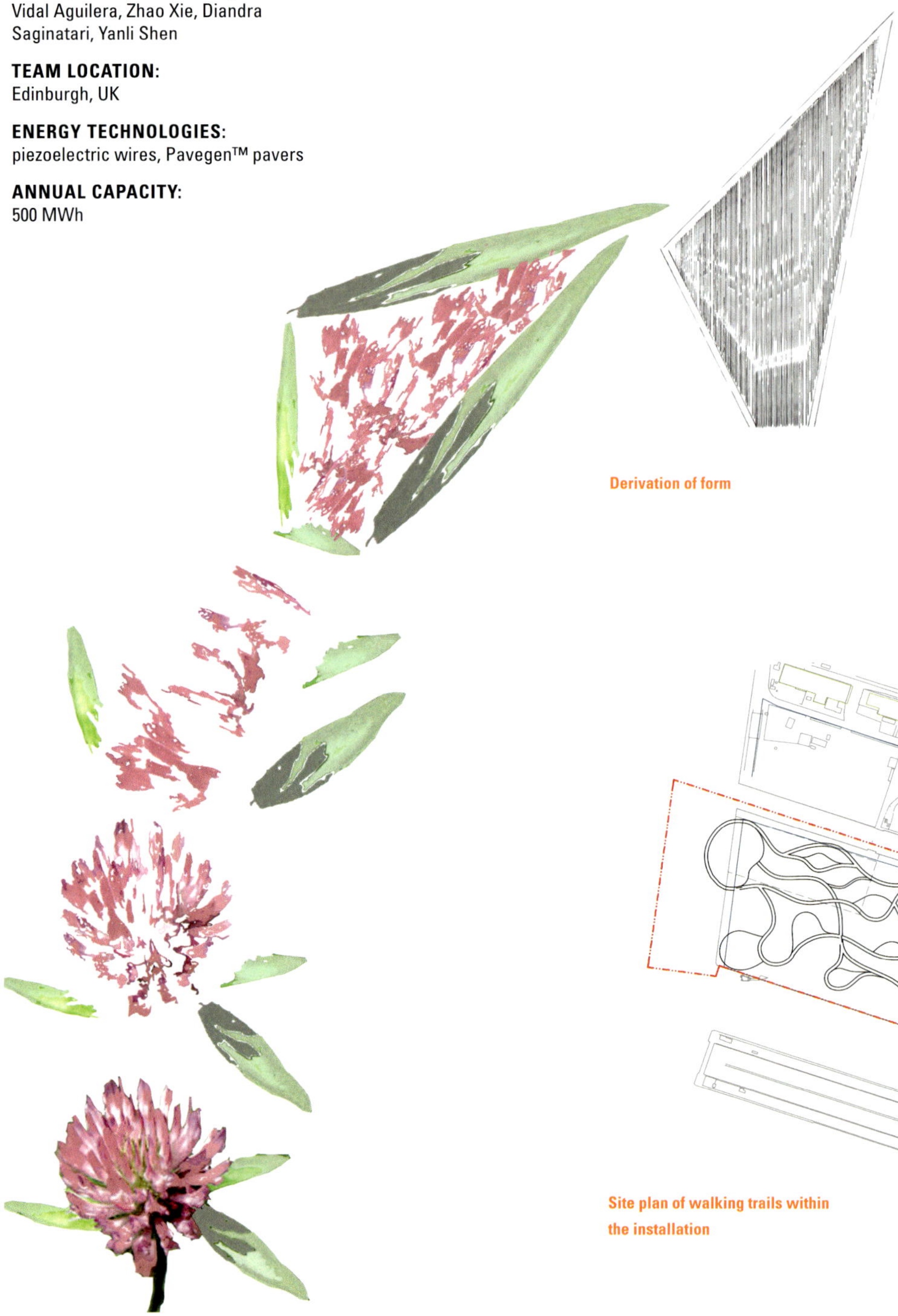

Derivation of form

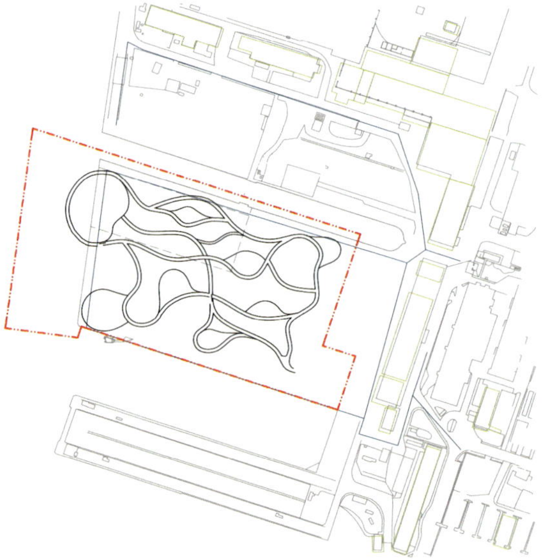

Site plan of walking trails within the installation

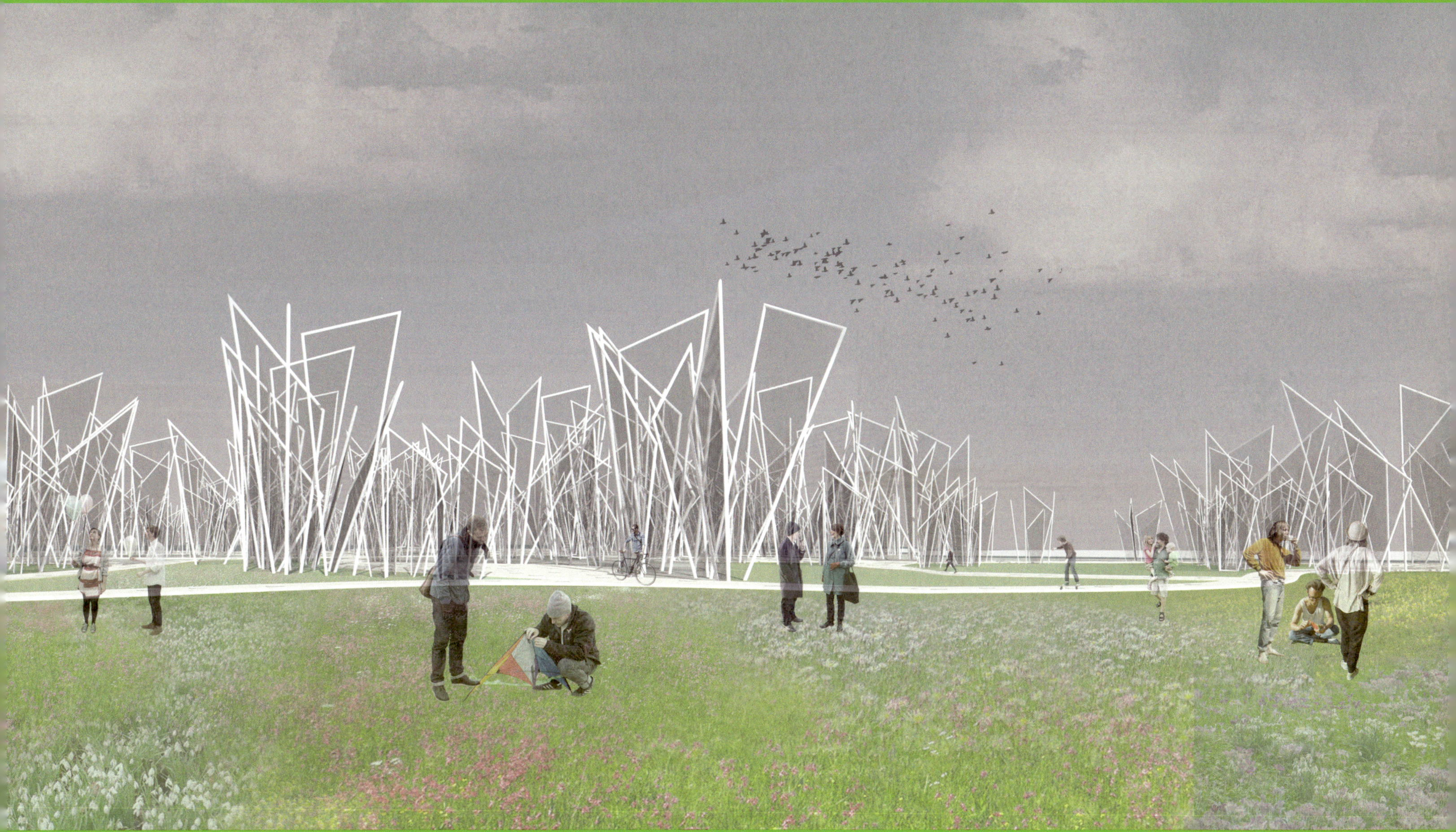

Refkløver creates a place of symbiosis between people and the landscape. A system of active walkways will connect passive nodes, which are guarded by the sculptural forms, making visitors feel like a small organism within the epicenter of the red clover.

Sails:
An Homage to Seafaring

Sun, wind, and sea are the constant companions of the seafarer. Whether centuries past or in the present, sailors have always known that they will encounter them on their voyage. In fact, they rely on them. The importance of seafaring for Denmark and Copenhagen throughout history has inspired the conceptual sails in this installation.

The 2,930 m^2 of conductive nets are supported on several masts or posts. Energy generation occurs on pieces of photovoltaic cloth attached to this grid and at the connection point to the masts where tension is translated into electricity through piezoelectric discs. The net itself is a grid composed of loose electrical conductors inside flexible plastic tubes that carry the tension. The generation occurs on pieces of fabric (one meter square) that are attached to the net on its diagonal to optimize collection through the conductive net while allowing maximum movement.

The fabric is composed of four types of interwoven layers: the Sphelar® mesh, the piezoelectric fibers, the fabric, and the coating. The Sphelar® mesh is a grid of conductors that connects three-dimensional, light-capturing, micro-spherical solar cells. They are less dependent on heat and direct light, and their spherical shape allows them to capture residual reflected light from all angles. This gives the possibility of photovoltaic electric generation without the need of a fixed system or solar tracking.

TEAM:
Laura Ruiz

TEAM LOCATION:
Caracas, Venezuela

ENERGY TECHNOLOGIES:
Sphelar® photovoltaic, piezoelectric

ANNUAL CAPACITY:
750 MWh

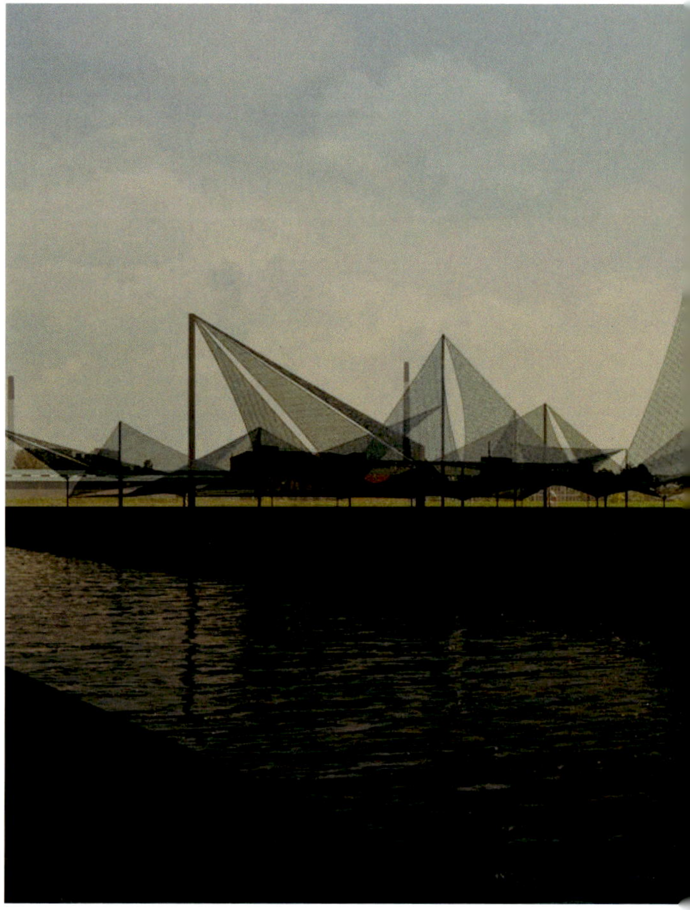

The metaphor of *Sails* pays homage to Danish history, to the hard working sailors who passed through Copenhagen, and to the ability of humans to harness the power of nature without destroying it.

View from across the harbor

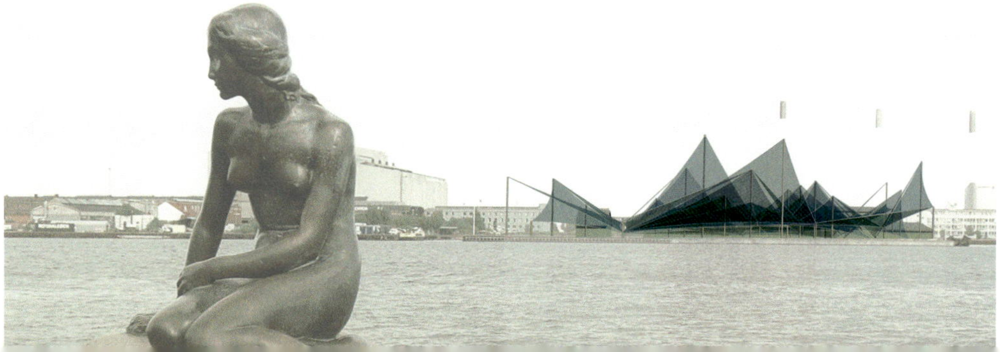

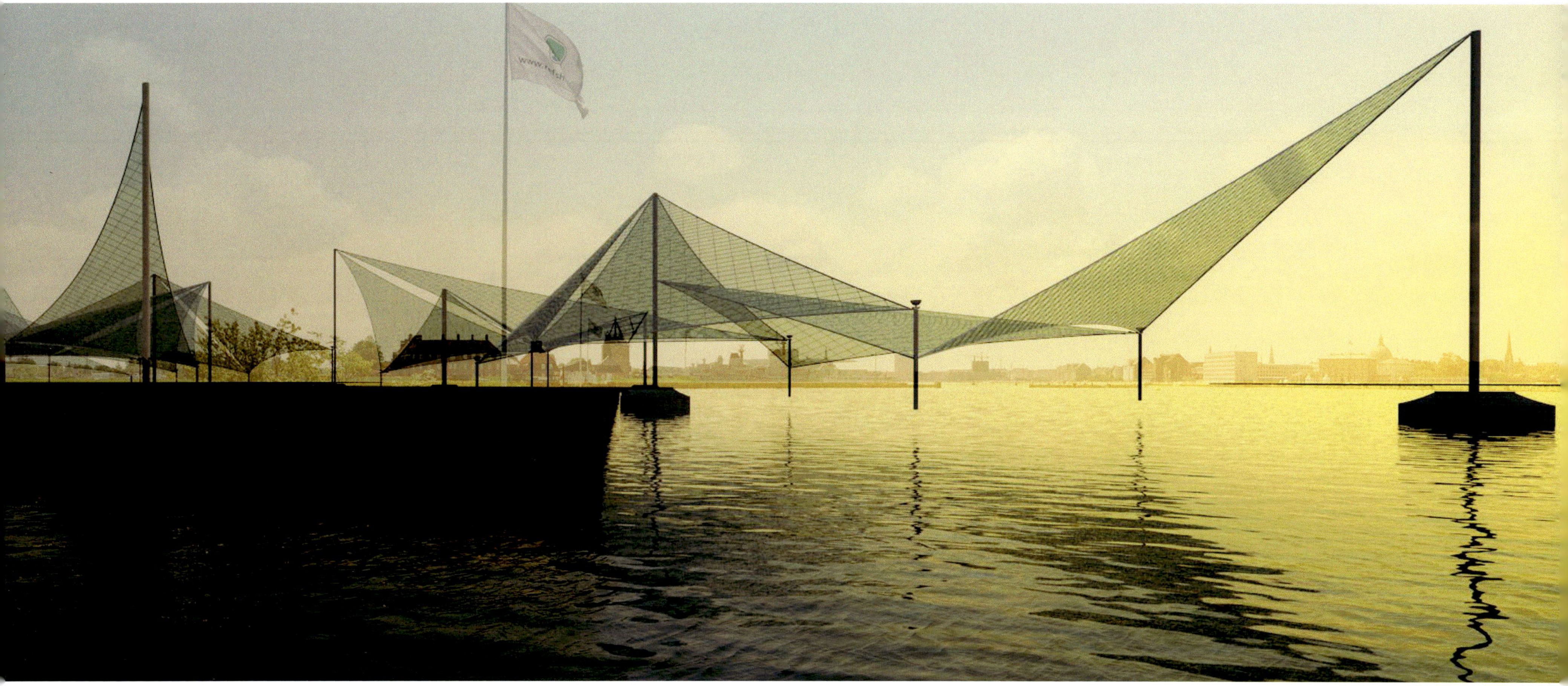

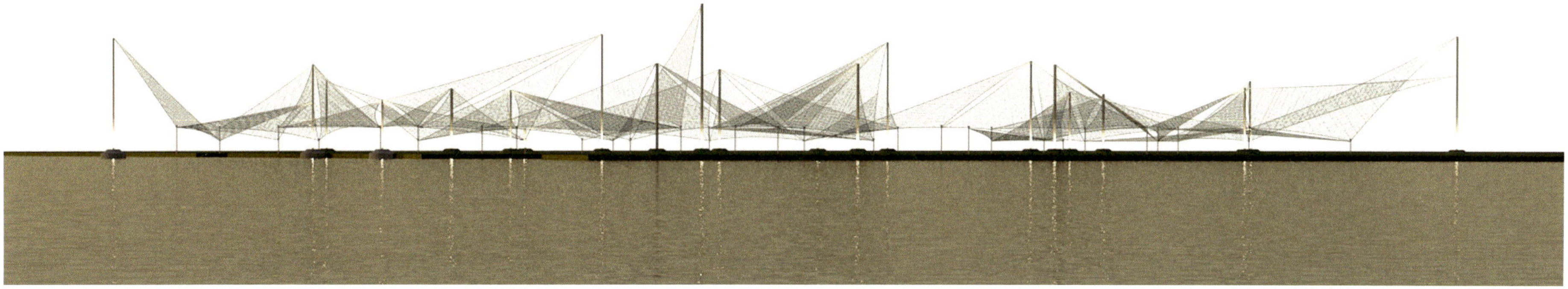

Site section

Regatta Fields

Regatta Fields defines a familiar yet otherworldly space that harmoniously complements and enhances its surroundings. The design site is configured as a gradual transition between land and sea. Fields are broken by paths evoking the geography of Copenhagen and Scandinavian coastal environments. The 60 wind turbines introduced in *Regatta Fields* have dimensions comparable to those of medium-sized trees to engage the human scale and to establish a playful appearance. Sails inspire their shape while their colors and structures subtly recall the patterns of kites and spinning tops.

The design of this new wind turbine is based on the Savonius vertical axis type. The shape is optimized, enlarging the top to catch stronger winds and tapering at the base where winds are not providing any relevant contribution. The minimal footprint further emphasizes the relationship of the turbines to the human scale, while not obstructing the view and allowing people to come close and walk around them.

Vertical axis wind turbines are considerably quieter than the commonly used horizontal axis type. The wind turbines specifically designed for *Regatta Fields* improve this feature due to their light structure and ethereal materiality. The generator and the gearbox are placed underneath the hills on which the turbines rotate.

The structure of the turbines is made of aluminum, chosen for its lightness and resistance to corrosion. The sailcloth wrapping the structure is made of Polyethylene Terephthalate (PET). Both materials are highly recyclable.

Energy-harvesting paving paths lead to the winter baths, which allow visitors to swim year-round in the city center of Copenhagen under a PET structure.

TEAM:
Riccardo Mariano, Salvatore Maraniello

TEAM LOCATION:
Berlin, Germany

ENERGY TECHNOLOGIES:
vertical axis wind turbines,
Pavegen™ pavers

ANNUAL CAPACITY:
500 MWh

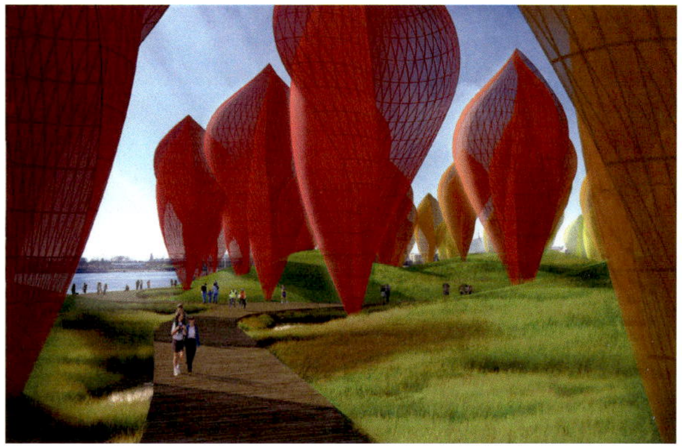

Visitors walk along the promenade.

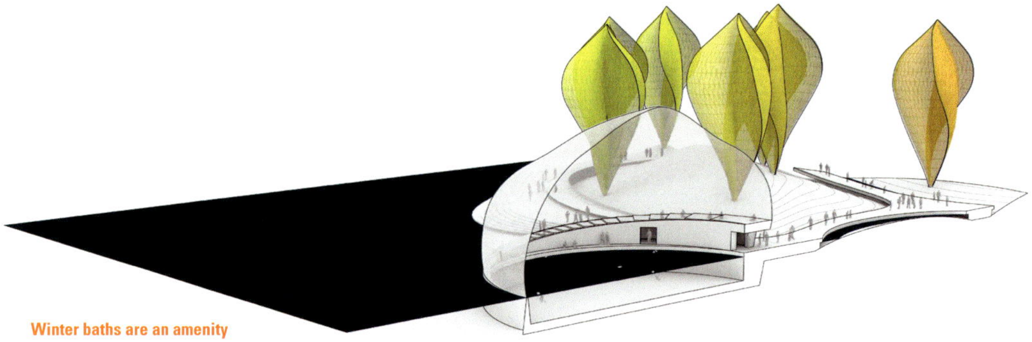

Winter baths are an amenity within the site.

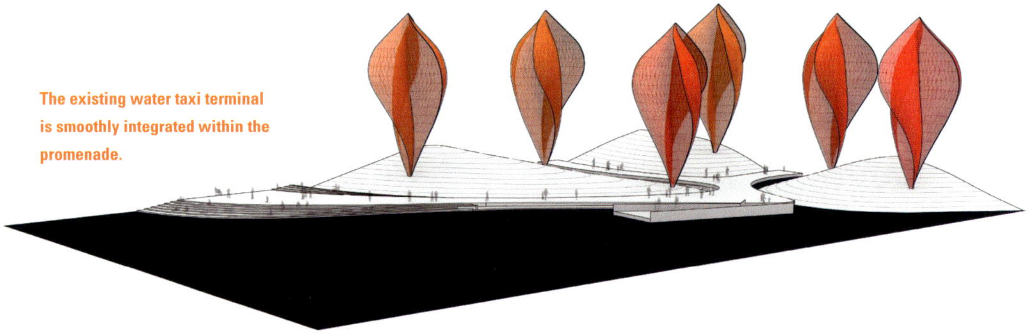

The existing water taxi terminal is smoothly integrated within the promenade.

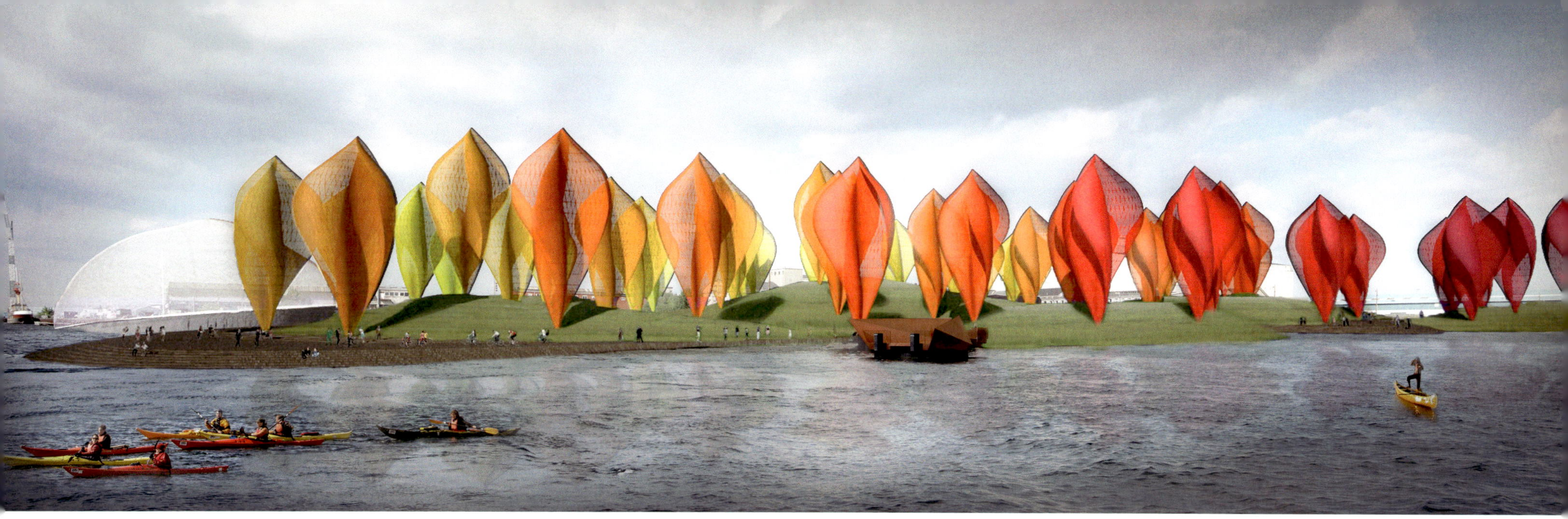

Regatta Fields defines a familiar yet otherworldly space that harmoniously complements and enhances its surroundings. The iteration of a specially designed wind turbine is the key feature of the installation. The site hosts 60 wind turbines with an output power ranging from 245 kW to 724 kW.

Plan view

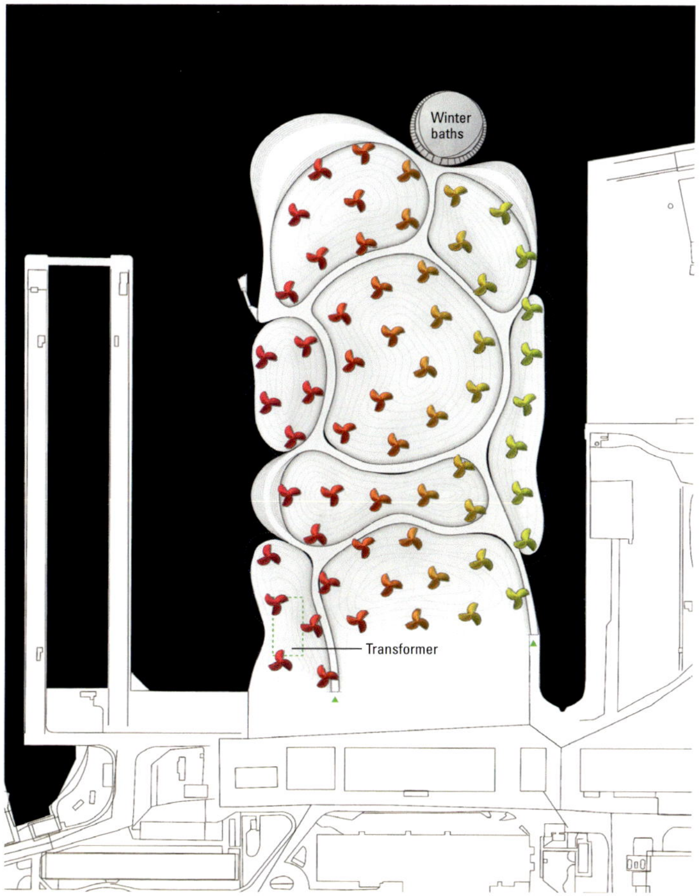

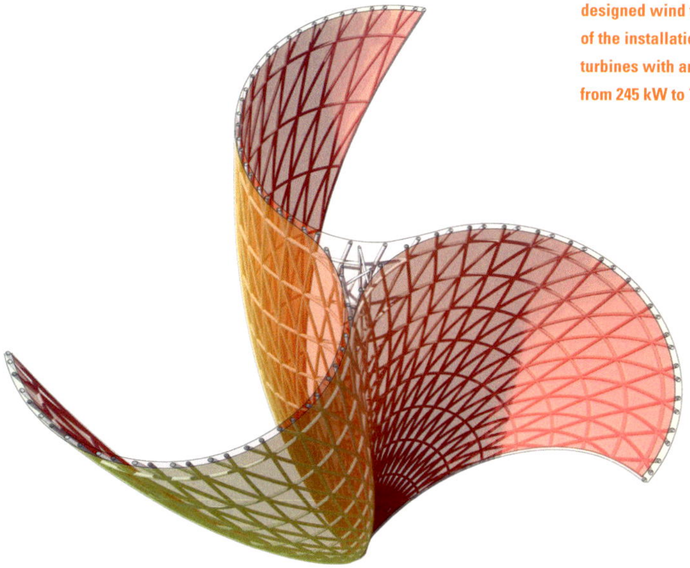

The transparency of the form connects the inner structural elements with the overall shape while encouraging a renegotiated perception of the surroundings.

Frame

TEAM:
Aucant Jordan, Marlhoux Lola, Aucant Mélissande

TEAM LOCATION:
Paris, France; Berlin, Germany

ENERGY TECHNOLOGIES:
vertical axis wind turbines

ANNUAL CAPACITY:
270 MWh

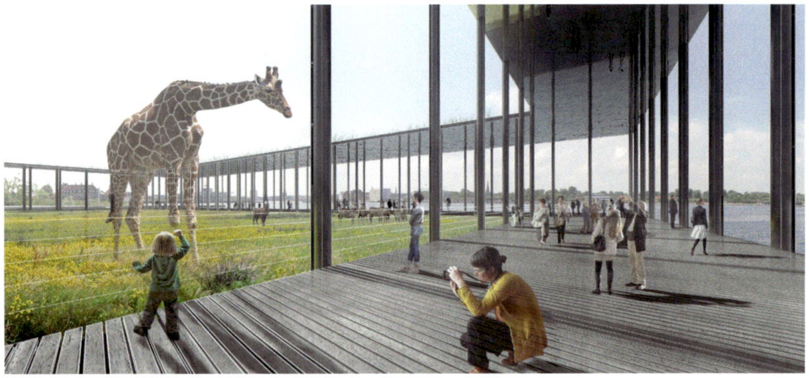

An empty interior encourages the installation to be alive. Various events and recreational activities, such as concerts, carnivals, fairs, and a circus could be organized on site.

Frame is a cloister on a metropolitan scale, designed for a powerful site that demands emphasis and protection. It is a mirror to the horizon, a temple of energy, a wind field, an exhibition space, a green roof, and a space-on-demand for pre-established uses.

The installation is built on the edges of the site in an offset 11 meters deep, leaving the rest of the site completely free for open use. Inside, landscaping will help clean the soil via phytoremediation.

Frame is designed so that cultural institutions can propose events (e.g., art exhibitions or boat displays) and recreational activities (e.g., concerts, performances, or a circus).

Six vertical axis Darrieus turbines are seamlessly integrated into each of the structural columns. These 12-meter tall "pillars of wind" carry a green roof that transforms the 220 days of annual rainfall into reusable filtered water while emphasizing the horizontality of the installation and becoming a new element on Copenhagen's skyline.

The style of the framework is intended to be as simple as possible and is comprised of steel and recycled metal from the deconstruction of neighboring port warehouses. The floor around the promenade is a subtle reminder of the decking of ships. The underside of the roof structure is a mirror that plays with the reflections of the water—returning the image of everything below.

Ice skating in the winter

Concerts and festivals in the autumn

Garden exhibit in the summer

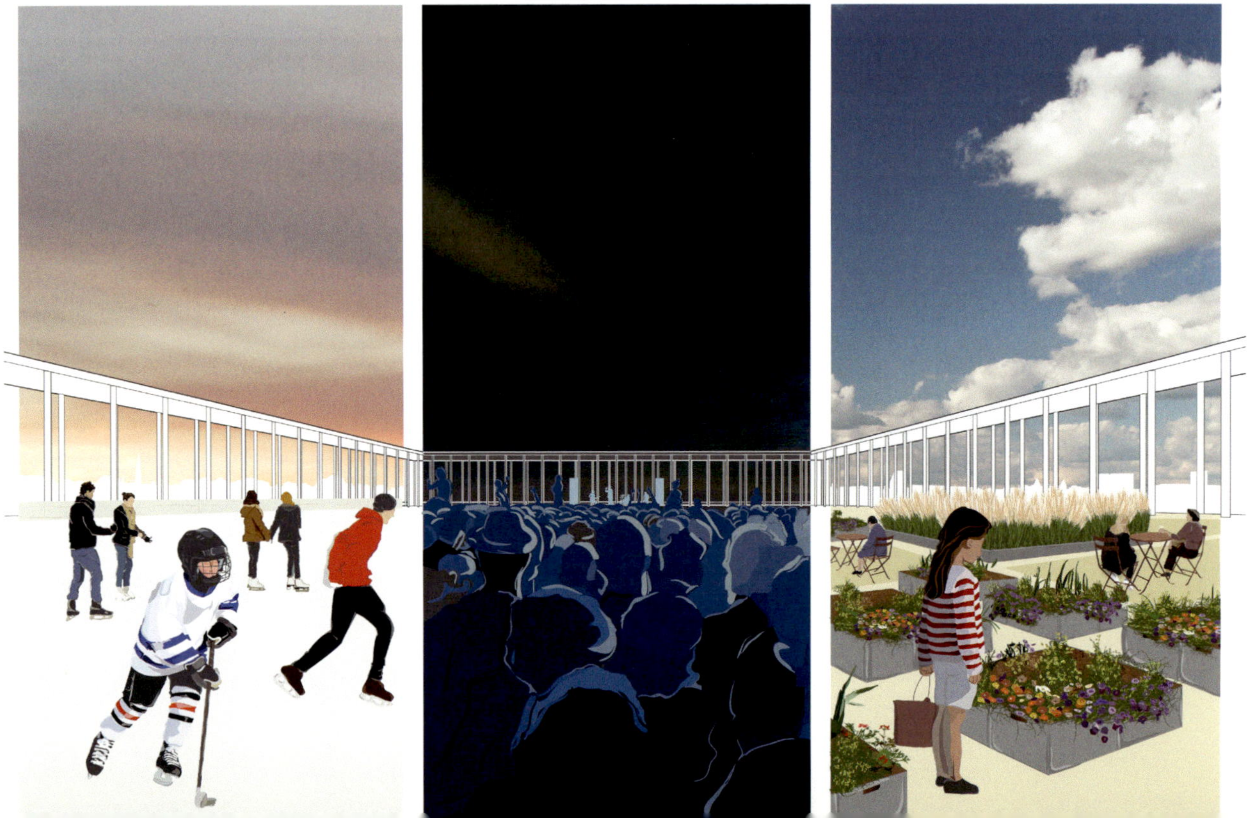

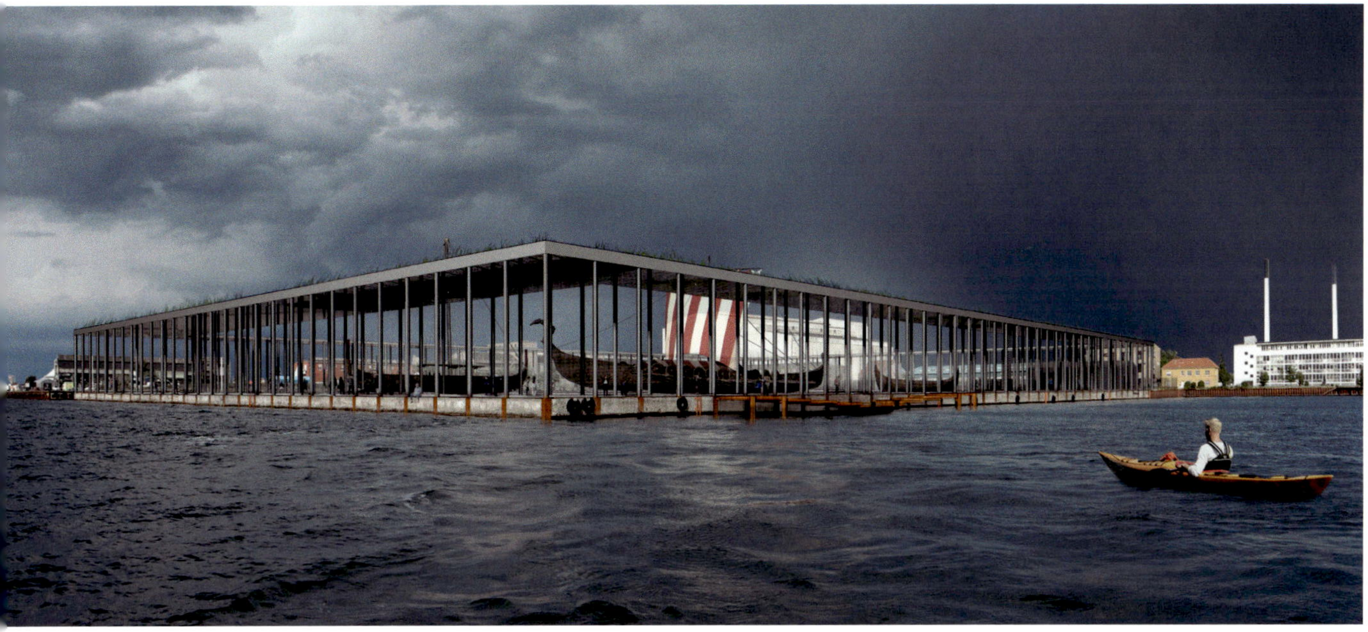

A Viking longship exhibition at Refshaleøen is just one of the many activities that can occur within the *Frame*.

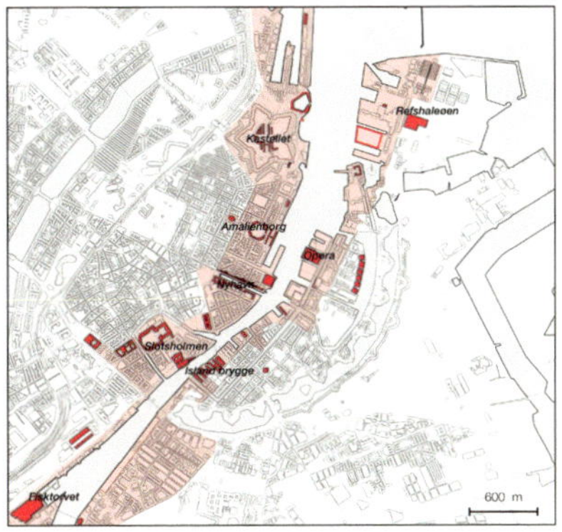

Context 1: The waterfront of Copenhagen

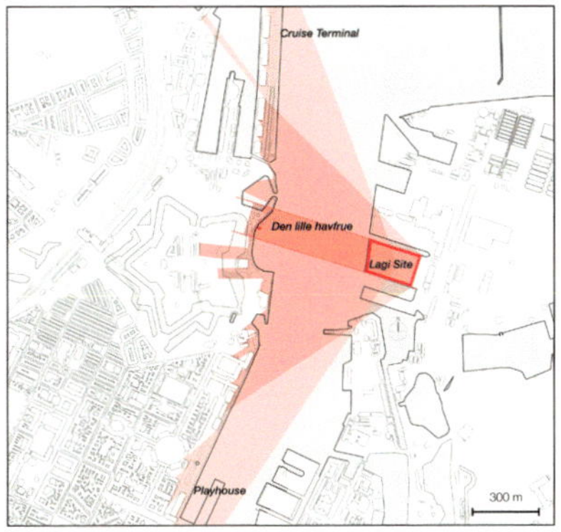

Context 2: The opposite bank

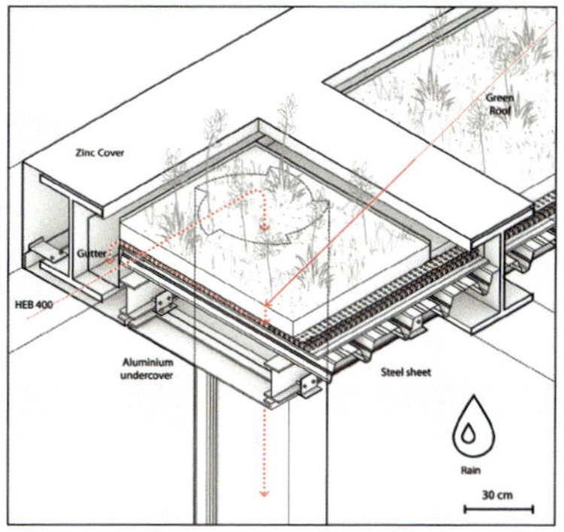

Detail of roof and pole

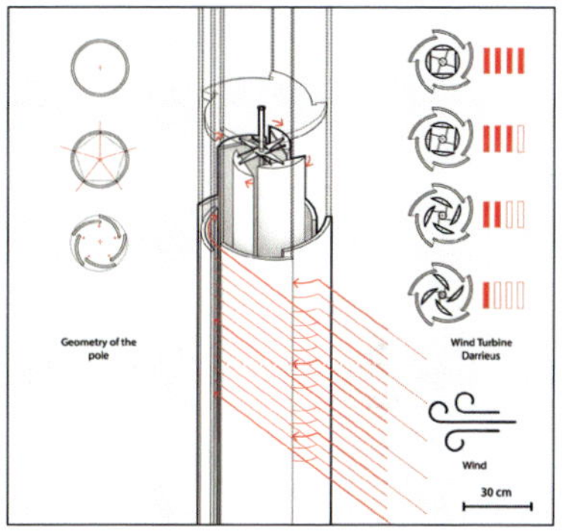

Detail of wind turbine and pole

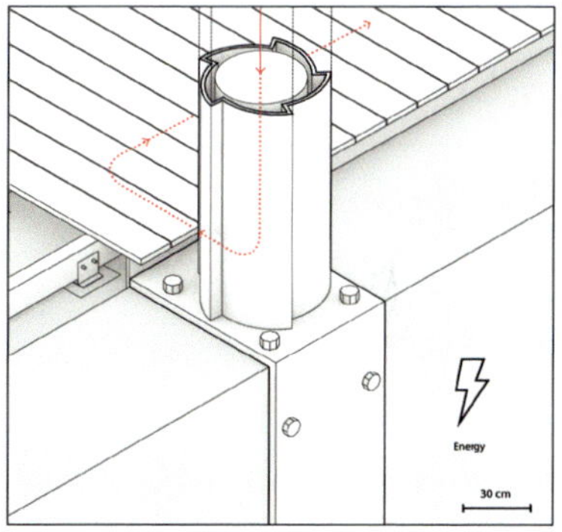

Detail of pole at the quay

The structure is composed of three basic elements of architecture in their simplest representation: 1. The floor is a subtle reminder of a ship's deck. This material delimits the promenade in the peripheric area of the plot. 2. Poles on the outer perimeter rest on the edge of the dock. They are hybrid in the sense that they support the roof and also provide electrical generation by way of six vertical Darrieus turbines in each pole. 3. The roof is the final element and emphasizes the horizontality of the project. Its underside is a plane mirror playing with the reflections of the water and everything below. Its surface is vegetated in order to retain rainwater, slowly filter it, and channel it to an underground reserve for reuse.

Cloud

TEAM:
Laura Maria Gonzalez,
Amir-Behan Jahanbin

TEAM LOCATION:
Pittsburgh (PA), USA

ENERGY TECHNOLOGIES:
flash geothermal with carbon nanotubes

ANNUAL CAPACITY:
75,000 MWh

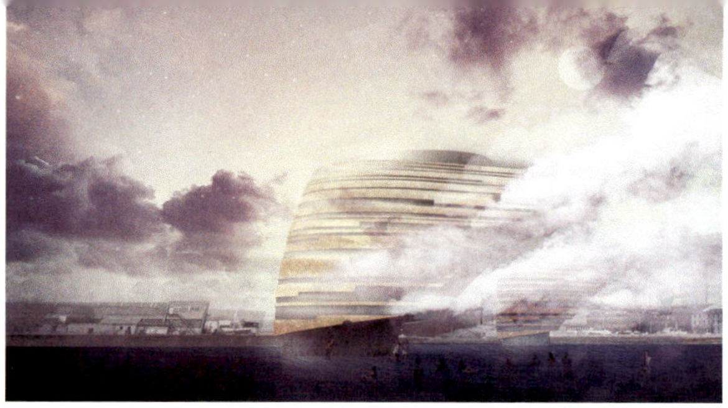

A series of three steaming sculptures in the landscape of Refshaleøen give rise to a cleansing ritual of body and environment.

Cloud is a revelation of transient states through steam. The installation brings earth and water together to establish a cleansing ritual for both the environment and the body as one, connecting people and nature in a dynamic system of ever-changing currents and flows.

Three sculptures make use of a well drilled into the ground that brings hot water to the surface. Because the geothermal resources below Copenhagen are generally not hot enough to generate steam directly, the hot water is mixed with carbon nanoparticles. The physical properties of carbon nanoparticles allow them to absorb heat more efficiently, creating pockets of steam at a molecular scale. Once sprayed onto surrounding water from the shallow pool, flash steam is created to run an electric turbine and heat a bathhouse.

Cloud embraces traditional bathing customs as visitors immerse themselves in cool water before entering a steam room. Oscillating steam turbines are visible as people slip into the steam rooms within the sculptures. The contemplative interior space provides an intimate experience between people and technology.

Cloud is a purifier of body and environment. As people use the sculpture to cleanse their bodies, the sculpture cleanses the surrounding water of impurities, rendering it valuable for the environment and ecosystems of Refshaleøen. The purification process also generates the power, since it is the same steam that powers the turbines for energy. The result is a compelling relationship between people, technology, and the environment.

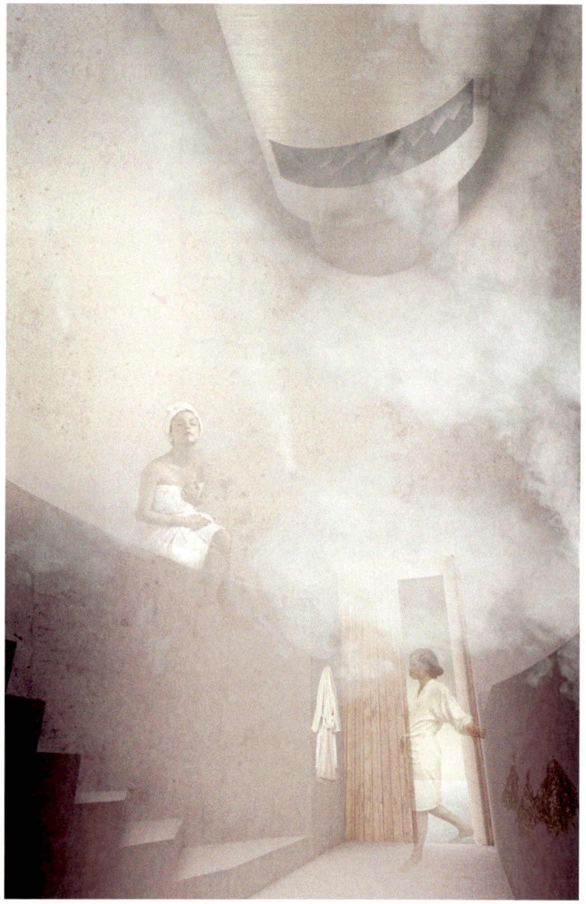

The experience inside of *Cloud*

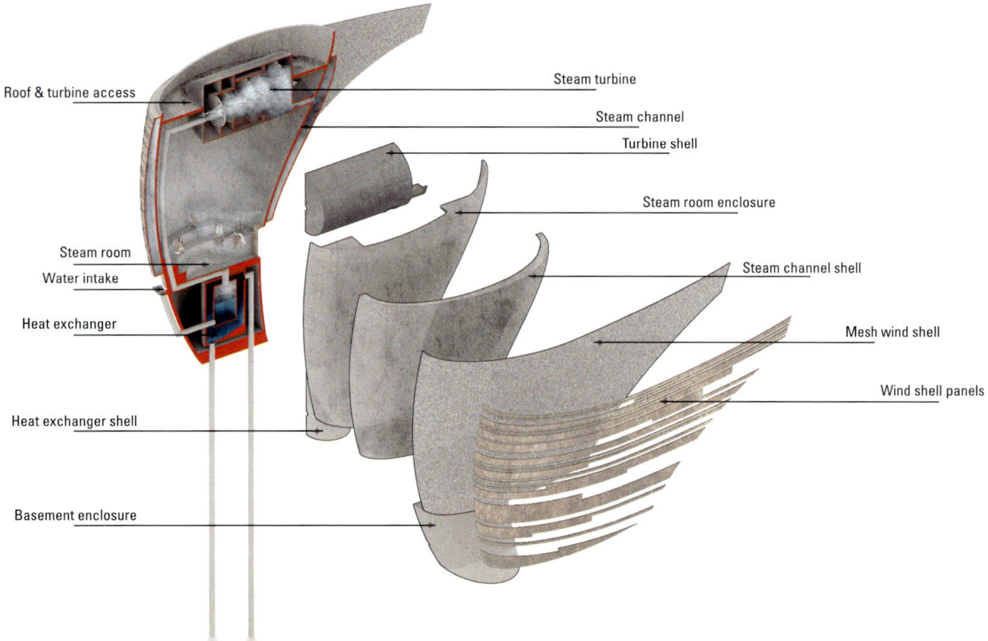

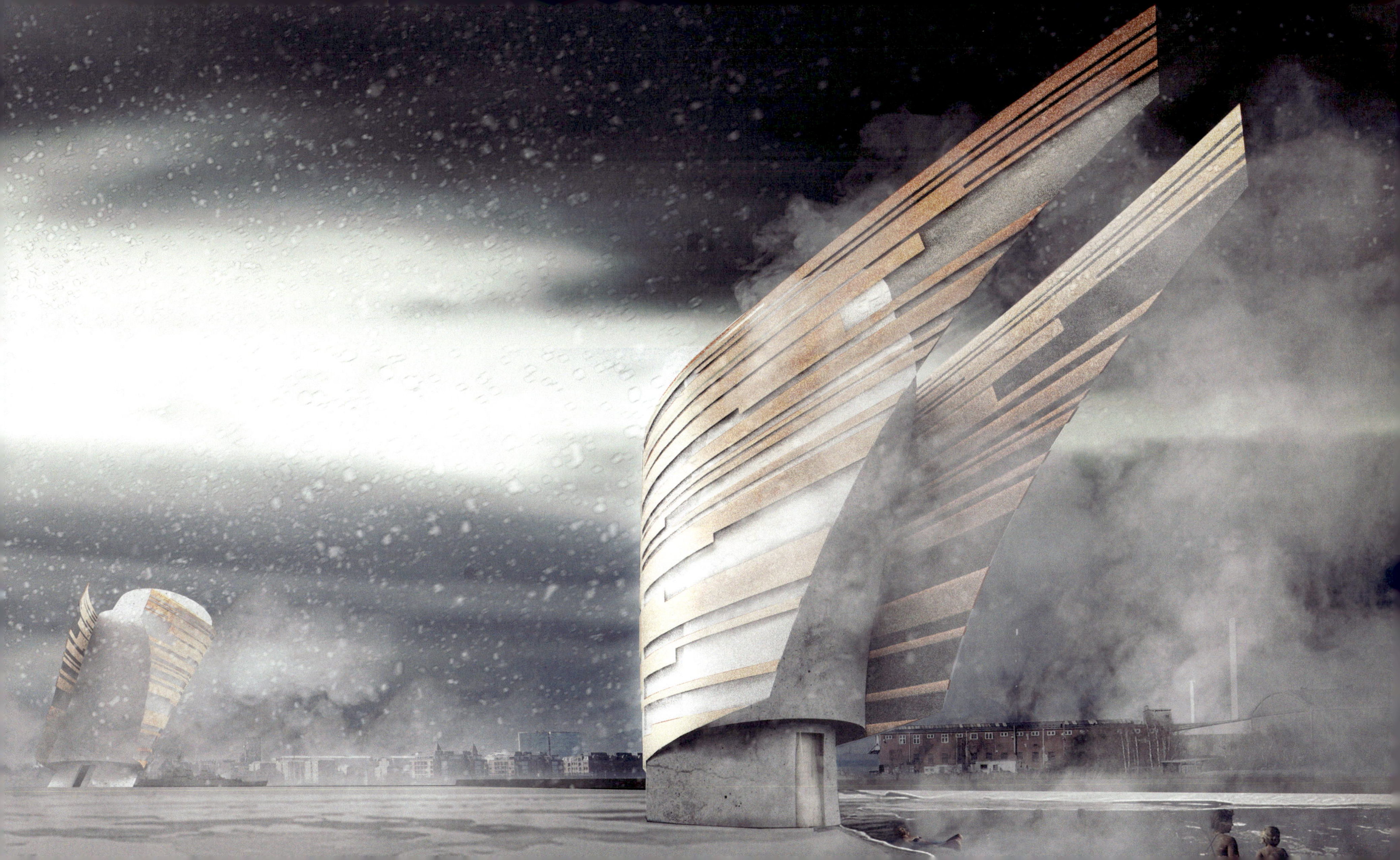

100*100*100 Wind Tower

*100*100*100 Wind Tower* is a 100-meter cube consisting of a myriad of small windmills that can be seen across the Copenhagen harbor.

Overlapping frames rise up, giving the appearance of an ordered morning mist. This huge volume is empty at first. It is for the citizens of Denmark to populate the tower with individual contributions of spherical, vertical axis wind turbines that are available in three sizes: 600 mm, 1200 mm, and 2400 mm. Similar to a crowdsourcing model, a family can contribute a small-size windmill, and a company can purchase a large-size windmill.

This community of turbines will offset the energy use of the individual investors and help Copenhagen achieve its Carbon Neutral 2025 plan. Once the tower is completely filled with turbines, its output will rival that of a conventional wind turbine. By 2025, the tower will contain the maximum number of turbines from the citizens of Copenhagen. The result will symbolize Copenhagen's commitment to a sustainable future.

Between the double walls of the artwork, a gentle ramp extends from the ground to the summit. An observatory at the top offers a view around the City of Copenhagen. A nighttime glow at the top of the tower represents the amount of electricity produced by the artwork. Days with strong winds will produce a brighter glow than days with faint wind.

The turbines are similar to units developed by the Japanese Environmental Promotion Association. Because of its small size, *Wind Tower* can operate with wind speeds as low as 2–6m/s. Its spherical shape enables it to react to the wind from all sides.

TEAM:
Mitsuhiro Wada, Takanori Ishii

TEAM LOCATION:
Brussels, Belgium

ENERGY TECHNOLOGIES:
vertical axis wind turbines

ANNUAL CAPACITY:
30,000 MWh

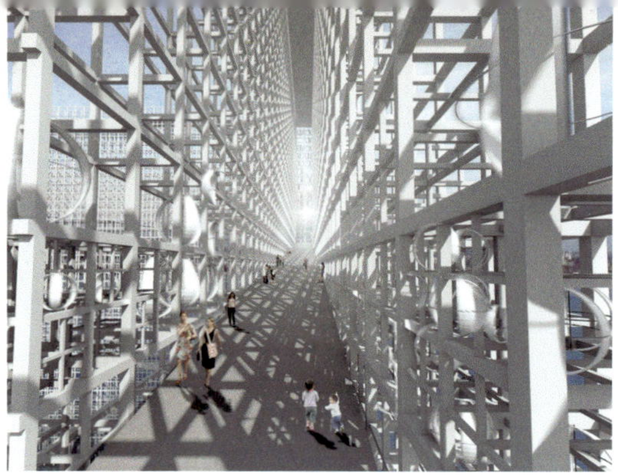

View of the visitor access ramp inside of the installation

Citizens can enjoy the spectacle of their collective clean energy generation.

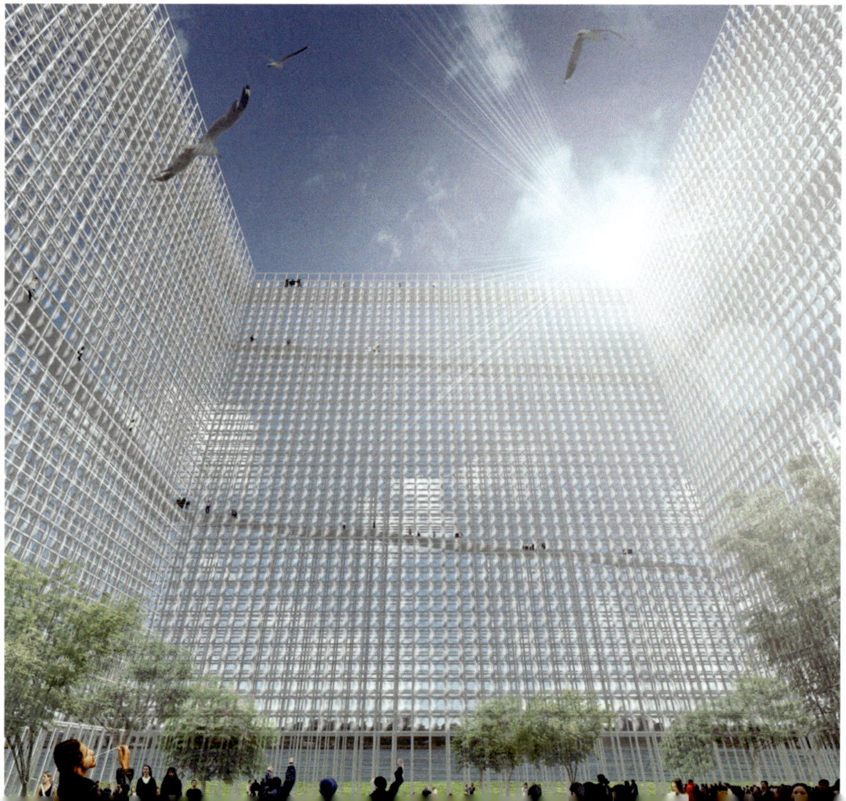

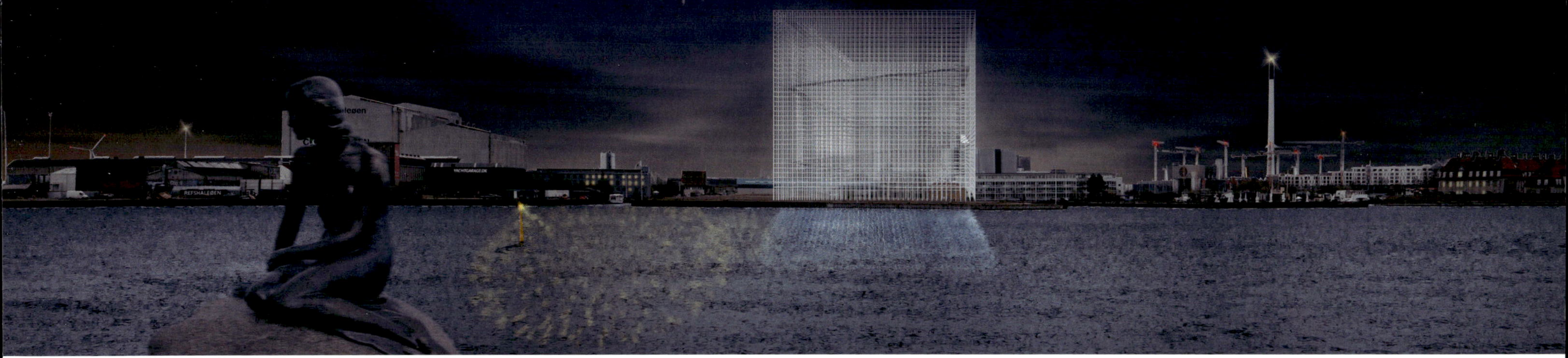

The artwork shines at night in proportion to its production of electricity.

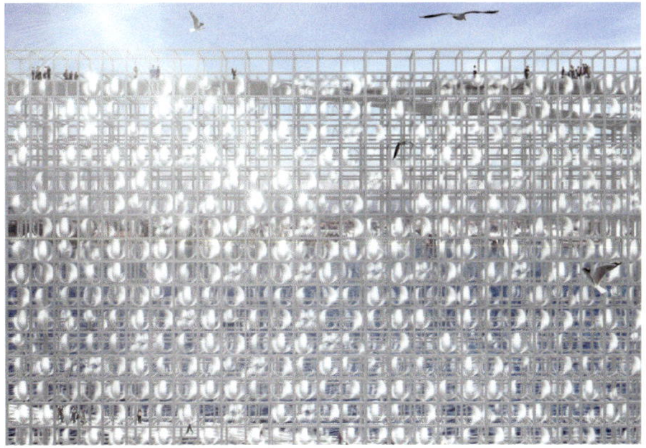

If completed by 2025, the tower would be a symbol of Denmark's commitment to renewable energy.

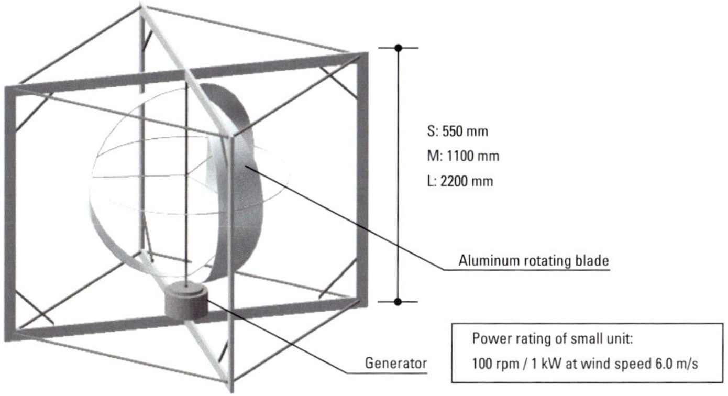

Spherical electronic generation unit

The Kyst

TEAM:
Janka Paulovics, Annika Janthur

TEAM LOCATION:
Berlin, Germany

ENERGY TECHNOLOGIES:
piezoelectric discs and fibers

ANNUAL CAPACITY:
500 MWh

The Kyst is positioned between the poles of the natural, technical, and cultural history of the site, and the visionary all-encompassing and sustainable impetus of Copenhagen.

The form of the installation is inspired by the "phragmites" or reeds that are a very common part of Denmark's coastal flora. Just like a field of reeds in a natural setting, *The Kyst* energy park consists of many densely placed stems in a wide swath along the water's edge. They draw energy from their surroundings: the wind, the sea, the rain, and from their interaction with humans. Scaling the stems up to the size of a tree places the visitor in a fairy tale environment, like *Alice in Wonderland*.

The Kyst consists of 1,400 single stems that merge into a sculptural landscape installation. The installation occupies the water and the land at the same time, placing one-third of the stems in water and two-thirds on land. Concentrating the stems at the edge will have the effect of an iridescent installation with a moiré of alternating views as the beholder approaches the site.

As an ode to nature and a calligraphic element, the stems have leaf-like structures added to the stems to harvest energy as they move with the wind, water, or rain. Even more importantly, the leaves enable and encourage interaction by visitors.

Piezoelectric Fiber Composites (PFCs) are placed inside the hollow structure of the stems to collect the energy from the leaves and swaying stems. The highest output can be generated when the stems catch their own resonance frequency and vibrate like flagpoles in the wind.

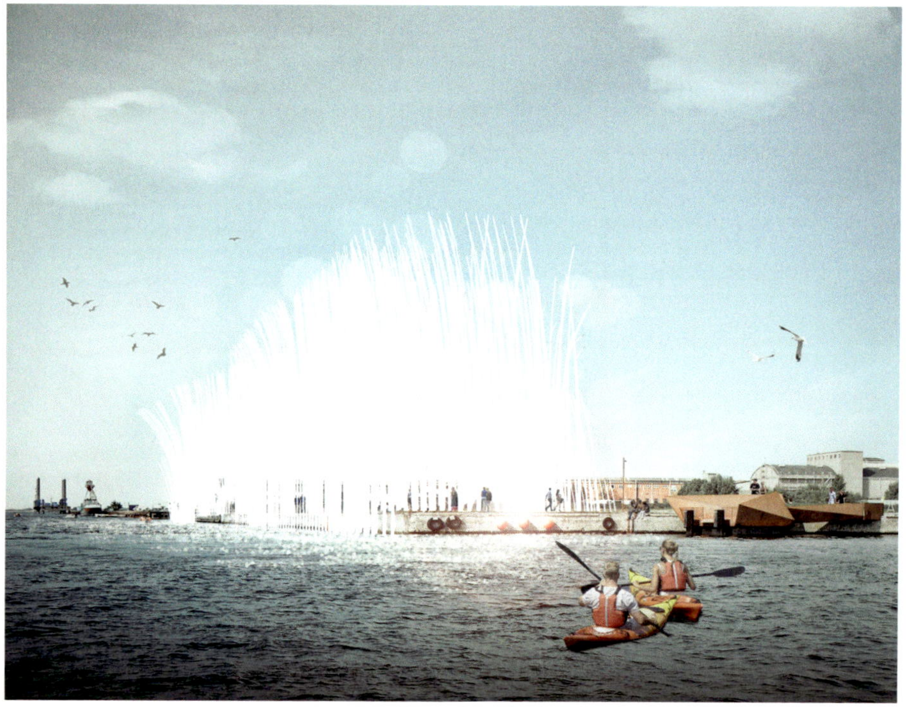

The conoid energy-generating stems come in four different sizes: the tallest with a diameter of 25 cm at the ground level and a height of 25 m, ranging down to the shortest with only 3 m in height and 5 cm in diameter.

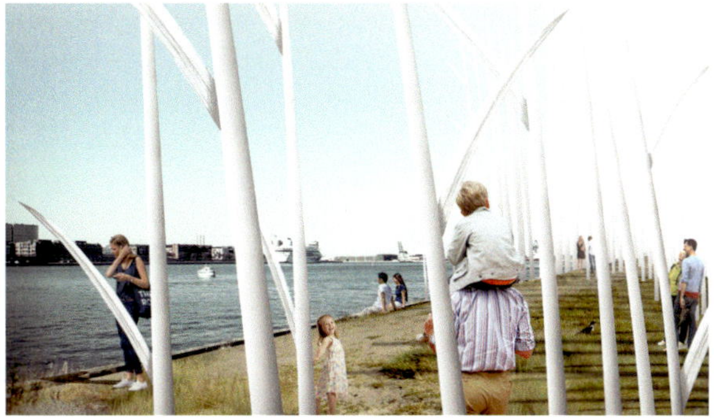

Just like the elemental forces, the visitor contributes to the power output. Through immediate physical interaction, the installation raises awareness about clean energy production and the constructed beauty that corresponds with this cultural achievement.

Each stem will be upholstered with a three-dimensionally woven piezoelectric fiber, which absorbs kinetic energy. This technology is currently being developed by the University of Bolton's Institute for Materials Research and Innovation.

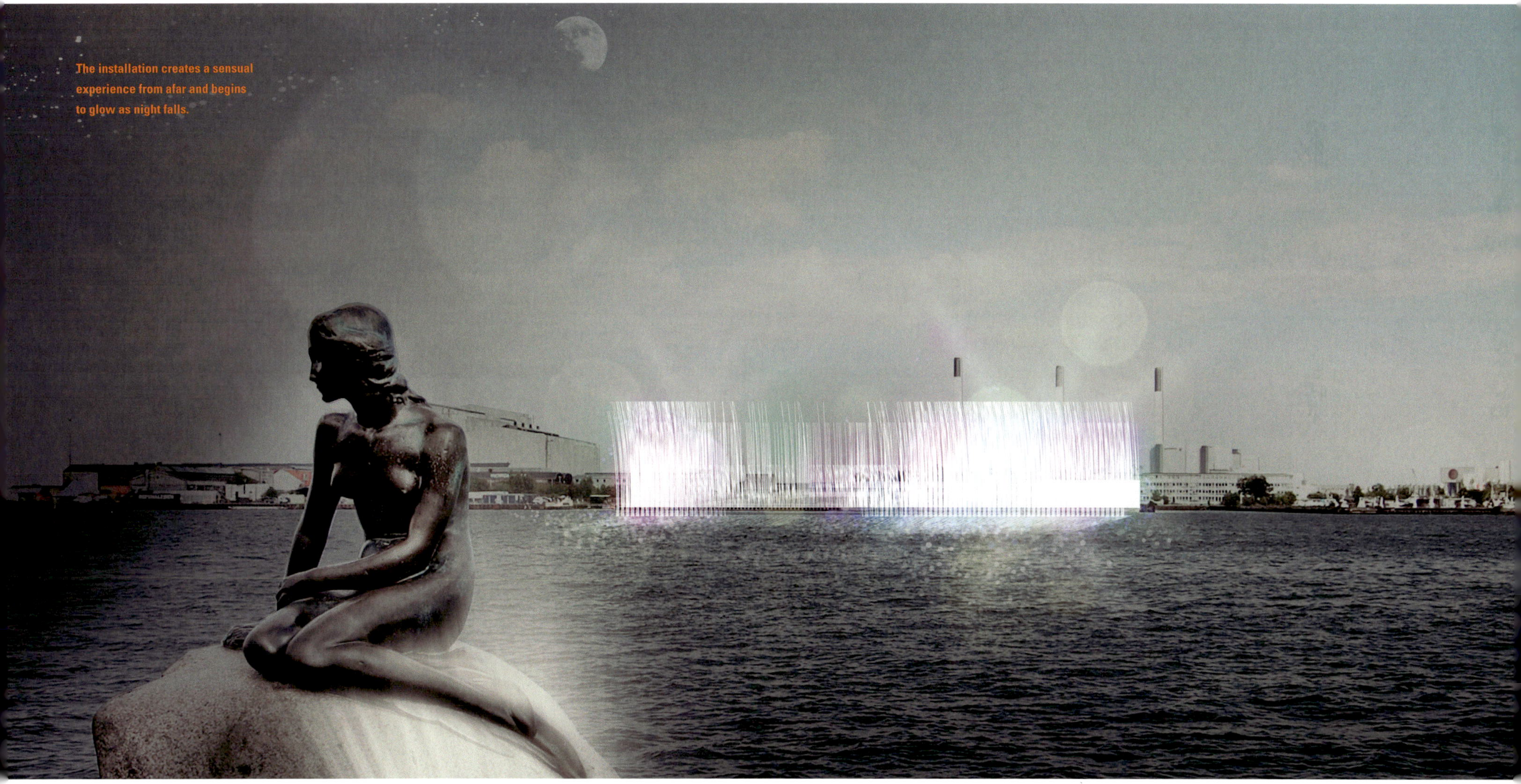

The installation creates a sensual experience from afar and begins to glow as night falls.

Floating Forest:

Between the Trees

Inspired by the movement of boats, the trees of *Floating Forest* are designed to sway in the water, moved by the force of waves and wind. Additionally, artificial elements work like a plant to accumulate carbon dioxide and give back oxygen.

The artificial trees that make up the installation interact with the sea and with a new dock. As they are activated, they produce energy while they clean the surrounding air, capturing CO_2 and releasing oxygen.

A suspended path runs through the installation, linking the water taxi stop with the rest of the design site. Natural areas, such as sand dunes, planted outcrops, and urban gardens intermingle with artificial elements, like the suspended path and a terraced square for outdoor events. An artificial pool is carved from the existing bank for water sports and open-air activities.

"Floating Poles" produce energy with their oscillatory movement. One system in the design harvests kinetic energy while a second system captures the hydraulic thrust to run a small turbine from compressed air. "Land Poles" provide energy storage and transmission functions. On-site displays indicate the level of energy that is produced and the amount of CO_2 removed from the air in the park by the artwork.

TEAM:
Massimiliano Matera, Annalisa Pilati

TEAM LOCATION:
Rome, Italy

ENERGY TECHNOLOGIES:
kinetic harvesting, compressed air turbines

ANNUAL CAPACITY:
164 MWh

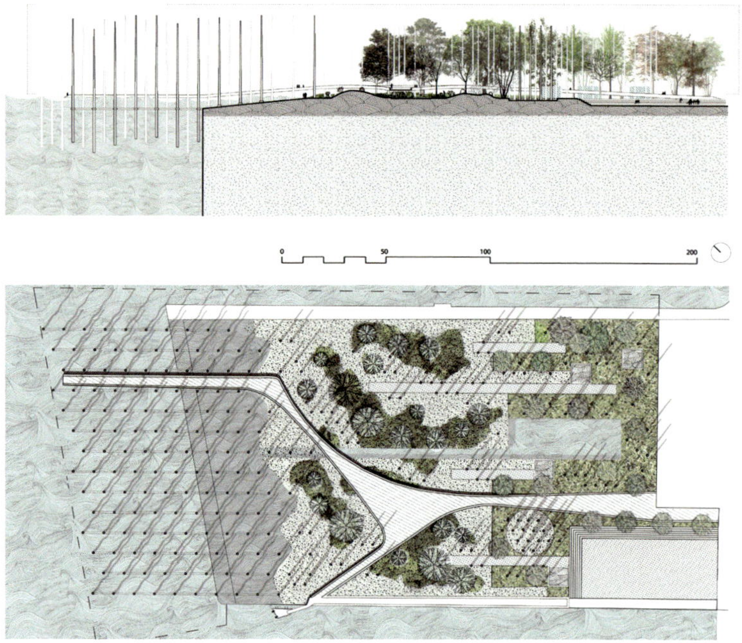

Longitudinal section and site plan

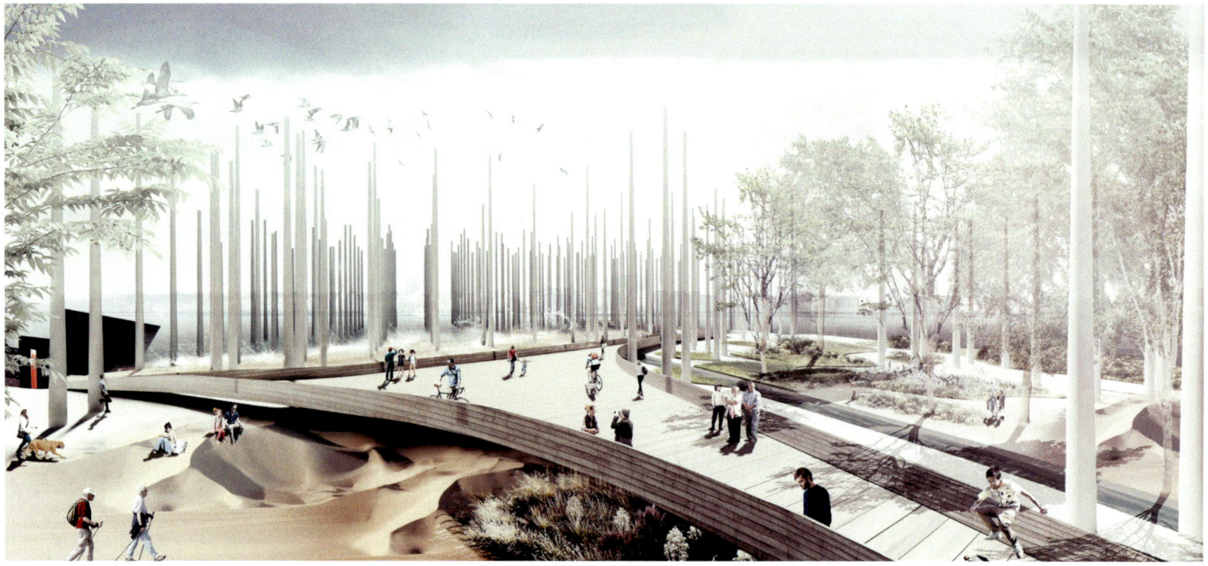

There are three types of poles in *Floating Forest*: Floating Poles, Water Poles, and Land Poles. The rotary movement of the turbine at the base of the Water Poles and Floating Poles generate energy by harnessing the force of the water on the beach. All pole types provide an air purification system, similar in function to that of a natural forest.

1. Mobile arm; 2. Main structure connected to the dock; 3. Rotor; 4. Entry air holes; 5. Valve; 6. Space dedicated to the movement of the piston (fixed); 7. Piston; 8. Compressed air chamber; 9. Valve; 10. Turbine section; 11. Generator; 12. Carbon dioxide accumulation section; 13. Purification section (clean air exit)

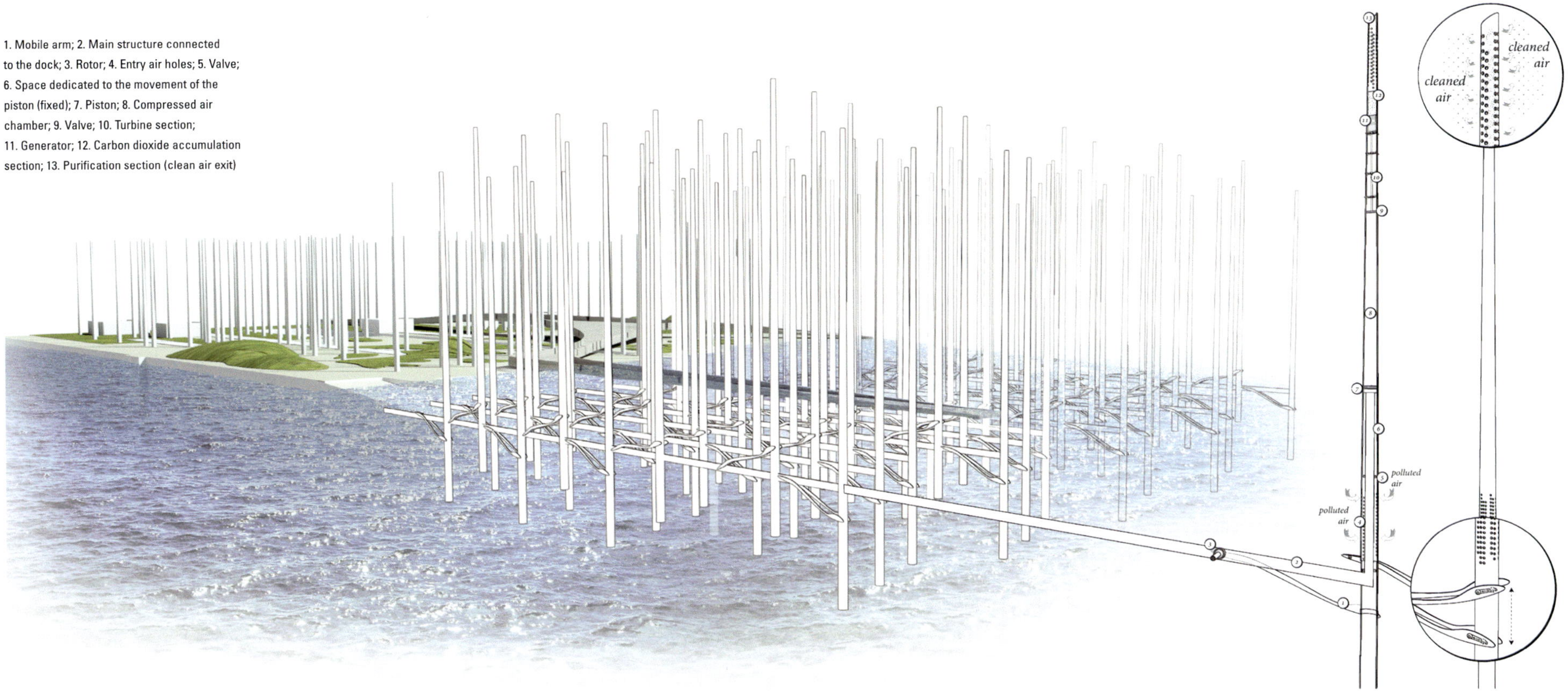

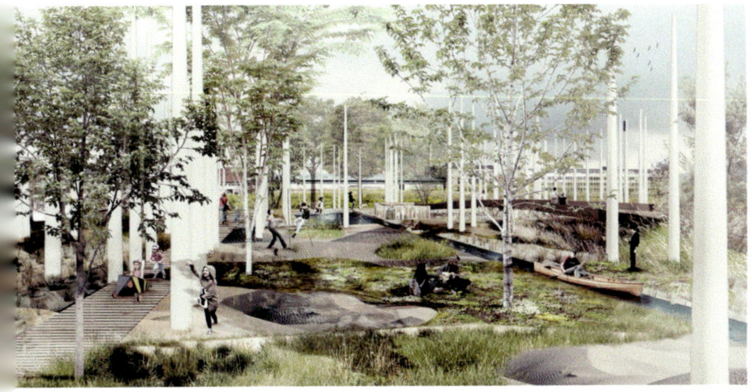

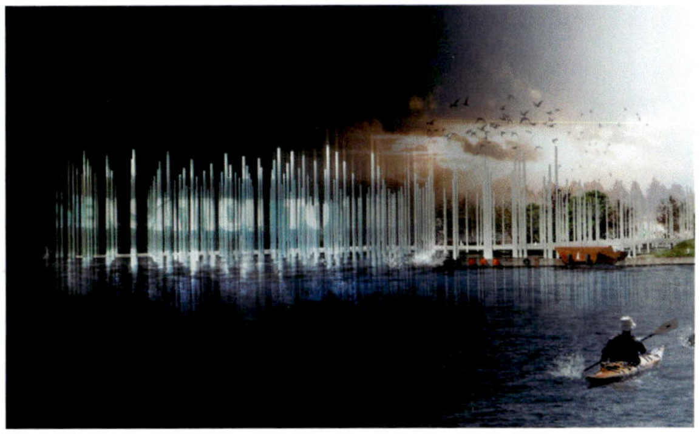

Natural and artificial elements cooperate to improve the environment.

Golden Roots

TEAM:
Ronny Zschörper, Franziska Adler

TEAM LOCATION:
Münster, Germany

ENERGY TECHNOLOGIES:
biomass, piezoelectric paving

ANNUAL CAPACITY:
52 MWh

Golden Roots contrasts the urban environment of Copenhagen with the experience of unspoiled nature, suggesting a childhood walk through golden fields of grain. The installation recalls Denmark as a sparsely populated, agrarian country, where its inhabitants maintained a close relationship with nature. A system of paths and bridges guides visitors through a constellation of crop circles, bringing the calm countryside to life in the city.

The experience of being enveloped by nature is achieved through circular fields of rye, which have a natural maximum height of two meters. Red poppies, so familiar in the region, provide a splash of color and biodiversity. The fields are periodically harvested to generate high-energy biomass as well as to provide materials for the construction of observation towers, which change every season. Rye bales coated with potassium silicate are used to create towers of up to 18 meters tall when maximum crop yields are achieved.

Paths and bridges allow visitors to walk within and to rise above the circles of rye. In addition to the energy production through biomass, each crop circle (connected by the paths) contains a piezoelectric floor, transforming the kinetic energy of people into electric energy.

A sea of rye is highlighted with red poppies in the summer months.

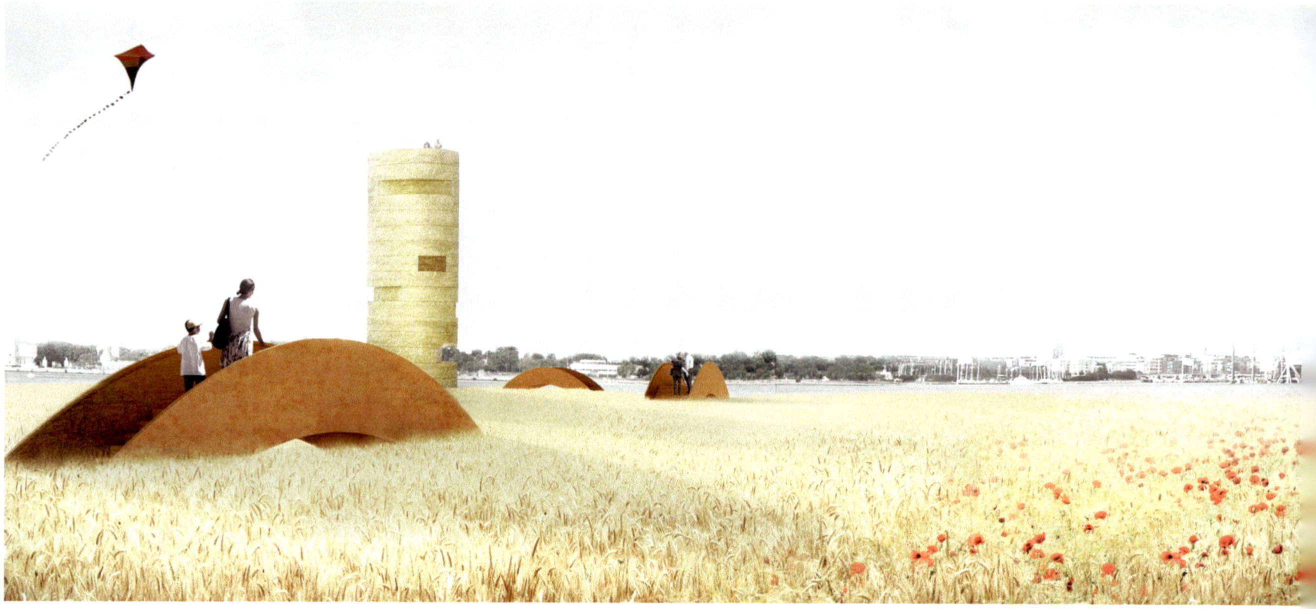

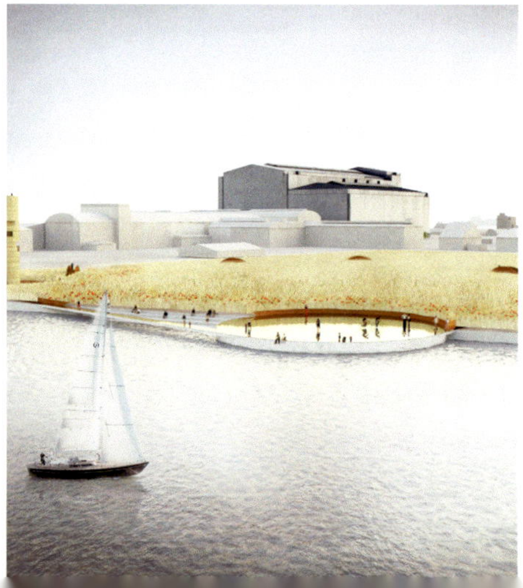

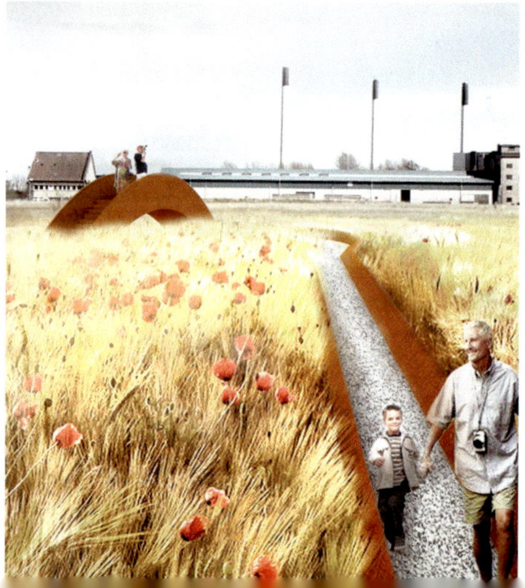

The path through a field of rye invites visitors to relax and reflect on the cycles of nature.

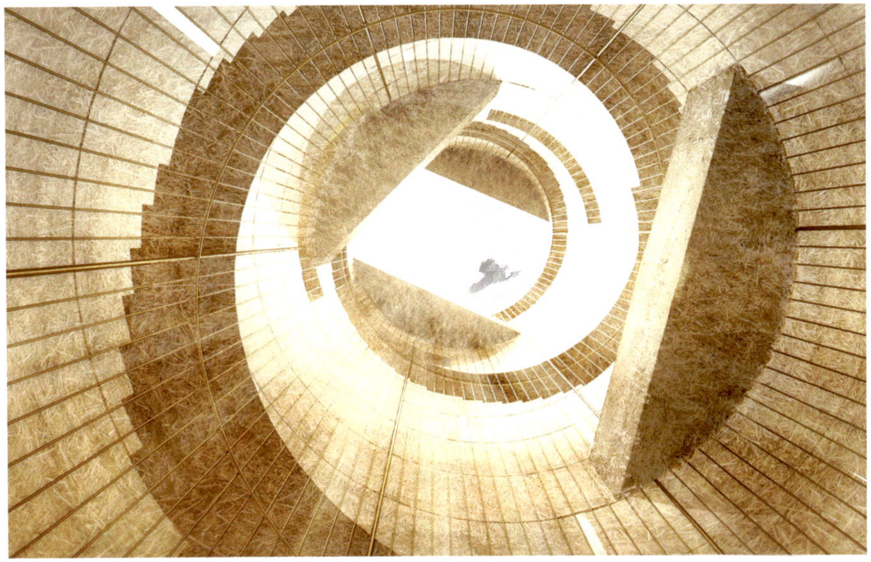

The tower grows through a modular system as an outcome of the harvest.

Crossing Bridges
Bridges and their underpasses provide barrier-free and unlimited access to the observation tower and the water's edge.

Water Platform
A direct water connection, additional entrance point, and a place to relax with a view of the city.

Observation Tower
The tower provides an additional view point with an altitude of 18 meters above the ground.

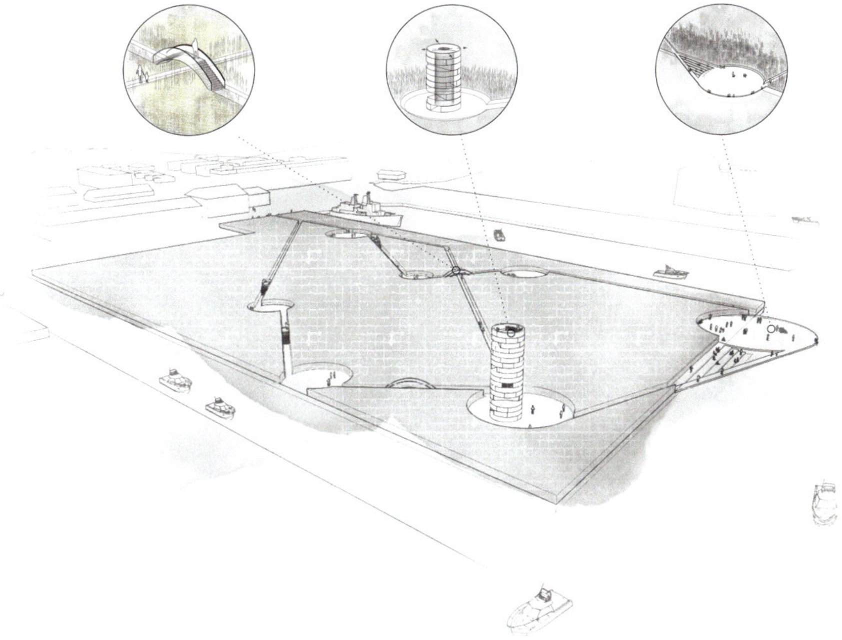

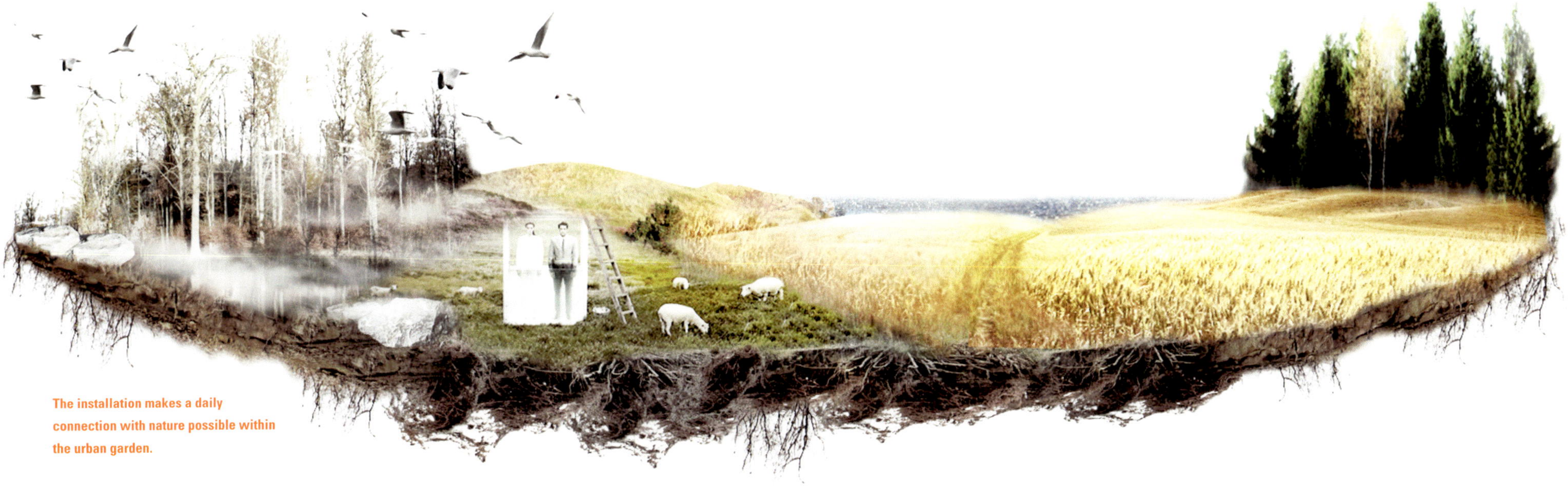

The installation makes a daily connection with nature possible within the urban garden.

Kinergic

TEAM:
Hector Paredes Gutierrez,
Jorge Arias Olaiz

TEAM LOCATION:
Barcelona, Spain

ENERGY TECHNOLOGIES:
piezoelectric toroids

ANNUAL CAPACITY:
300 MWh

Inspiration for *Kinergic* comes from the great strength and unpredictable temperament of the wind, the constant and tireless motion of the water, and the ever-changing and ambitious goals of humans.

The *Kinergic* structure invites people to interact with it: playing with the filaments, walking up and down, and transforming motion into electricity. Visitors have a new and productive relationship with the energy that they will later consume.

Kinergic also offers activity spaces, such as an urban garden and two amphitheaters. The goal is to generate a continuous flow of people throughout the year, their kinetic forays contributing to a healthier planet.

Electricity is generated on the nodes of a 60-meter tall wooden structure made out of laminated structural wood beams set in a grid module of 10 meters square. There are two types of nodes: wind and water. Metallic hemispheres hold eighteen 4.5-meter long high-density EPE foam wind filaments. In each filament, a flexible inner shaft houses 2.5 meters of piezoelectric toroid generators and has an aerodynamic tip that facilitates the movement. The water filament has a plastic floating device that transmits the wave's movement like a buoy.

The wind nodes are distributed throughout the structure, while the water nodes are located in just the first two rows, touching the water with special floating devices that transmit the wave's movement to the piezoelectric pieces of the filaments.

There is an LED light at the tip of each filament that shines according to the amount of energy produced, creating a highly visible landmark on the site.

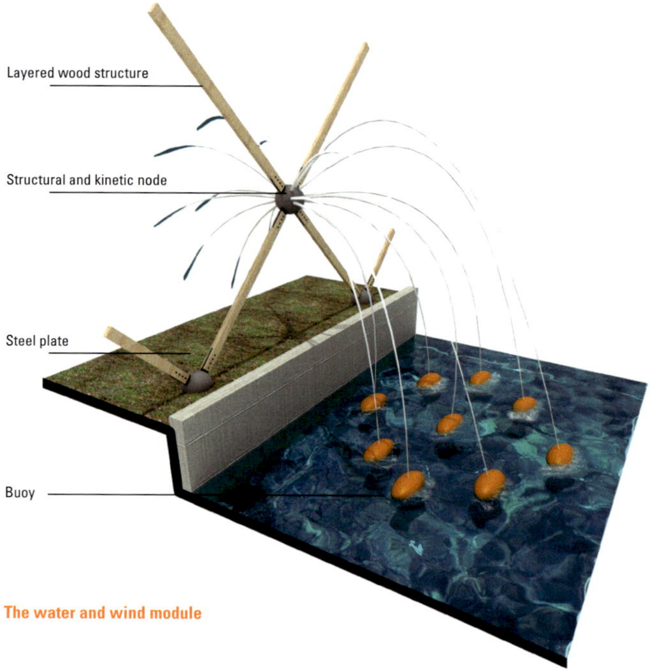

The water and wind module

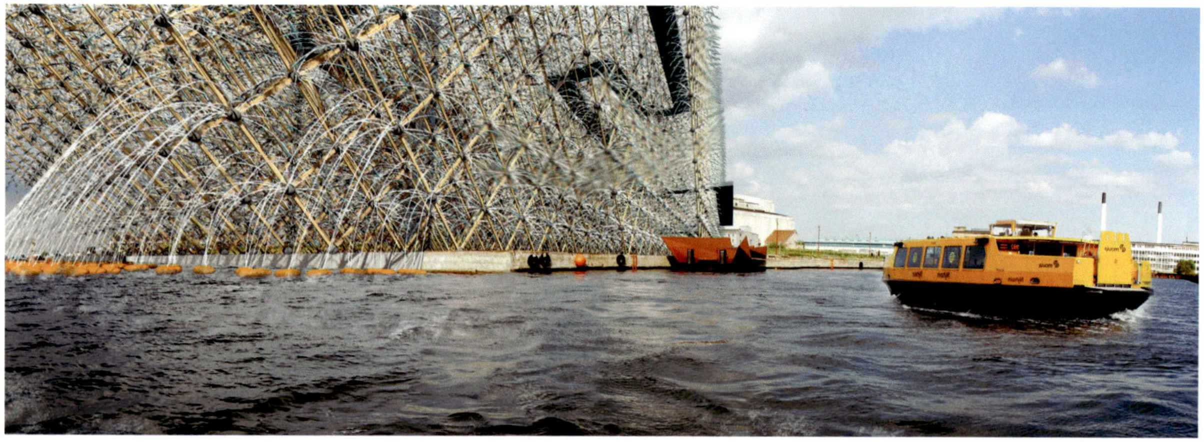
Copenhagen harbor bus nearing the dock at the edge of the installation

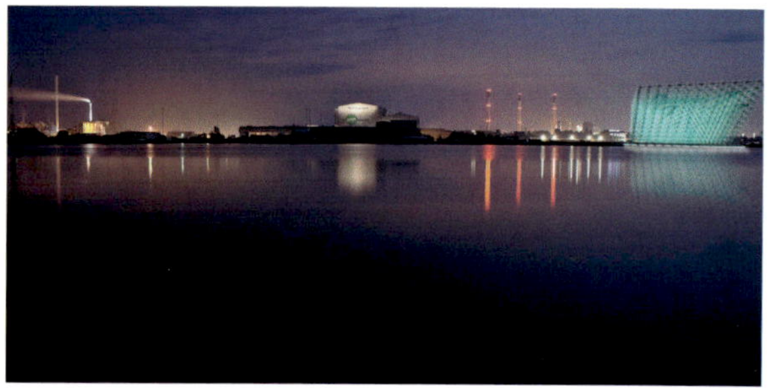
Night view from across the harbor

Background photograph by Johan Biilmann

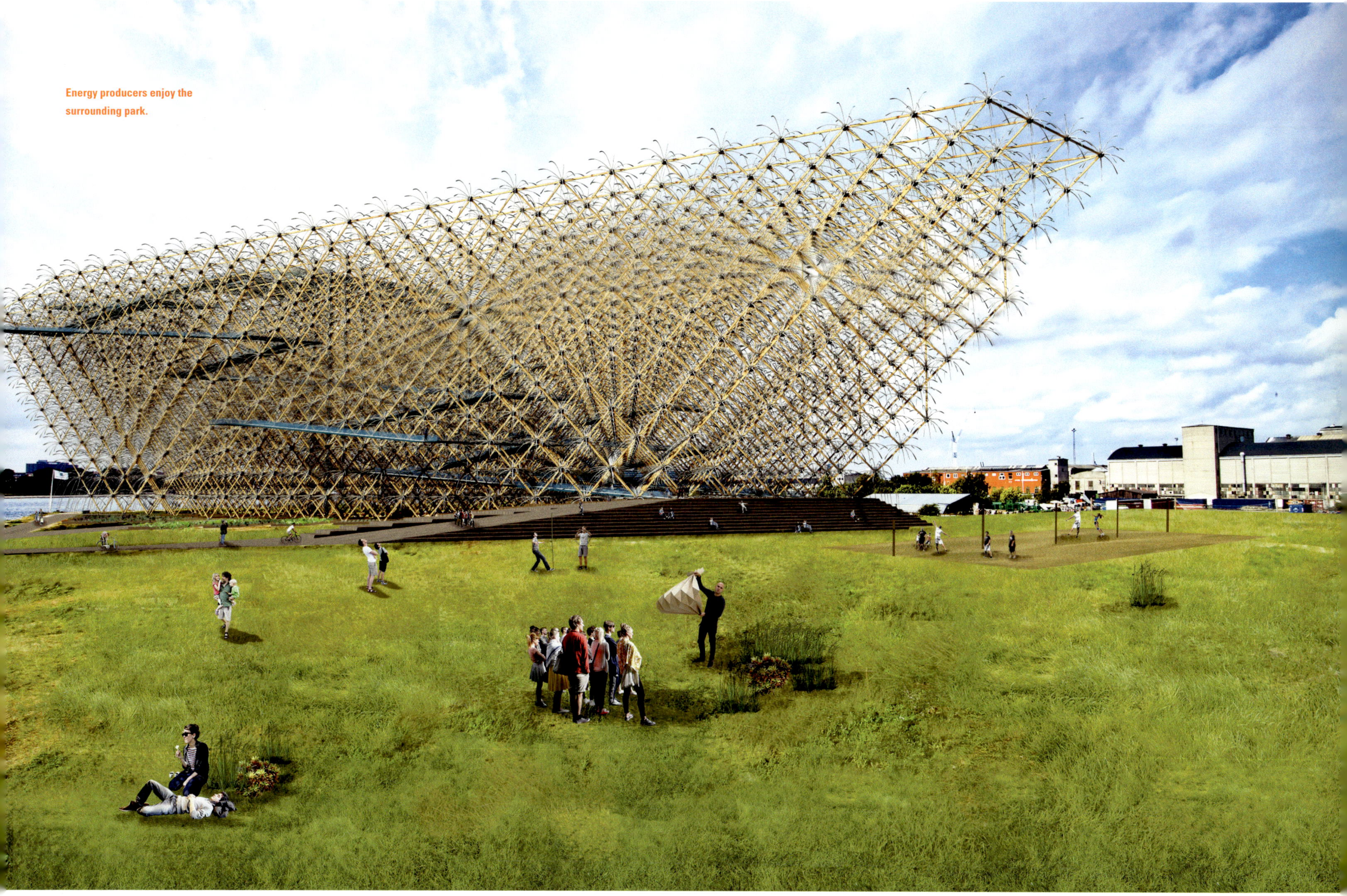

Energy producers enjoy the surrounding park.

Light Guard //

Know your flood risk

Light Guard is a land art generator for both energy generation and flood-prevention risk. It is a dynamic structure that responds to the design site through climate prevention.

The installation consists of a series of 137 metal columns spaced one meter wide and measuring 35 meters in height. Each series of two columns support an autonomous system of alternating photovoltaic panels and a system of plates with bright colored LEDs.

Glowing from light blue to deep red, the LED panels serve the city as a warning mechanism for flood risk and prevention. Ranging from normal risk to heightened risk to evacuation due to imminent threat, *Light Guard* helps to keep the citizens of Copenhagen safe from rising sea levels.

The installation takes advantage of the area in front of the design site while preserving the site itself as a space for various events. A footbridge connects Refshaleøen to *Light Guard*, inviting visitors to pause and enjoy the lookout facing the sea.

TEAM:
Vincent Trarieux

TEAM LOCATION:
Brive-la-Gaillarde, France

ENERGY TECHNOLOGIES:
photovoltaic panels

ANNUAL CAPACITY:
1,250 MWh

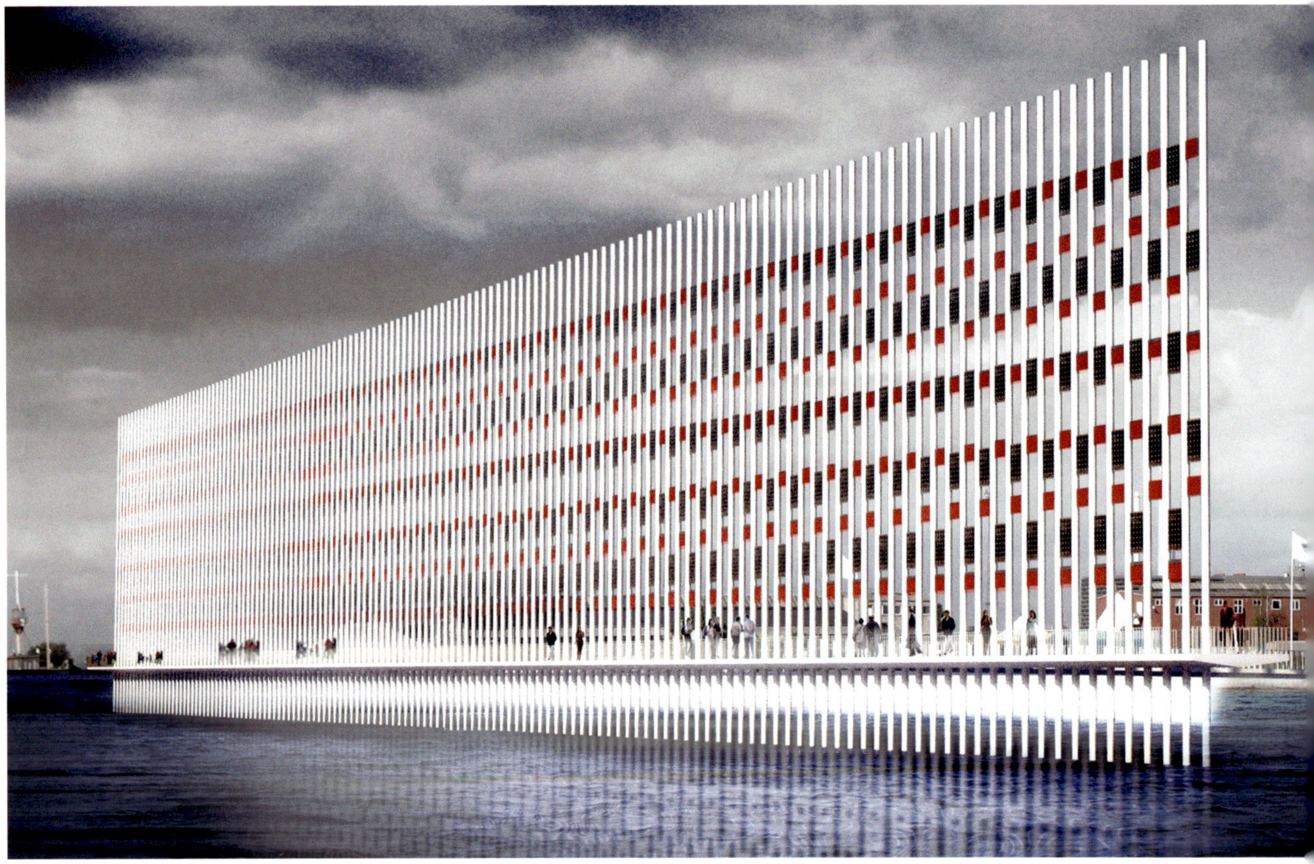

The pattern of colors signifies the level of flood risk. The red and black pattern above is associated with imminent risk of devastating flood activity.

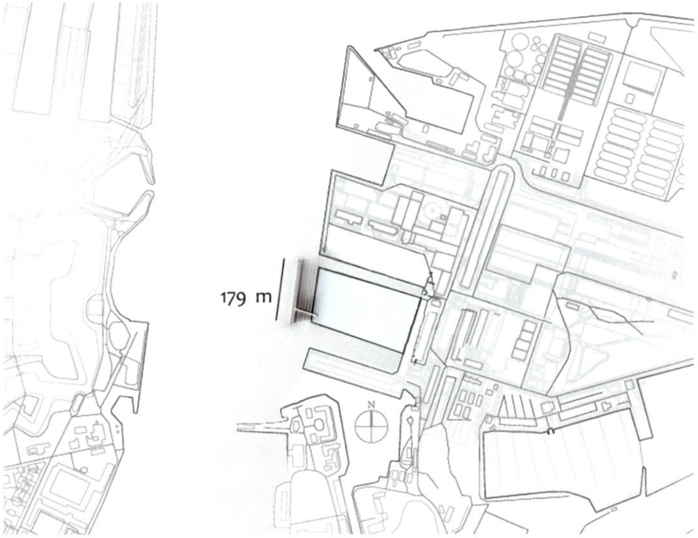

Plan view. The modular facade creates a new public space just offshore.

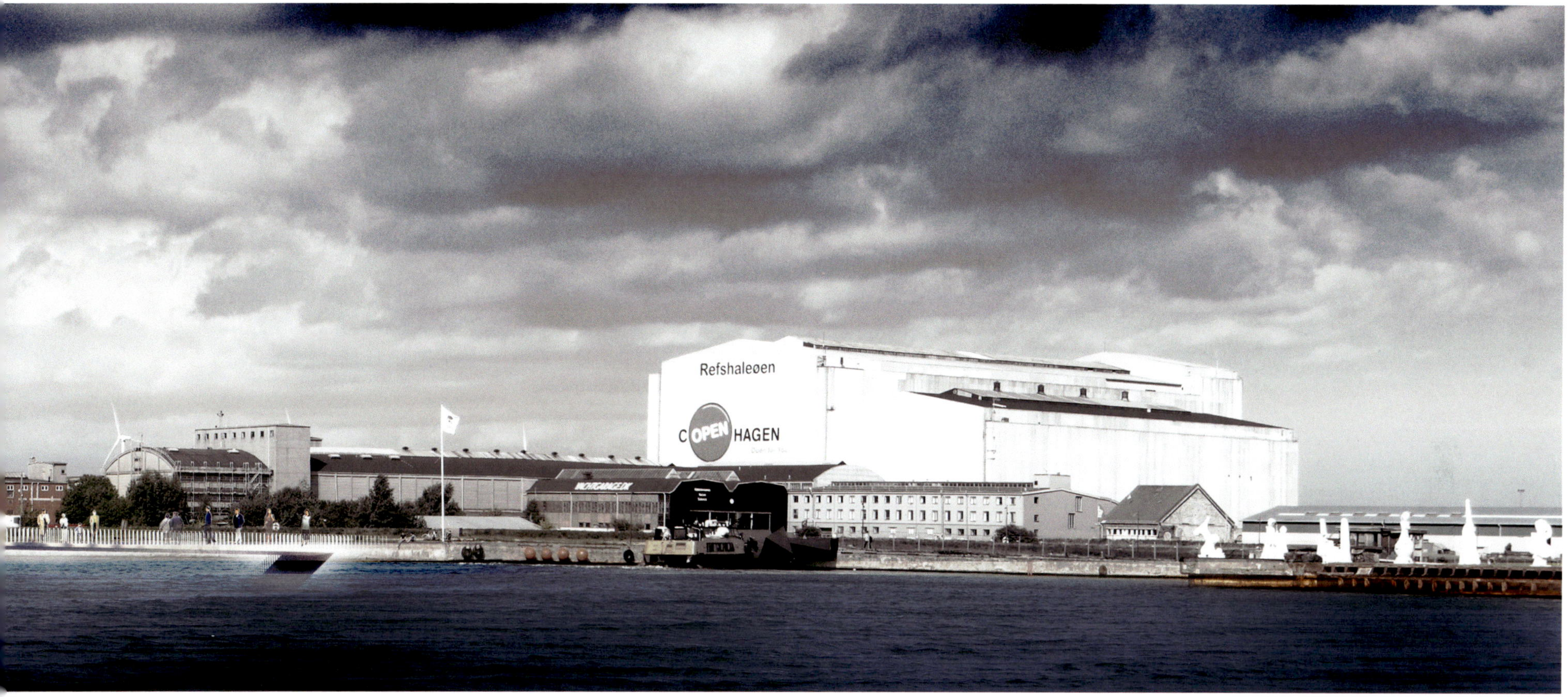

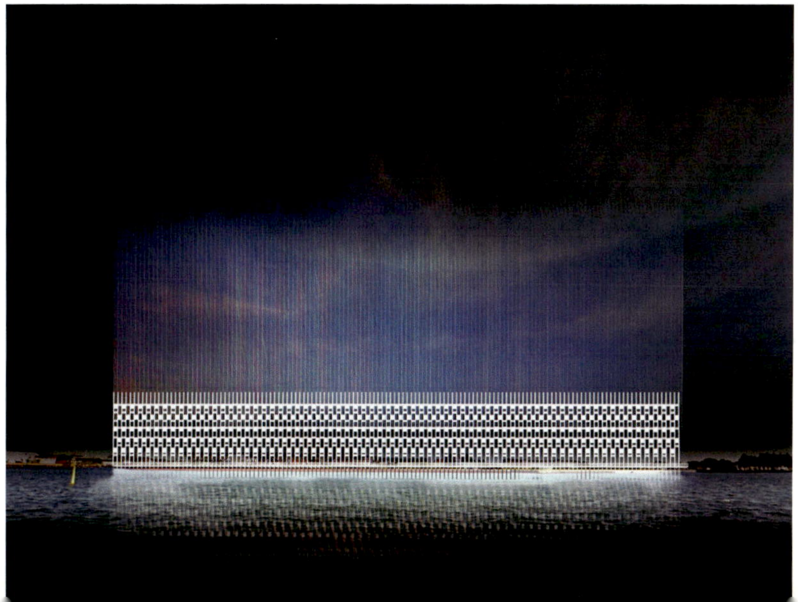

The lighting signal indicates the flood warning level to those across the Copenhagen harbor.

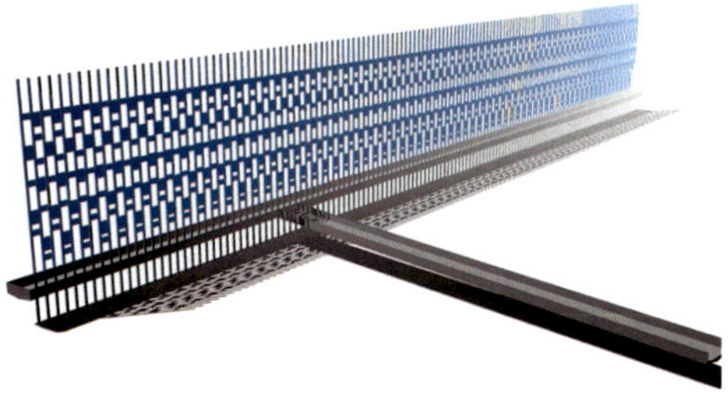

Light Guard is an autonomous lighting and signal device.

A Place for a Chocolate Ice Cream

TEAM:
Juan David Ramirez, Diana Marcela Manrique, Bao Bao Huan, Rasmus Johansen

TEAM LOCATION:
Cali, Colombia

ENERGY TECHNOLOGIES:
photovoltaic panels

ANNUAL CAPACITY:
3,500 MWh

A Place for a Chocolate Ice Cream creates a modular grid of retractable, energy-generating pavilions, inspired by Danish Rococo-style mosaics—three-dimensional sculptures derived from two-dimensional drawings. The energy pavilions change between day and night, revealing new appearances for every occasion. Daytime reflections are replaced by nighttime glow.

The experience begins at the site entrance, where a dense mass of pavilions slowly decreases in number as visitors continue west into the park until they reach an open green space near the canal. The placement is based on a hexagonal grid, giving the installation the flexibility to be reorganized and redesigned according to the conditions of the site.

Each modular pavilion can unfold its six petals to expose 93.53 square meters of solar surface (for a total of 11,000 m^2). The petals shift to follow the sun throughout the day. The solar panels incorporate MPPT (maximum power point tracking) controllers that sample the output of the cell and selectively disable those that are not collecting enough energy. This ensures that the pavilion is receiving as much charge as possible for any particular orientation.

The solar petals can close like a flower, permitting the project to exist in two very different states. During the day, the retractable arms are open, permitting maximum solar energy harvesting and providing a 65-percent shade to the ground plane for events.

At night, the petals can be retracted quickly, using a simple, solar-powered hydraulic system. The spatial experience changes completely, transforming the site from a covered space to an open-air park in minutes, revealing views towards the harbor.

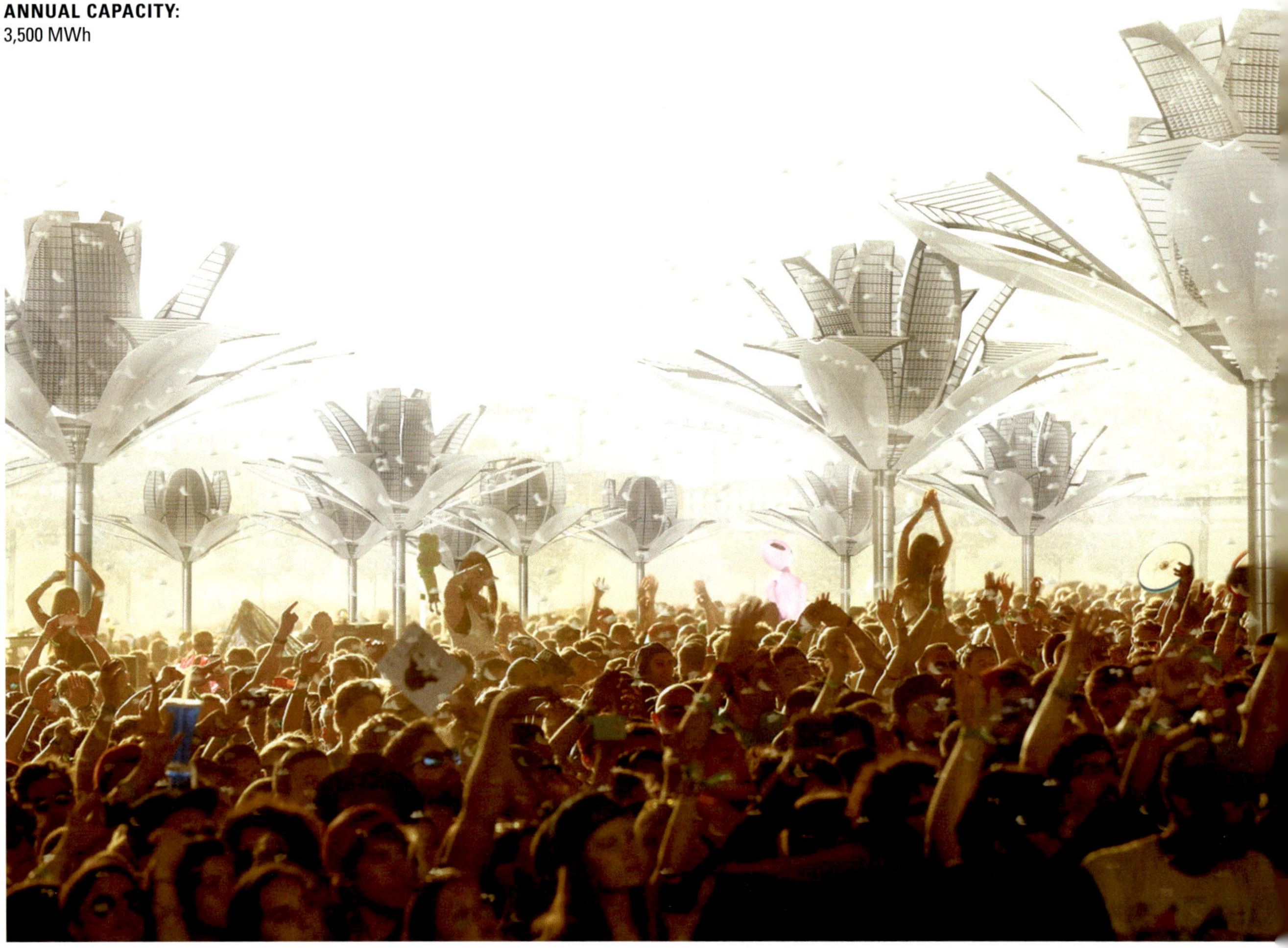

The pavilion permits a full spatial experience. The intervention area can transform from a covered public space to an open air park in just minutes, opening the views towards the harbor and creating an interactive mediated outdoor space.

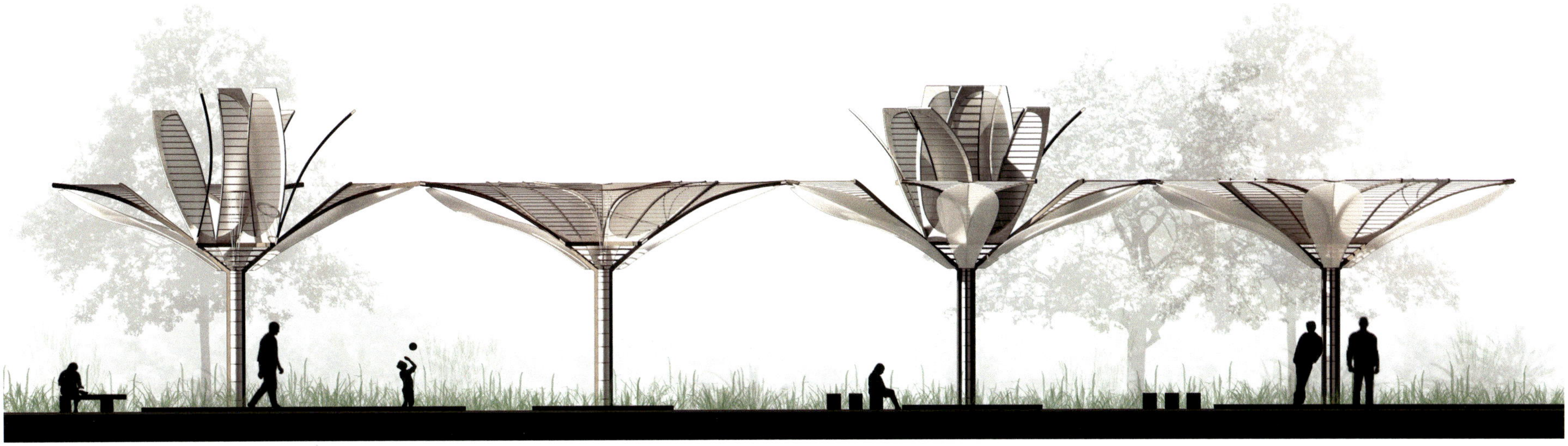

The solar-shades close like a flower, permitting the project to exist in two different states. During the day, the retractable arms are open for maximum solar energy harvesting and create a soft continuity in the space.

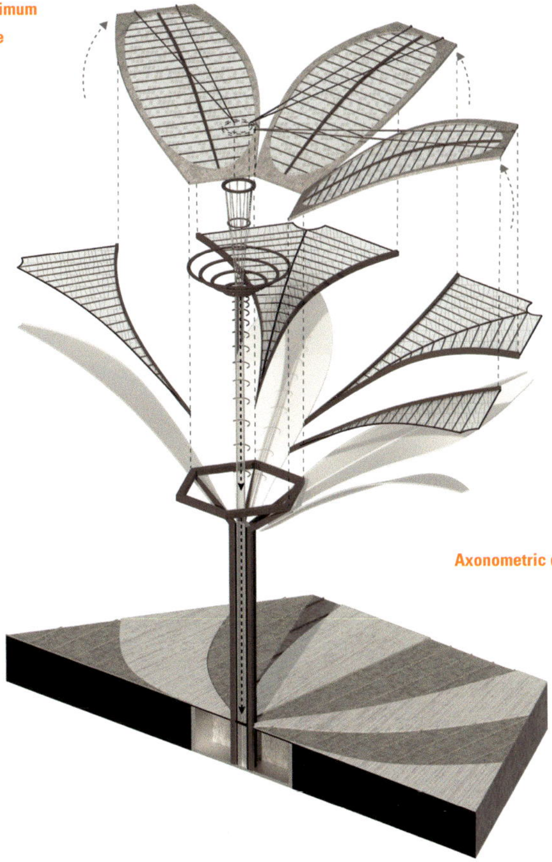

Axonometric diagram

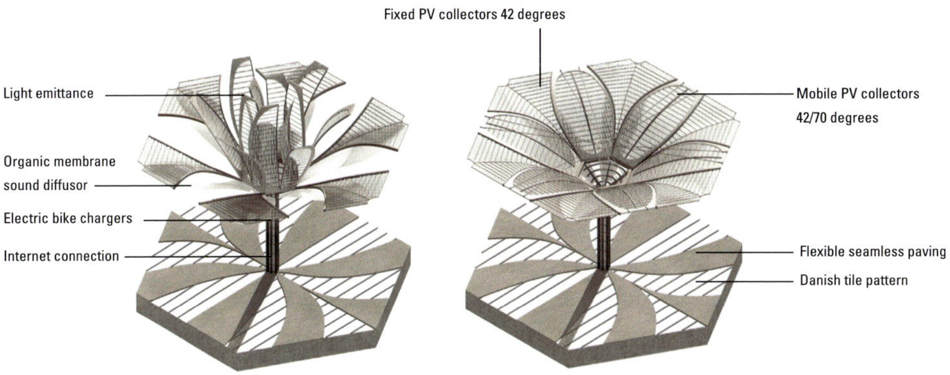

Retracted sun-shades

Open sun-shades

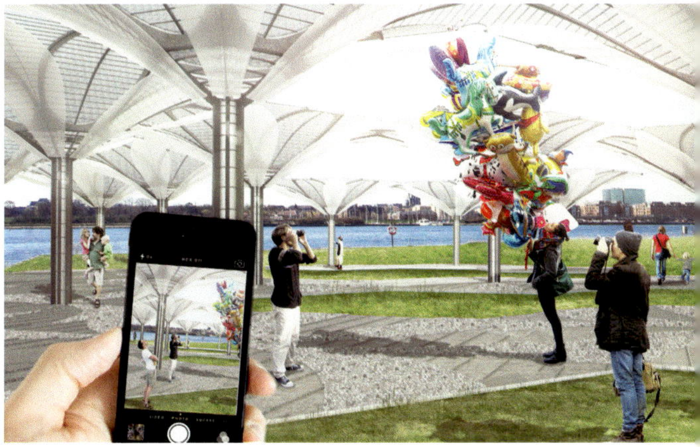

Lightfoam

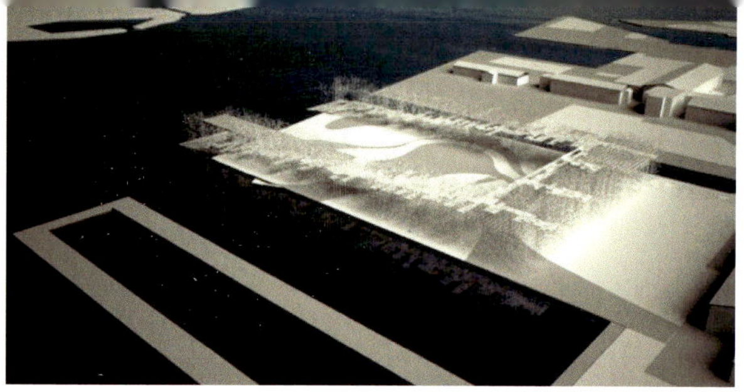

Bird's-eye view model

Long poles standing on artificial wind slopes continually move like a field of reeds according to the direction of wind. Daylight reflections and dim lighting at night invite visitors to imagine a scene from the Little Mermaid.

Lightfoam is an array of 2,400 polished, white bending poles that harvest the energy of the wind with piezoelectric generators. The 20-meter high poles naturally respond to any change in wind direction. Each is embedded with stacks of piezoelectric discs set between rigid backup plates. To increase the intensity of wind, these piezoelectric structures protrude through a platform shaped to form a valley. The framework of poles is free to bend and respond to all forces, such as wind or people. The more they bend, the more electricity they generate.

At night, LED light sources inside each pole allow the structure to glow or dim depending on the amount of energy they generate. As the wind moves, the fiber rods softly sway to create a dynamic effect and memorable scenery.

The site is transformed into a public park where children can shift or climb between poles, people can ride bicycles, or walk along vantage points on the slope. As they are enjoying the park, their activities generate additional energy through the poles and pavements.

As families come together on the playground, this active and energetic atmosphere will influence the surrounding island.

TEAM:
ByungEon Song, SooHyun Kang, John Song

TEAM LOCATION:
Seoul, South Korea

ENERGY TECHNOLOGIES:
piezoelectric discs

ANNUAL CAPACITY:
1,400 MWh

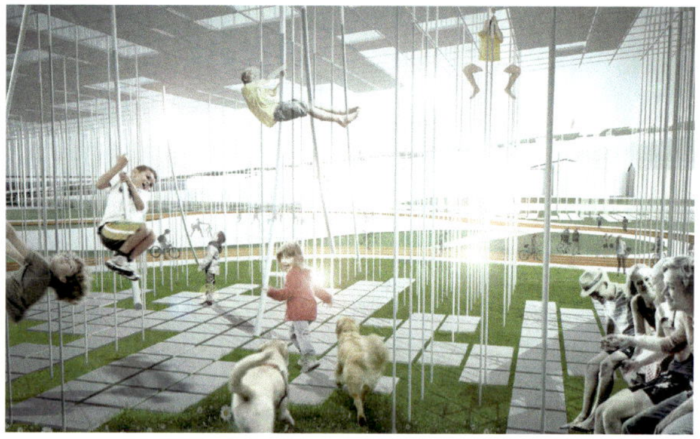

Lightfoam creates an interactive space. Children can sway or climb inside poles and people can ride bicycles or walk alone along vantage points on the slope.

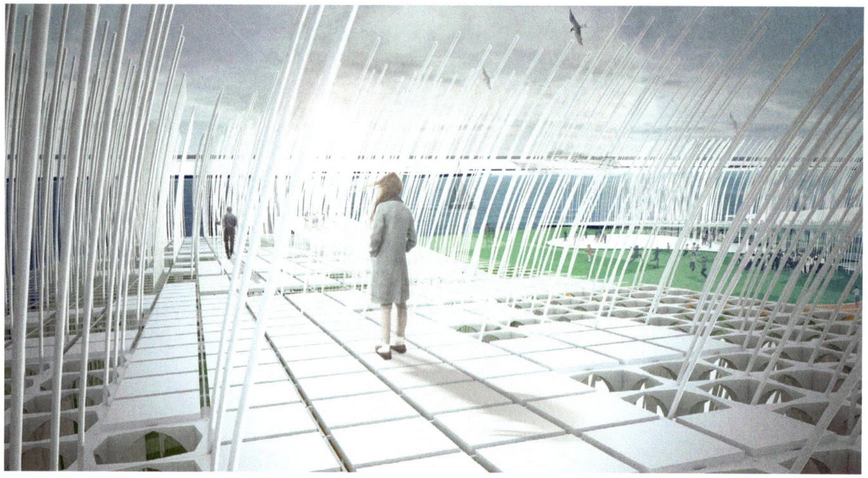

Visitors can walk along a path between the poles and gaze out over the harbor.

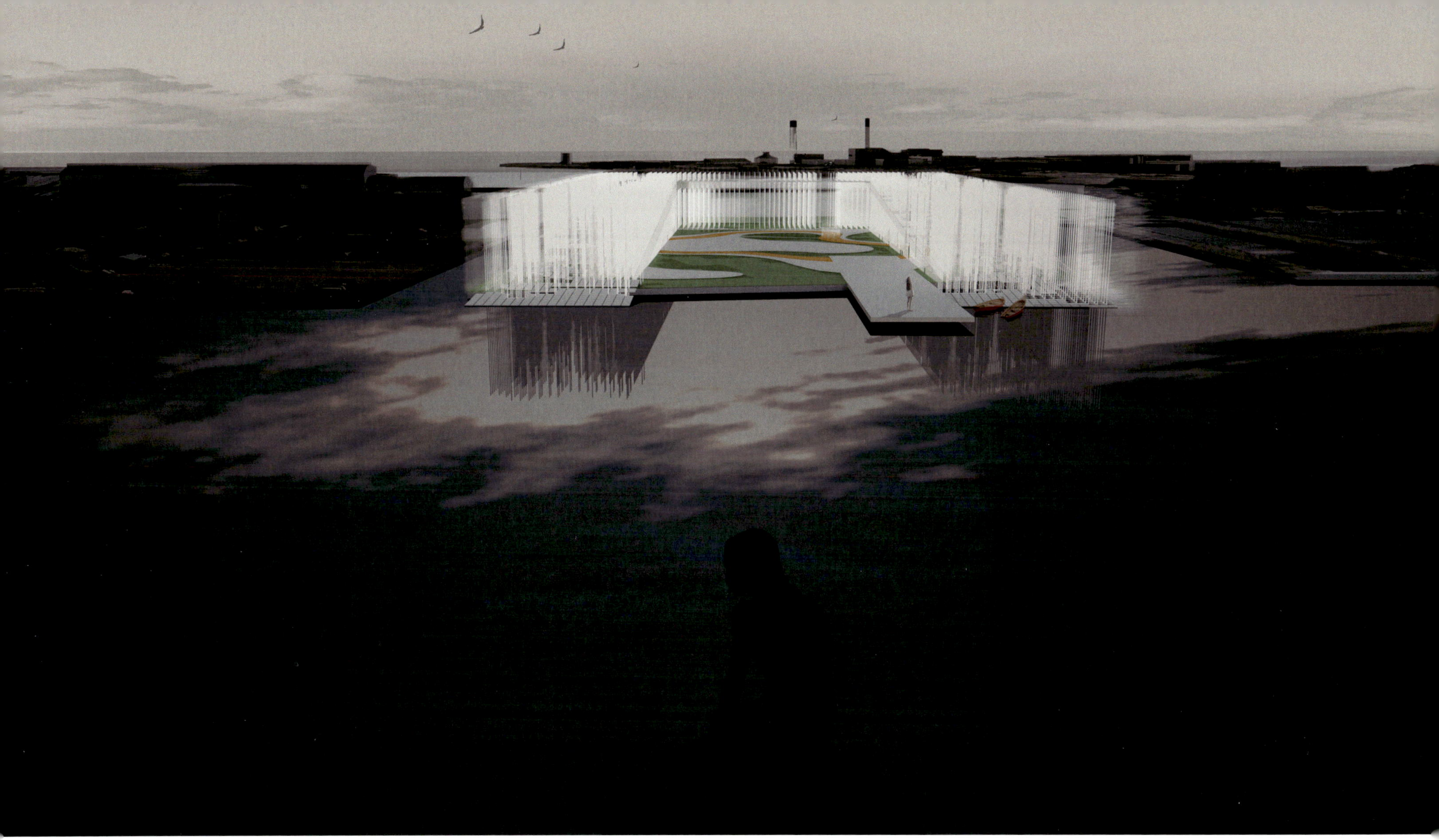

Entering from the harbor

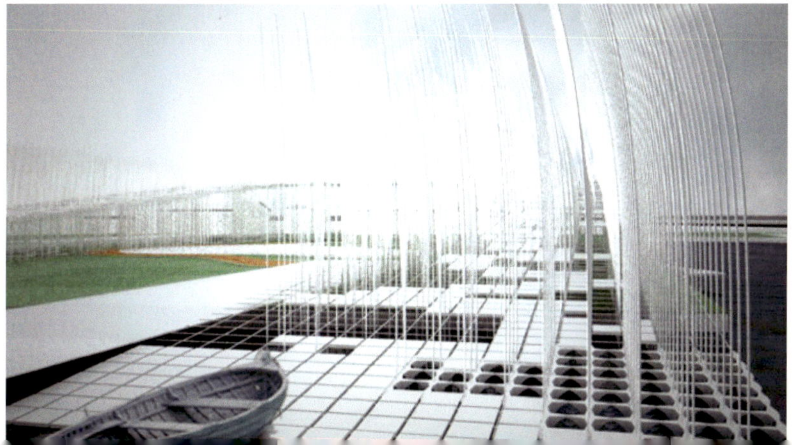

Celebrating the Sun

TEAM:
Sturla Sandsdalen, Knut Helge Teppan

TEAM LOCATION:
Oslo, Norway

ENERGY TECHNOLOGIES:
heliostatic photovoltaic panels

ANNUAL CAPACITY:
20 MWh

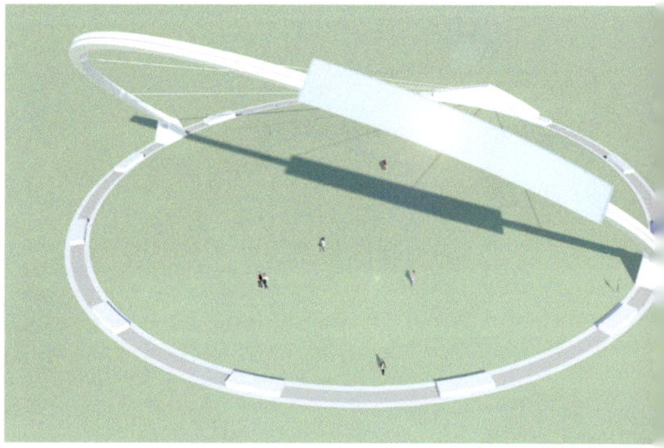

Visitors can experience the movements and changing geometry of *Celebrating the Sun*, with slow changes during the day and subtle adjustments during the year. An almost meditative motion is achieved, where the installation slowly adjusts to the sun's movement across the sky.

The sun is the origin of all life and energy on our planet. It has spawned myths, awe, and wonder throughout history and defines every day and every year of every person on our planet.

Celebrating the Sun is an art installation that connects us to the sun by reflecting its movements through the sky. Visitors can experience the slow movements and changing geometry of the installation with slow changes during the day and subtle adjustments during the year.

The installation consists of a large structural arch along the length of which moves a solar panel trolley, driven by an electric engine and powered by the solar panel itself. The movement of the trolley is a three-dimensional representation of the sun's path across the sky over the course of a day. Tension cables tilt the entire arch up or down slightly each day to maintain the seasonal rhythm and match the height of the sun in the sky. All electric components and mechanisms are hidden inside the arch or in installations on the ground, achieving a clean and elegant expression.

The arch has a radius of 30 meters, making it clearly visible on the other side of the water. It will reach its highest angle of 57.71° on June 21. From there, it will slowly lower itself to 10.96° on December 21.

The arch and its solar panel are fitted with LED lights to mark the times and days related to the sun and the seasons.

The area surrounding the installation is used for urban gardening. Horticulturalists will be assisted in their seasonal planning with daily reference to the angle of the artwork—a solar clock of sorts.

The arch spans from east to west and is placed on the center axis of Kastellet. Together, they form a relationship through their strict east-west orientation, building a new bond between the two sides of the water. The installation gives an ambience of a time before technology and development caused humans to live at arm's length from nature.

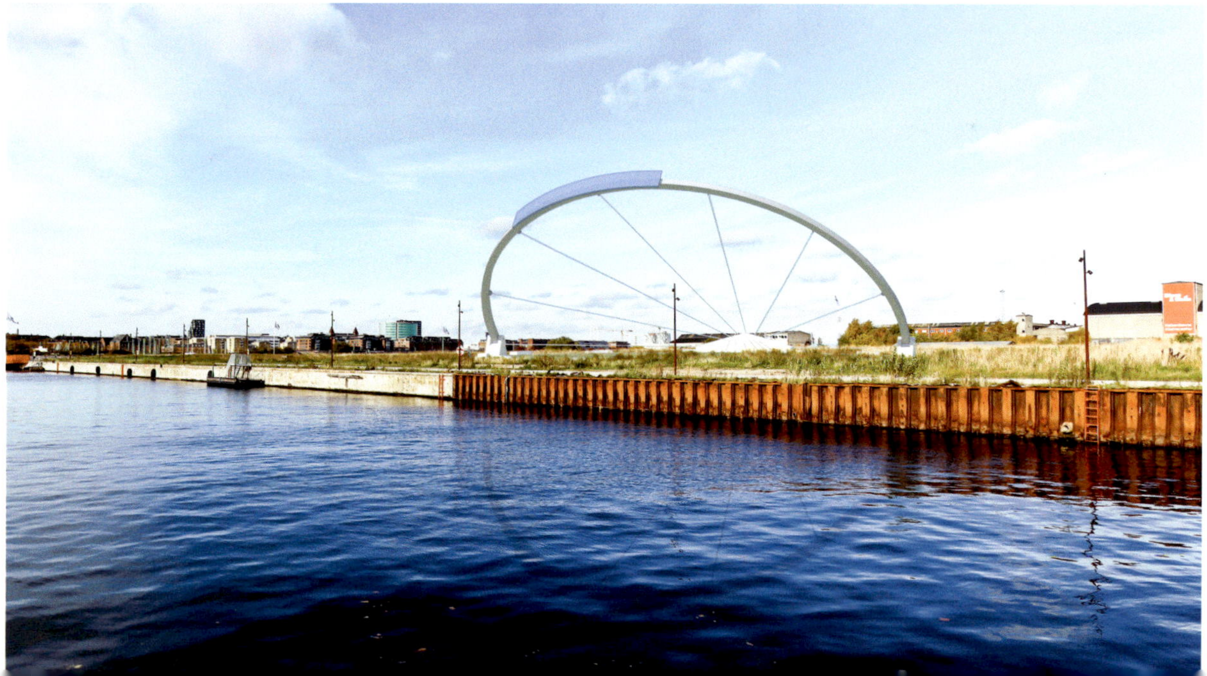

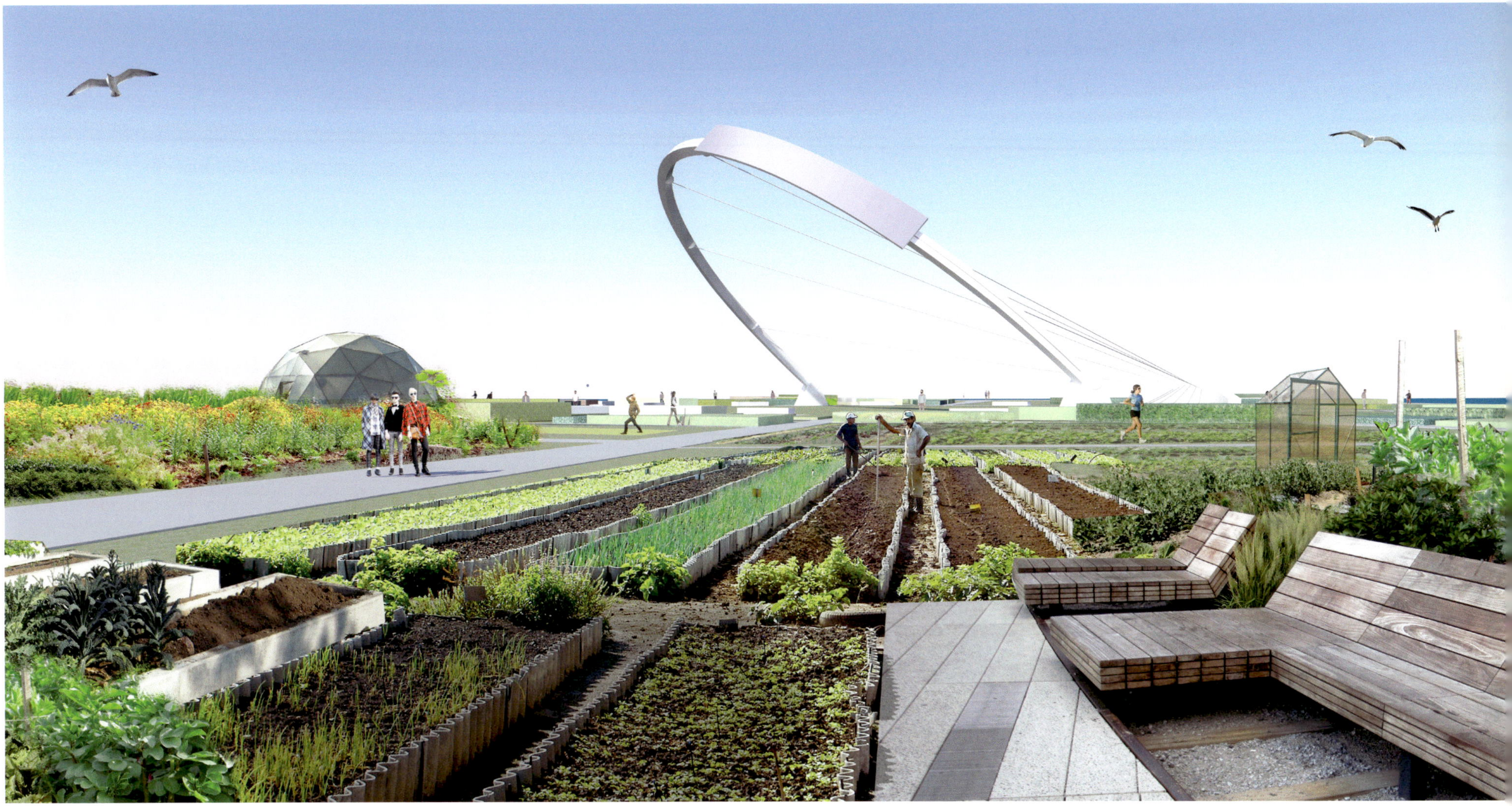

Agriculture is a stimulating and engaging activity where the citizens can participate, develop, and shape the area surrounding the artwork.

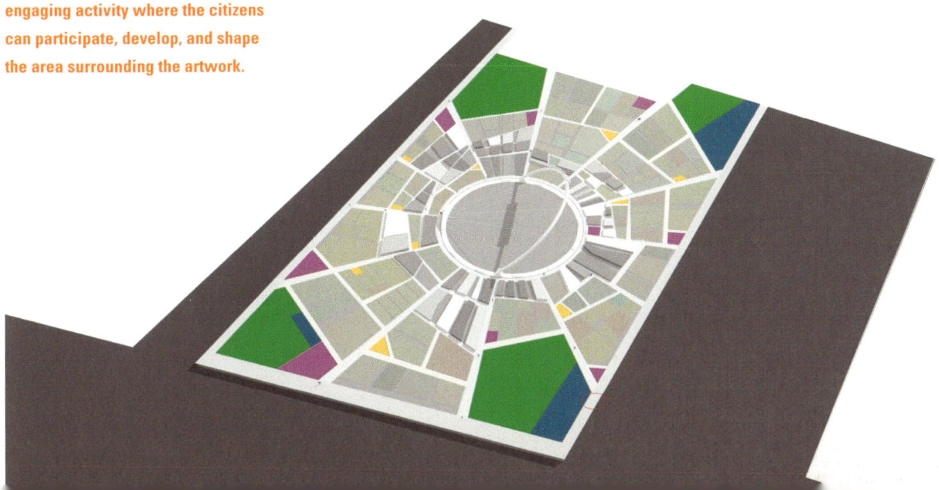

Aerial view shows the layout of the urban farming plots, arranged radially with public walking paths in between.

Tree of Knowledge

TEAM:
Boguslaw Barnas, Rafal Barnas, Omar Gharaibeh, Natalia Jejer

TEAM LOCATION:
Krakow, Poland

ENERGY TECHNOLOGIES:
piezoelectric toroids, wind microturbine

ANNUAL CAPACITY:
468 MWh

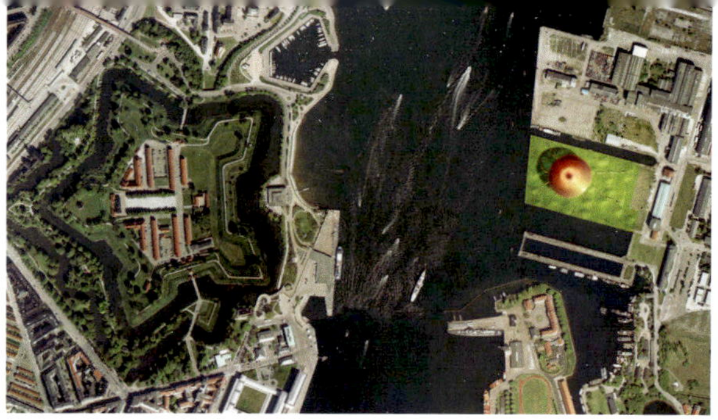

The legacy of Hans Christian Andersen spans the water creating a visual and thematic bridge between the two land masses.

Tree of Knowledge draws on motifs found in the fairy tale "The Garden of Paradise" by Hans Christian Andersen. The electricity-generating sculpture is "The Tree of Knowledge" while "The Four Winds of the World" are the source of renewable energy.

The sculpture's roots wind their way through a wild and mystical landscape, evoking an otherworldly image that is in contrast to the industrial surroundings of Refshaleøen. The organic shape of *Tree of Knowledge* stands out amid the rigid forms of chimneys, turbines, and warehouses. Upon stepping into the garden, the visitor is transported from the urban reality of Copenhagen into a world of fairy tales. The tree is a symbol of nature and knowledge, and the aim of this artificial landscape is to promote contemplation on humankind's impact on nature.

At the top of the tower is a small-scale wind turbine that captures air as it moves through the roots and into the tower by way of the stack effect (warm ground air naturally rises up to mix with the cooler air at the top of the 70-meter tower). Additionally, the steel structure holds two large industrial balloons, the surfaces of which are covered with PVC pipes containing millions of piezoelectric components. The wind and the mechanical stress from the balloon movements will stimulate the piezoelectric elements to produce electricity.

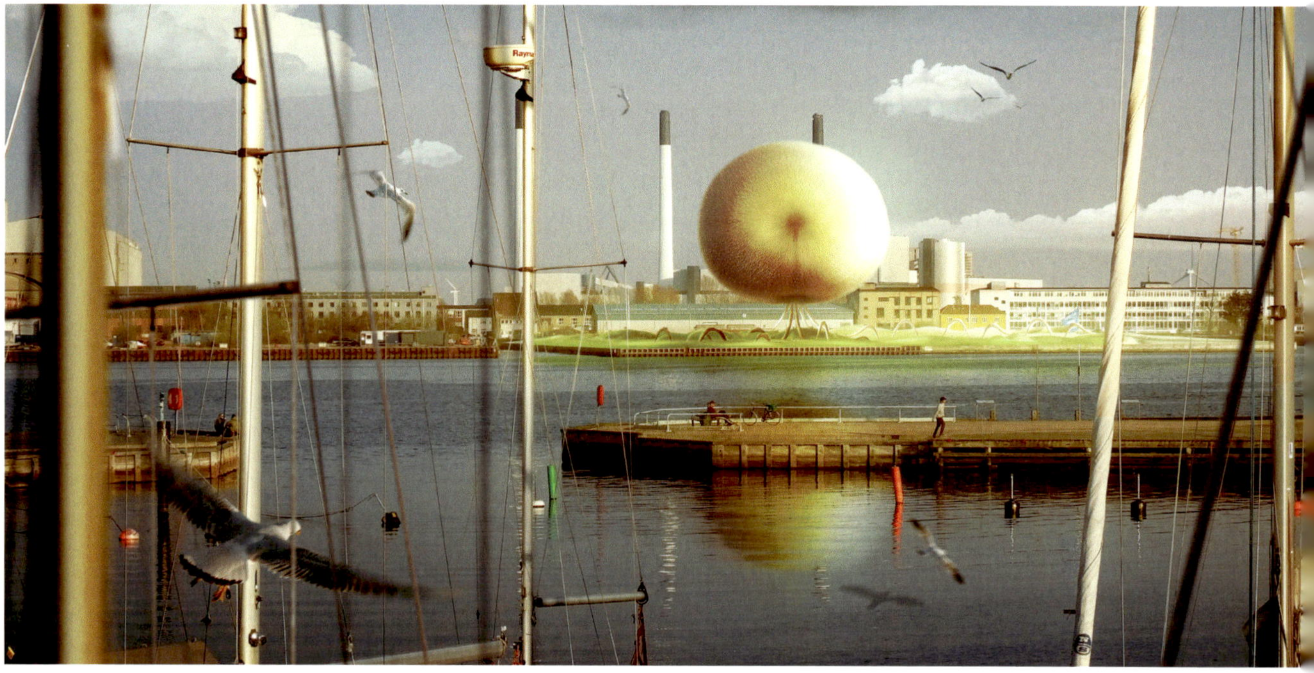

The sculpture and undulating garden contrast with the industrial surroundings of Refshaleøen.

"In the center of the hall grew a large tree, with luxuriantly hanging branches. Golden apples large and small hung like oranges among the leaves. This was the Tree of Knowledge."

—Hans Christian Andersen

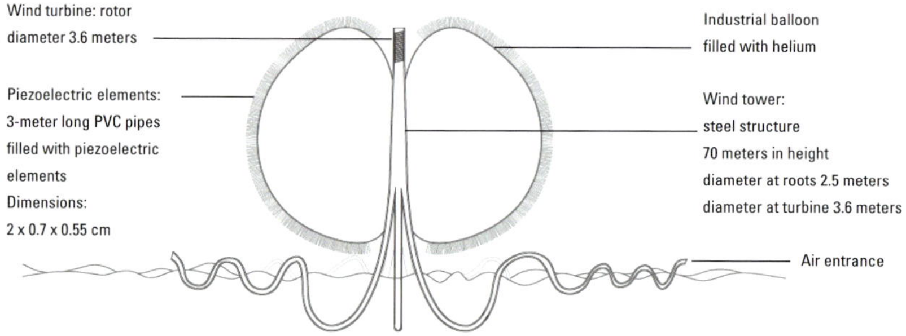

Windbrator

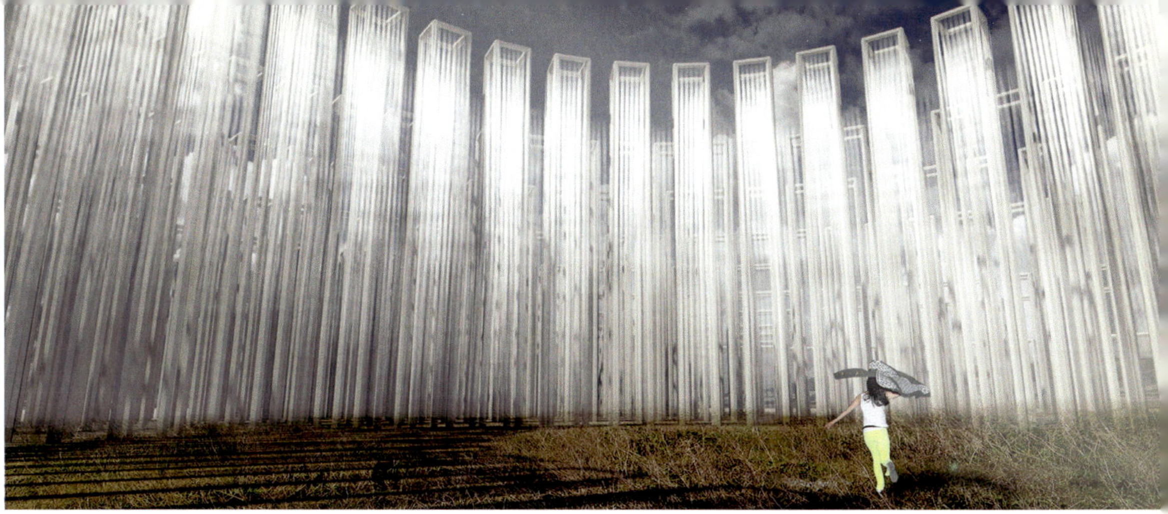

When the piezoballs shake, the kinetic force is captured and converted into electric power. The electric power is sent to the source through the conductive cable made of carbon fiber reinforced resin.

Windbrator is a work of land art for the regenerative metropolis, providing electricity to the city through the natural force of wind. The key aim is to generate renewable energy from the vibrations caused by the infinite, yet ever-changing, supply of wind.

Kinetic force is captured and turned into electric power by the use of piezoelectric ceramic discs that are arrayed around 220,000 piezoballs—the primary module of the artwork. The structural housing consists of 552 sectional 3-meter square quadrate poles that vary in height from 15–35 meters.

The towers are arranged in a plan to reflect the prevailing wind direction at the site throughout the four seasons. At the center is a serene and monumental public gathering space.

Piezoballs—each 30 centimeters in diameter with a transparent outer skin—adorn the top 10 meters of each pole to harvest electric power. Vibration sensors light up LEDs that function as a seasonal calendar as well as a nighttime spectacle.

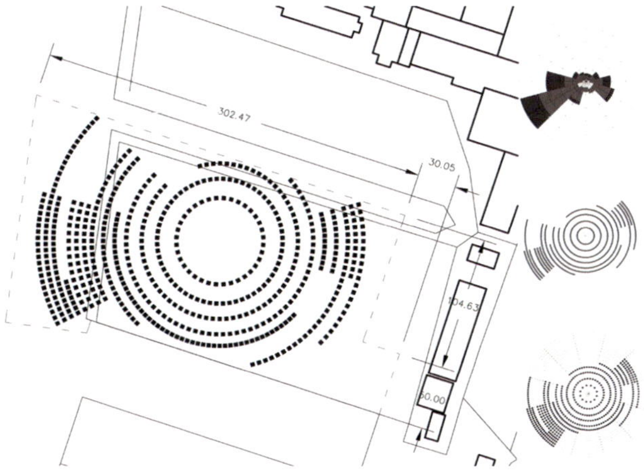

Site plan

TEAM:
Chanin Sheeranangsu,
Phatra Wongsantimeth

TEAM LOCATION:
Bangkok, Thailand

ENERGY TECHNOLOGIES:
piezoelectric ceramic discs

ANNUAL CAPACITY:
11 MWh

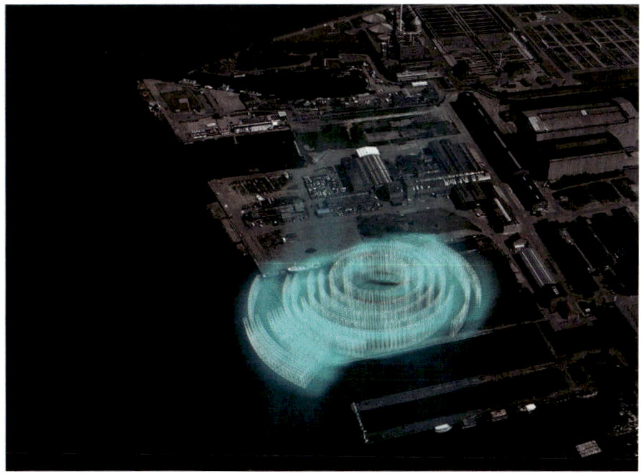

At night when the wind moves through the piezoballs, the vibration sensor activates the LED lights.

Sun & Wind Plaza

TEAM:
Michał Kowalski,
Aleksandra Kaszubska

TEAM LOCATION:
Warsaw, Poland

ENERGY TECHNOLOGIES:
Onyx™ solar panels

ANNUAL CAPACITY:
2,165 MWh

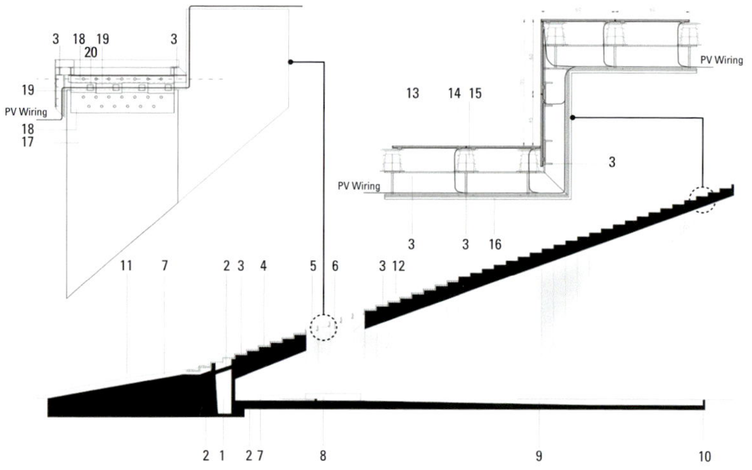

1. Transformer room
2. Reinforced concrete structure
3. Steel structure
4. PV floor panels
5. Voids
6. Blades/wind wings
7. Ground embankment
8. Walkable water lamps and PV
9. New water basin
10. Retaining wall
11. Terrain lamps and PV
12. Glass balustrade
13. Walkable PV floor panels
14. Junction box
15. System support member
16. Fiberglass panels, white finish
17. Fiberglass wind swinging panels
18. Load-bearing steel hinge plate
19. Load-bearing steel hinge
20. In-hinge generator

Sun & Wind Plaza is a multipurpose public plaza cantilevered above the Copenhagen harbor. Splendid views over the city and water are provided from the south-facing terraces fitted with walkable solar panel treads. The terraces are each 1.05 meters in height and 3 meters deep, connected by a system of ramps.

Visitors can move about the plaza or sit and observe how the mechanism works. Triangular windows in the terraces provide views down to the wind panels hanging above the water. The fluctuating movement of these huge panels is visible even from across the harbor.

The skin of the terraces comprises approximately 32,000 anti-slip floor panels of integrated photovoltaic Onyx™ panels. The panels will have a semi-transparent white surface. The south-facing orientation of the panels provides the optimum gain of solar energy, assisted by the reflection of the risers.

The west-facing wind panels are intended to respond to wind flow. They are made of triangular-shaped lightweight fiberglass and hung underneath the main steel structure. Hinge supports allow for swinging movement of the panels, which creates energy at kinetic harvesting devices within the hinges.

Visitors can walk below the cantilevered terrace to experience the scale of the wind panels and hop between floating triangles within the harbor waters.

Entering the site from the ferry or bus stop, visitors cross the city meadows, a place for picnics, sports, family gatherings, and walks, smoothly rising from the water towards the white and shining solar terraces.

Individuals can go underneath the terraces, where the wind moves the hanging triangular wings, for relief from the heat. Swimming in the open pool, sunk into the ground on the northeast side of the site, visitors can admire the play of shadows and light peeping through the triangular openings in the terraces.

Changes in the position of the sun play unpredictable tunes with the wind. This subtle mechanism produces energy for the city.

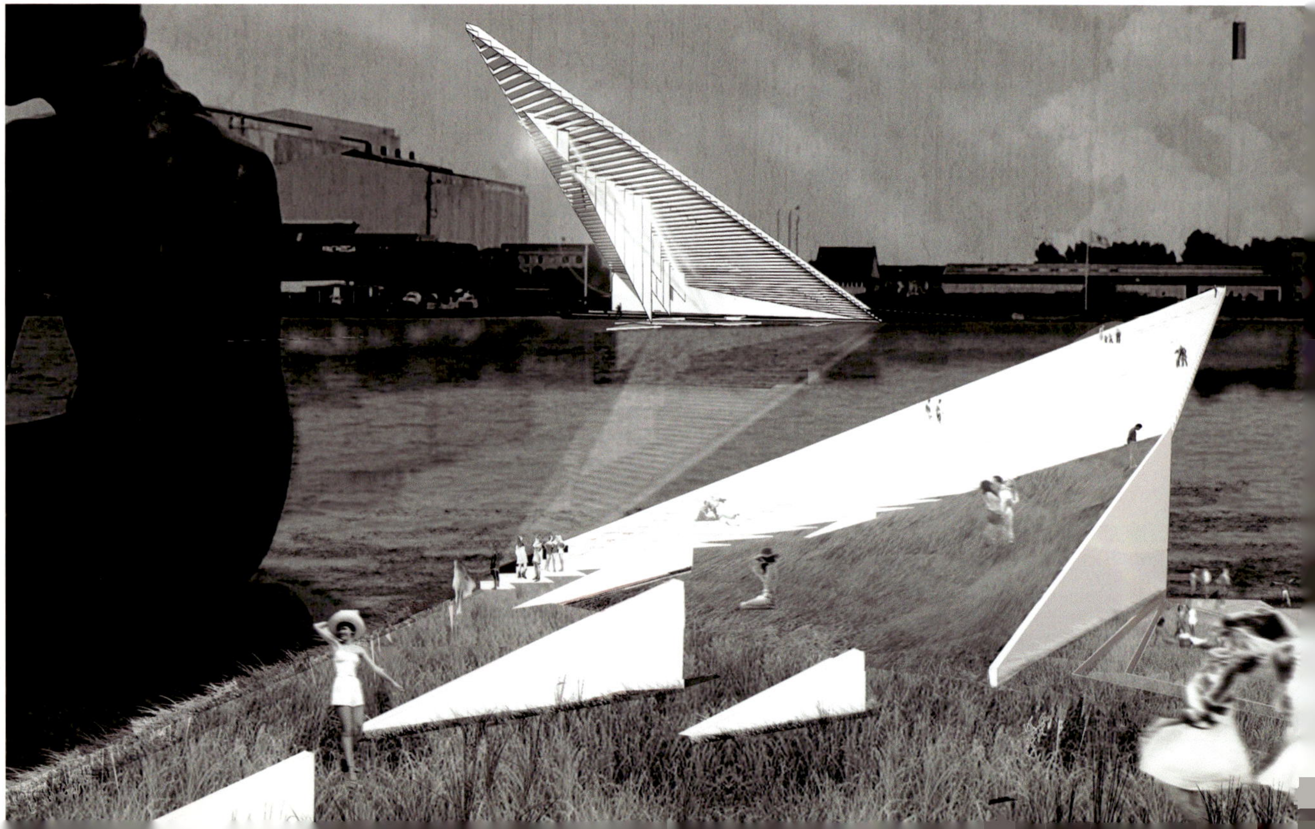

Rør

TEAM:
Evan Wakelin, Christopher Tron, Sylvia Zhou

TEAM LOCATION:
Toronto, Canada

ENERGY TECHNOLOGIES:
wind microturbines driven by the solar thermal expansion of air

ANNUAL CAPACITY:
20 MWh

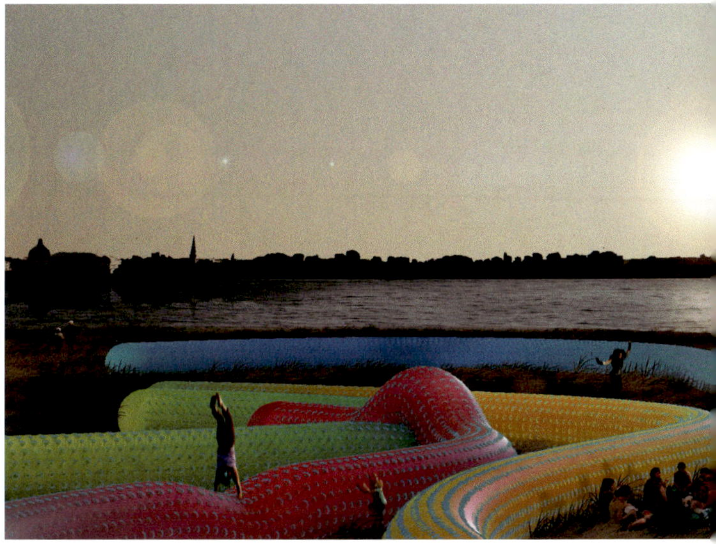

The structures consist of an inner and outer membrane. The space between these membranes is inflated with air to hold their form. This inflated structure houses an inner air chamber. Air passes in and out of this chamber through conical shafts. Each shaft is fitted with a microturbine to respond to pressure differences between the inner chamber and outside air.

Rør is an inflatable public art park that harnesses solar energy and invites visitors to actively participate in the generation of power. The greater the level of physical interaction with the structures, the more energy is generated. Using a lighthearted and lightweight approach, the multi-colored pneumatic tube park is a fun celebration of the environment, challenging the public to take a positive stance toward the pressing issues of climate change.

Climate change research estimates that Copenhagen will face increased and prolonged heat waves in the future as well as heavier rainfall. *Rør* responds to these future trends by covering only 10 percent of the existing green space to minimize heat island effect and to absorb all rainfall on site.

Rør takes advantage of future heat waves by using solar radiation as its primary power source. Its autonomous pneumatic structures generate energy through solar radiation and human interaction. This energy is harnessed using solar updraft tower technology, but rather than transfer heated air to centralized turbines, *Rør* incorporates an array of smaller turbines over the surface of the tubes. As visitors add their weight to the tubes, it increases the rate of escaping air through the microturbines.

The lightweight pneumatic tubes are easily moveable and the arrangement of the tubes is determined by the activities and requirements of the people—the visitors have complete control over how to create and use the space and are free to roll and rearrange the tubes to create new and unique spaces. The resulting pockets of space and pathways provide opportunities for community engagement, moments of play, and exploration.

Because *Rør* is a dynamic energy installation that can transform on demand, each visit to the park will be a new experience.

The electricity generated is conducted to an electrical transformer embedded at the end of each tube, which is then fed to the city grid.

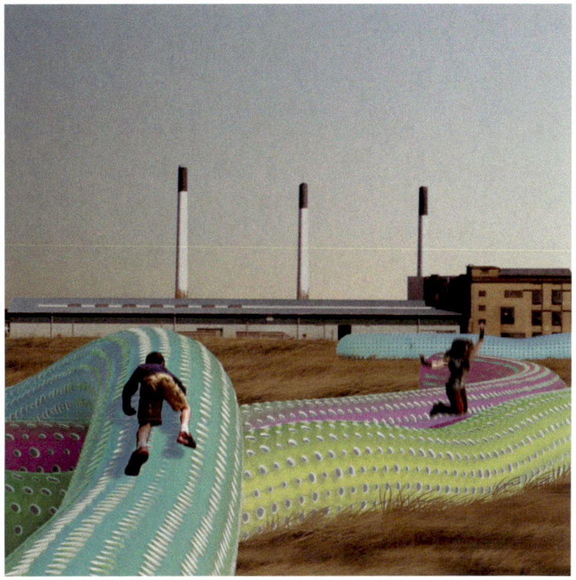

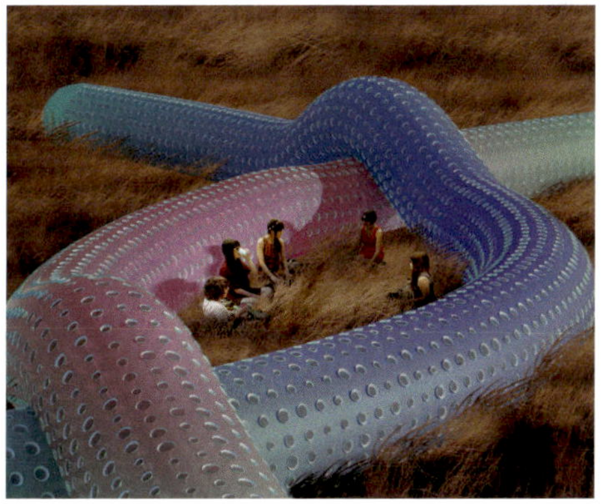

The process of air pressure differentiation can be accelerated through human interaction. Jumping on the tubes will rapidly increase and decrease pressure differences within the inner chamber.

Solar Lake

TEAM:
Michal Marcinov, Michaela Kesanova, Marek Kremen, Tomas Kristek, Karin Tajtiova, Andrej Moravek

TEAM LOCATION:
Bratislava, Slovak Republic

ENERGY TECHNOLOGIES:
photovoltaic panels,
piezoelectric generators

ANNUAL CAPACITY:
1,400 MWh

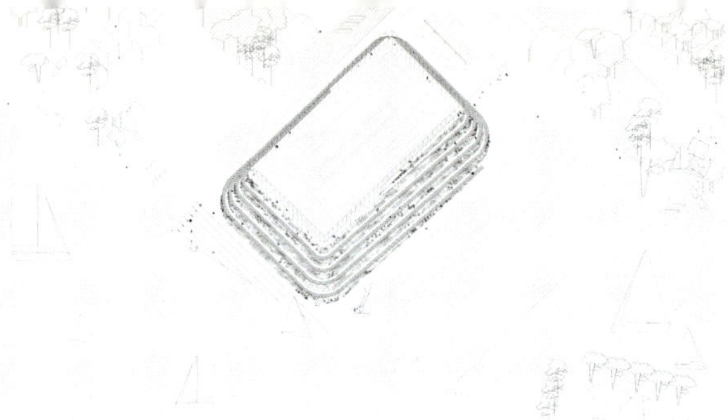

An earth embankment supports the reservoir construction while hiding a system of catacomb spaces inside. These spaces house the mechanical functions.

Solar Lake is a new, elevated landmark for Refshaleøen. Vertical gardens and a green spiral pathway comprise a three-dimensional landscape that culminates with an observatory on the top of a hill.

The hill itself is a structure that conceals a hidden reservoir, which fills as rain falls over the site and into its chambers. Before finding its way into the reservoir, the rainwater has a role to play in the visible face and surface of the installation.

Photovoltaic panels are arrayed below a shallow lake, the depth of which remains constant as overflow water finds its way to the 137,723 m^3 reservoir below, after passing through a set of filters. This type of walkable photovoltaic panel is an innovative solution for integrating solar energy, supporting 400 kilos in point load test. The cooling function of the water increases the efficiency of the panels so they can convert more energy compared to conventional solutions in open air.

A walking path around the perimeter of the hill provides access for visitors to reach the top, gaze across the *Solar Lake*, and perhaps wade through the ankle-deep waters. Along the walking path, piezoelectric tiles capture the energy of walking to add additional electricity to the overall system. Urban agricultural spiral terraces line the side of the path and take advantage of free rainwater irrigation from the reservoir within. The terraces can be rented for community gardening.

In the evening, the backlit panels under the lake deliver a spectacular feeling with their soft glow through the water.

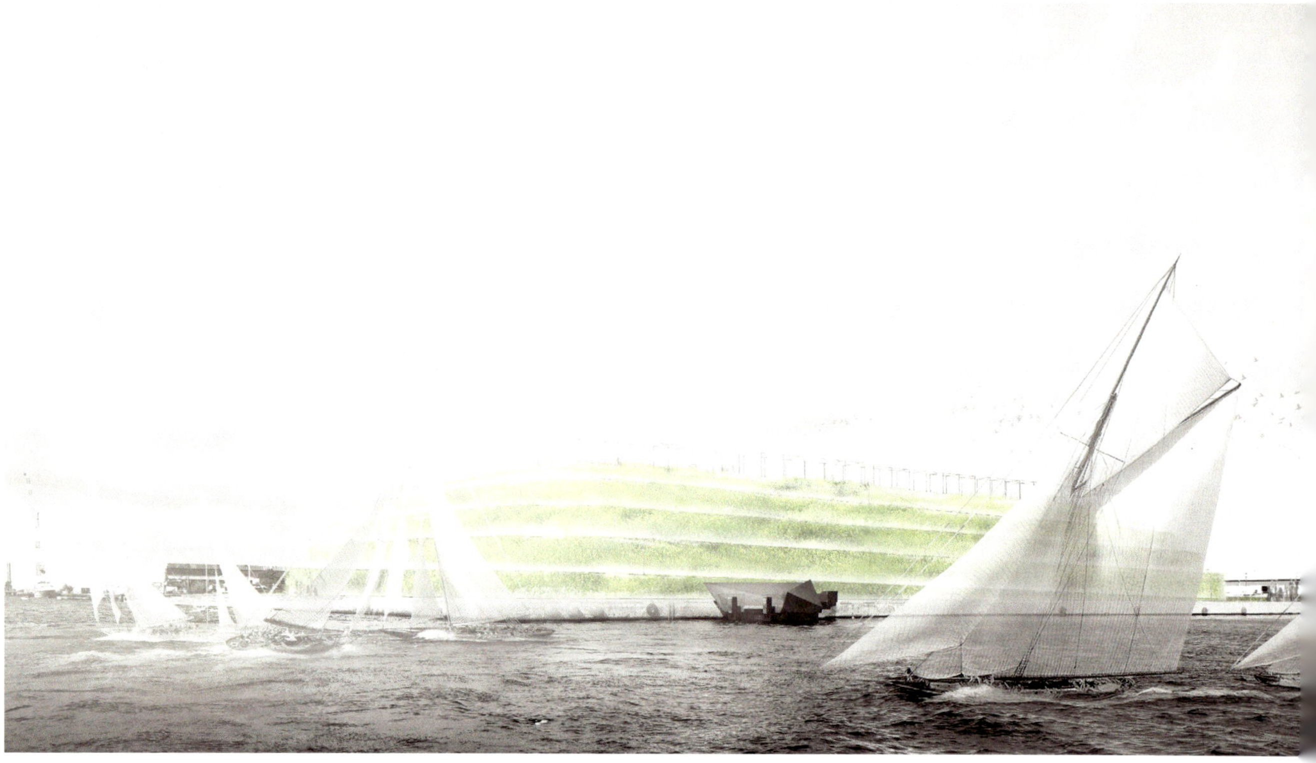

A reservoir is placed in the heart of the entire structure and is supported by a complex construction of vaults. One part of the collected water is used to irrigate the adjacent landscape. After filtering through the UV membranes, the remainder of the collected water is able to supply 634 households with utility water per year.

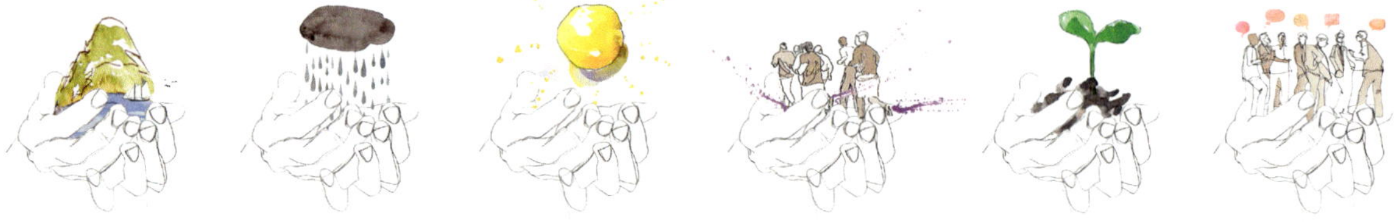

Energy Hive

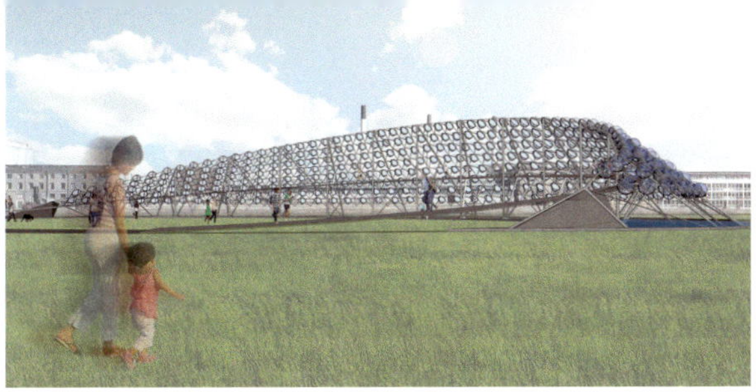

The harvesting of solar energy is exponentially multiplied by the collective orbs and stored within the two anchor pylons at each end of the hive arc. The anchor pylons are points of distribution for the collected energy.

Energy Hive is inspired by the natural system of bees working in choreographed synchronization to harvest and produce for the benefit of the collective. The installation metaphorically extracts the beehive structure, expressed as a sculptural arc of orbs, to construct a new interactive renewable energy hive.

The *Energy Hive* system harvests sunlight, tracking the radiant energy of the rays of the sun as it moves across the sky during the day. Structural pylons on either end of the hive arc also store and distribute the collected energy. The south-facing facade of the arc of orbs is oriented to maximize sun exposure for the site.

The 1.8-meter diameter orb unit uses a water-filled, acrylic-polymer spherical lens to concentrate sunlight on a small and highly efficient multi-junction photovoltaic cell. A dual-axis optical tracking pivot is integrated into the support structure enabling the lens to follow the movement of the sun.

At the ground level of *Energy Hive*, an education path traces the form of the hive through a ramp system that safely brings the viewer into close proximity to the arc of orbs and offers dynamic views of the city. A shallow reflecting pool provides a place to cool off during the warm seasons and to ice skate during winter months.

Energy Hive serves as a powerful energy generator, an arc of sculptural simplicity and beauty, a beacon of human ingenuity, and a space for human interaction and play.

TEAM:
Glenn LaRue Smith, Vinson J. Camacho

TEAM LOCATION:
Washington D.C., USA

ENERGY TECHNOLOGIES:
concentrated photovoltaic (CPV)

ANNUAL CAPACITY:
4,200 MWh

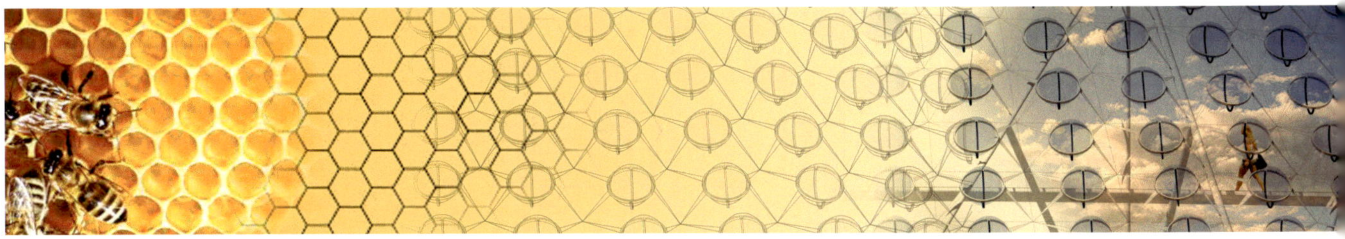

The inspiration stems from the transformation of a natural system to a man-made system.

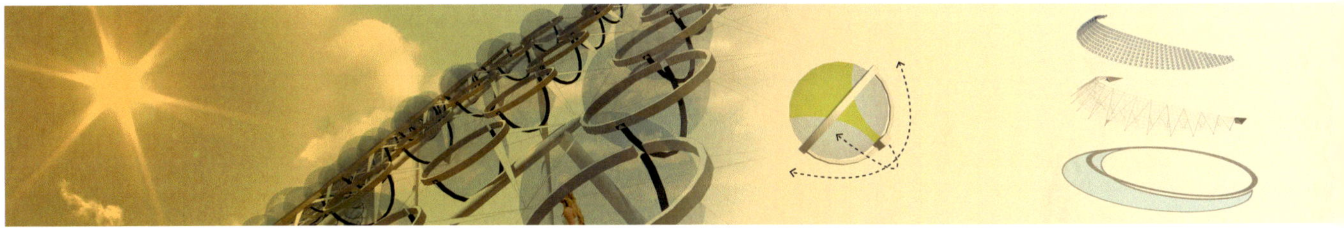

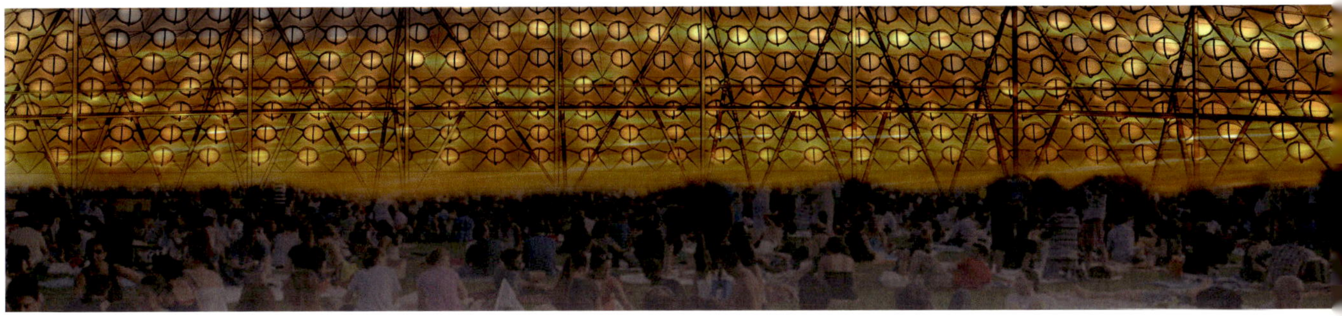

Energy Hive was inspired by the natural system of bees working in choreographed synchronization to harvest and produce for the benefit of the collective. The installation metaphorically extracts the beehive structure, expressed as a sculptural arc of orbs, to construct a new interactive renewable energy hive.

Blowing Horn

TEAM:
Hooman Tahvildar Akbary

TEAM LOCATION:
Tehran, Iran

ENERGY TECHNOLOGIES:
Windbelt™, compact wind acceleration turbine, multi-rotor wind turbine

ANNUAL CAPACITY:
500 MWh

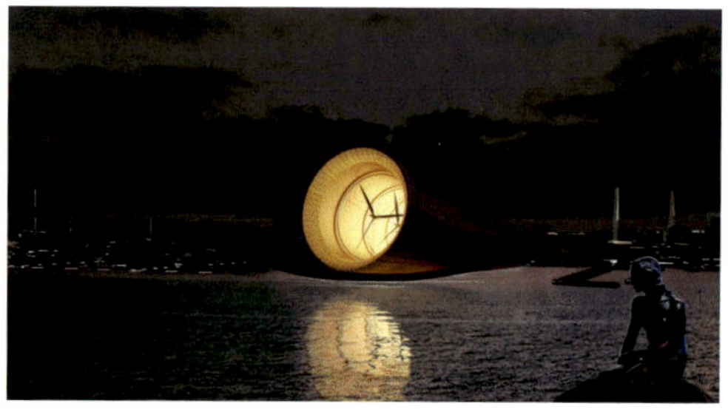

Nighttime view from across the harbor

This new energy monument is oriented toward the most powerful and prevailing winds on the site. Its form gives reference to the large horns that were used throughout history to call across long distances, but it also serves an important utility. By reducing the cone diameter from the mouth, the speed of the wind increases towards the narrower end of the horn—an application of the compact acceleration turbine lens that makes use of the venture effect.

Inside of *Blowing Horn* there are a series of wind rotors with a single drive shaft—a multi-rotor wind turbine invented by Doug Selsam. By placing multiple rotors onto a single shaft, the amount of energy that is generated on a single turbine is multiplied.

The ship form at the base of the monument is designed to act as a channel, which leads wind through a Windbelt™ array on the deck and the outer shell of this new golden horn for environmental energy production.

"Near the shores of the great Belt, which is one of the straits that connect the Cattegat with the Baltic, stands an old mansion with thick red walls. I know every stone of it, says the wind."

— *The Story of the Wind*, Hans Christian Andersen

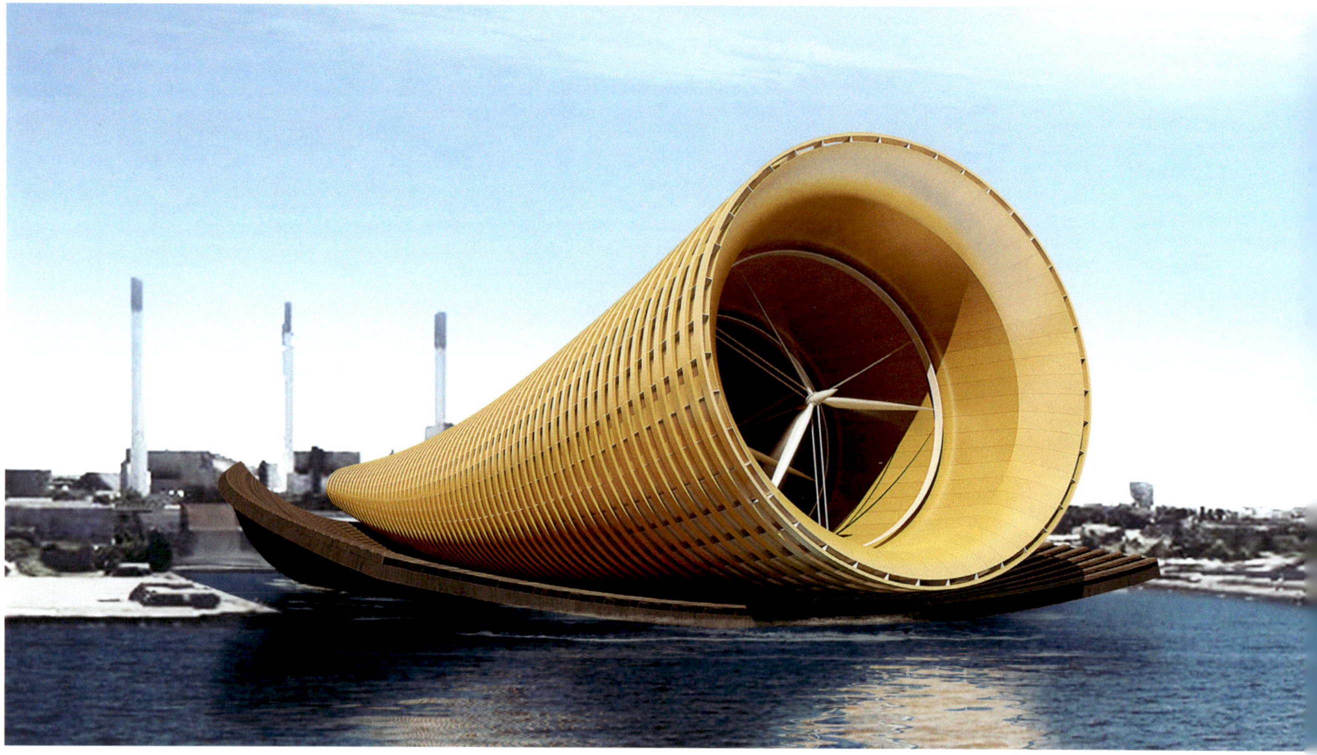

Blowing Horn incorporates three different techniques and methods:

1. Windbelt™ uses a belt secured between two fixed points set within a rectangular housing. The belt oscillates rapidly creating a rocking motion at the two ends. This motion is harnessed by small kinetic energy generation devices employing magnets at the ends of the belt, which move rapidly back and forth between metal coils.

2. Compact wind acceleration turbine uses a cone or series of cones to concentrate the wind and increase the velocity as it passes through the rotor area.

3. Selsam multirotor wind turbine uses multiple rotors on a single drive shaft.

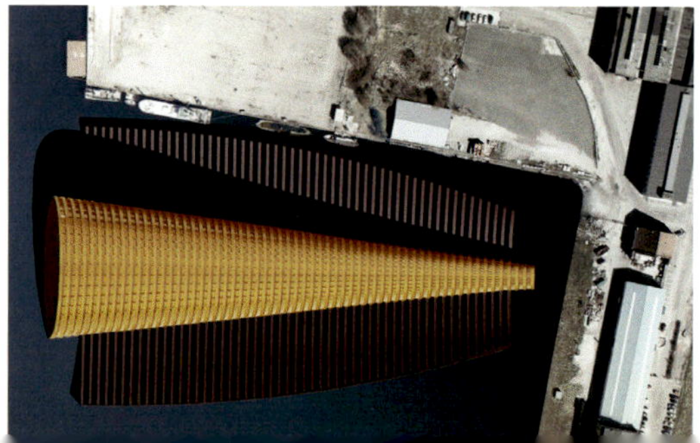

Site plan

#HelioTweet

"A short burst of inconsequential information."

—Jack Dorsey, the first tweet

TEAM:
Simon Nummy

TEAM LOCATION:
Dubai, UAE

ENERGY TECHNOLOGIES:
heliotropic photovoltaic panels

ANNUAL CAPACITY:
2,490 MWh

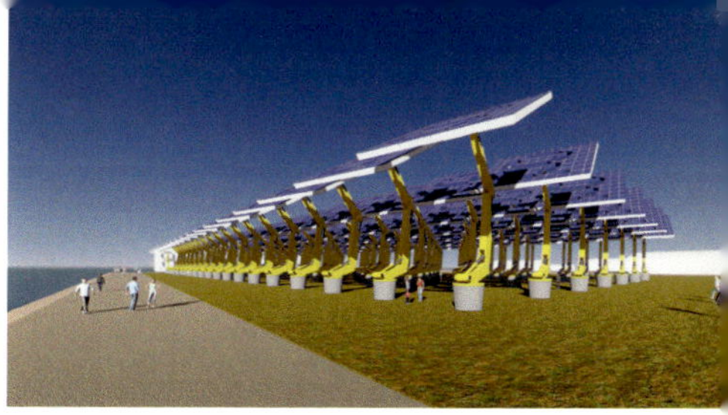

The balance of system (BoS) refers to all of the supporting devices that make the photovoltaic panel performance possible.

#HelioTweet is made up of 140 "characters," each with its own solar tracking (heliotropic) photovoltaic system. The PV panel is used to form a seven-segment display within a 7.5 meter x 9 meter module (67.5 m^2 of total PV surface per "character"). Sun tracking will maximize solar harvesting (22% increase on energy output compared to fixed panels), and the "characters" have the potential for many configurations and dynamic movements.

Laminated into each of the display panel glazed segments is a switching film—consisting of Polymer Dispersed Liquid Crystals (PDLCs), which is used to control the opacity of each of the segments. Light Tape™ is used to illuminate the segment when activated.

The feed from the installation's dedicated twitter account will be primarily focused on climate issues. The larger-than-life social media display is a canvas to explore many applications, functions, and outputs, such as atmospheric CO_2 concentration as demonstrated (401 ppm as of May 17, 2014).

The hydraulic arms of the "characters" are fabricated from repurposed earth-working excavator arms, bringing an industrial aesthetic appropriate to the site and creating a reference to resource extraction.

#HelioTweet has the potential to act as a reimagined "speaker's corner," an installation where analog and digital meet, encouraging freedom of thought and expression.

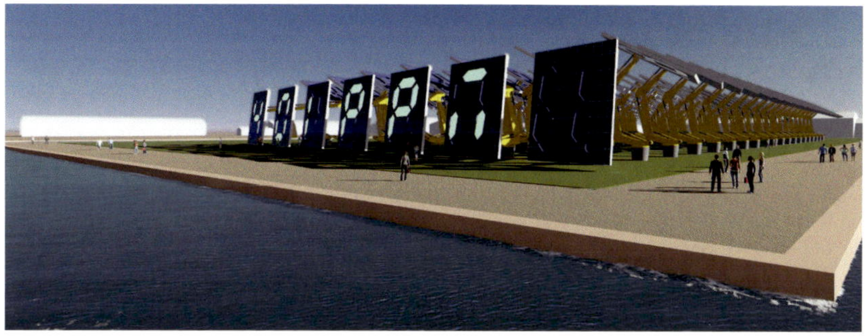

Sun tracking will maximize solar harvesting, while the "characters" have the potential for many configurations and dynamic movements.

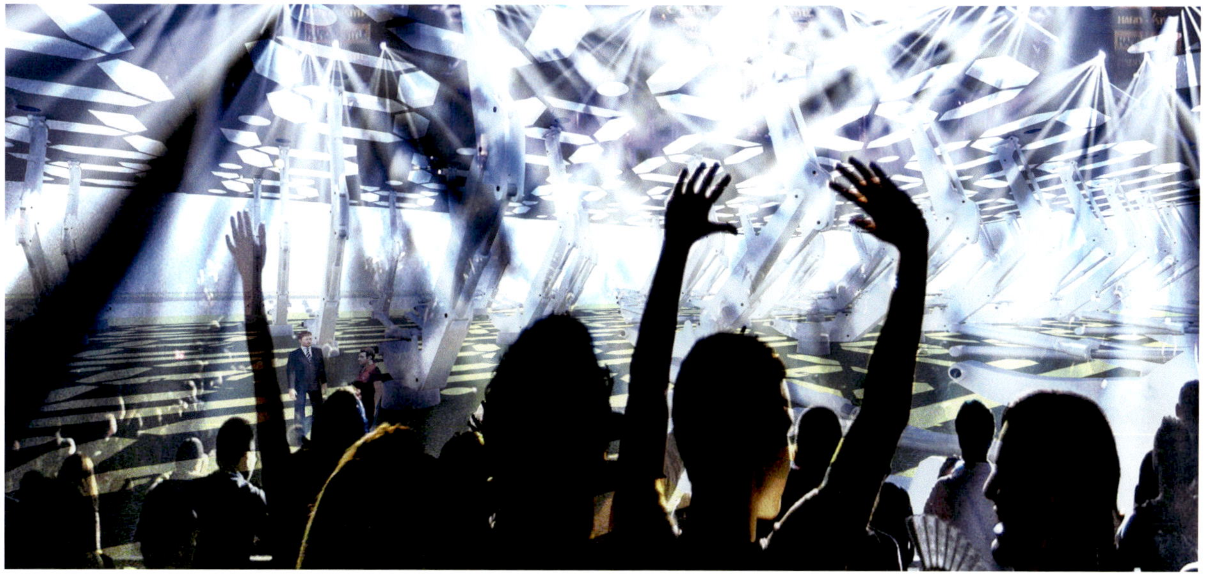

The artwork is an investigation into ways to inhabit the BoS space while generating utility and revenue from traditionally under-utilized aspects of a commercial-scale photovoltaic system.

Metamorphosis

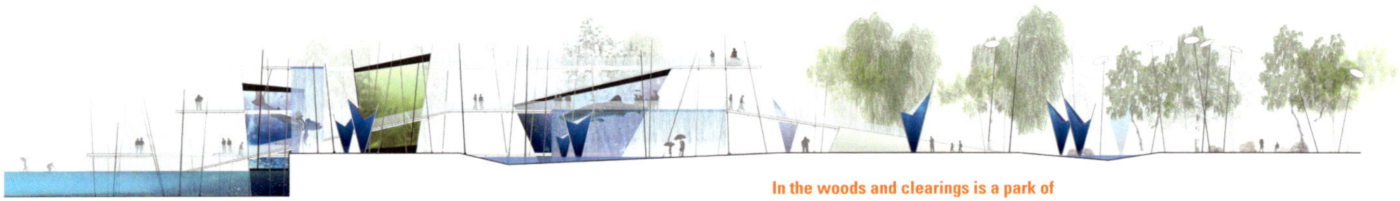

In the woods and clearings is a park of cultural innovation and experimentation.

The story of the Little Mermaid recounts a change that comes true in the water, on the ground, and then in the air—three states of nature that will be used for energy production in this new park for the Little Mermaid, opposite her statue across the water.

The chalice trees scattered in the park will collect rainwater and will supply the aquarium basins. Through osmosis, saline water from the sea and freshwater will be combined to produce energy. Organic thin-film trees will harness the power of the sun through artificial photosynthesis. Real trees, rooted deep in the ground, will help the polluted soil of the site to recover.

Pole trees are the main energy-production devices in the woods. With the breath of the wind, these free-swinging poles will channel kinetic energy to a generator placed in the soil at their base.

In the clearings of the park woods, visitors will encounter enchanted places from the Little Mermaid story, such as the aquarium where mermaids are projected inside the basins, and the prince's castle with a large ballroom, the floor of which converts the energy of human dance into electricity. Visitors will ride up the wind tower through energy-generating Windbelts™ by way of a compressed air lift to gaze down upon the energy woods and the harbor below.

TEAM:
Michele Galella, Pepe Barbieri, Michele D'Amico, Filomena Ricciardella, Francesco Ziccardi, Consuelo Nava, Raffaele Astorino

TEAM LOCATION:
Pescara, Italy

ENERGY TECHNOLOGIES:
osmotic generation, kinetic harvesting, organic photovoltaic thin-film (OPV), Windbelt™

ANNUAL CAPACITY:
498 MWh

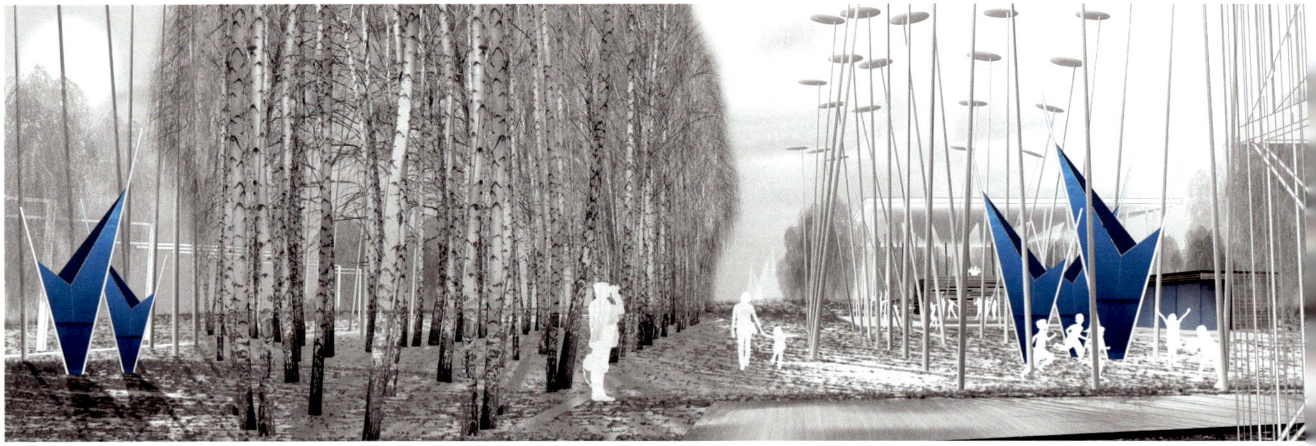

The walkway through the park is led by slightly raised routes, which give visitors the sense of a defined path. At the same time everyone is free to choose a different route and to explore hidden paths. The clearings are marked on the ground sometimes by benches and other times by walls or paved stripes. Finally, scattered in the woods, small wooden structures provide visitors with different services, such as cafés and shops.

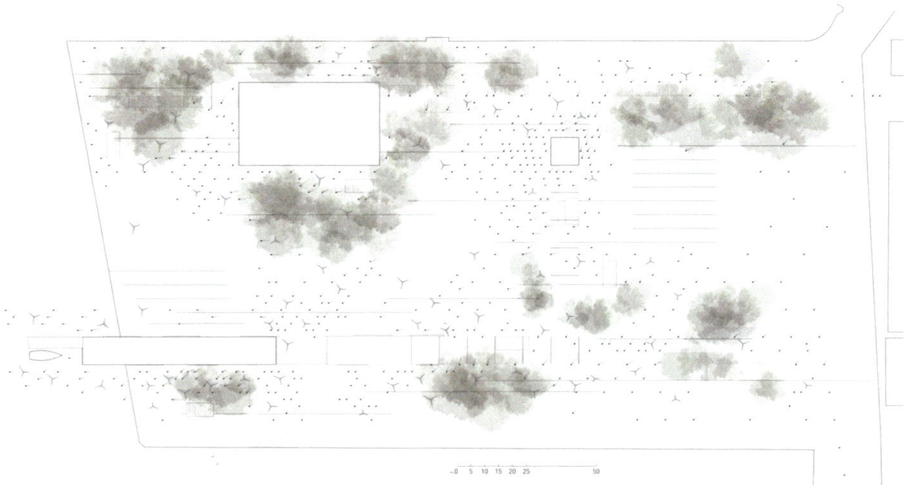

A renewable energy boat will carry visitors from the Little Mermaid statue to her park across the water.

Kinectart Generator

TEAM:
Mbaabu Stephen, Victor Basweti

TEAM LOCATION:
Nairobi, Kenya

ENERGY TECHNOLOGIES:
piezoelectric generators, compact wind acceleration turbines

ANNUAL CAPACITY:
260 MWh

The *Kinectart Generator* is a combination of a sea-wind farm (wind-tidal kinetic energy harnessed with offshore generators) and a windscape on the land (onshore wind generators). The windscape creates a dynamic, natural, undulating landscape that can be used as a public space for recreation and observation while generating electricity from wind turbines installed within its peaks.

The sea-wind generator is inspired by the soft movement of seaweed in water (and informed by the *Windstalk* entry to the 2010 Land Art Generator Initiative design competition). Two hundred sea-wind stalks, each 5 meters long, are anchored to a floating pad with a recycled aluminum and steel hull. The sea-wind stalks are made of carbon fiber reinforced resin poles, 30 centimeters in diameter at the base and 5 centimeters at the top. Within each hollow pole is a stack of piezoelectric ceramic discs that generate power through the bending of the stalks by wind and wave energy.

The windscape generators incorporate 4-meter diameter turbines situated to receive accelerated wind power by incorporating wind lens technology assisted by the valleys created by the landforms. Visitors to the installation can recreate in the valleys and walk the paths within them, while learning about the energy generation on their way to the sea-wind platforms and back again.

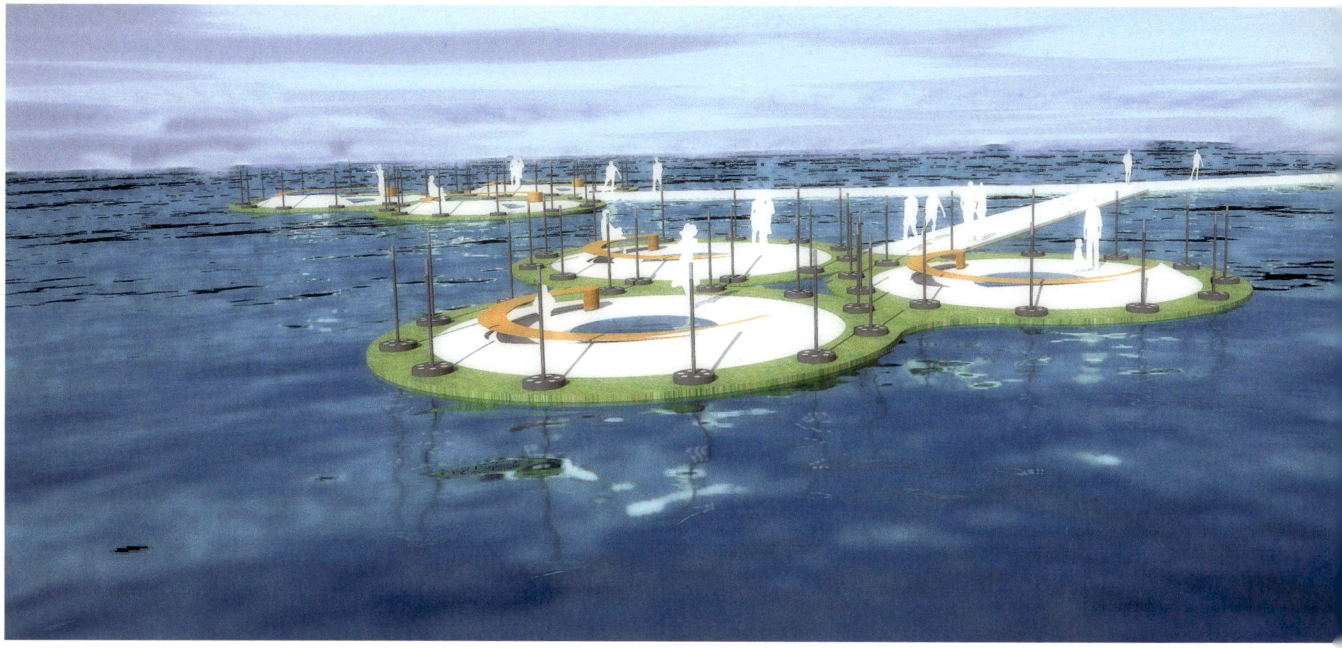

The windscape is a wind farm inspired by wave forms that create a dynamic, natural, and undulating landscape that can be used as a public space for recreation and observation while generating energy from wind turbines installed at its apex.

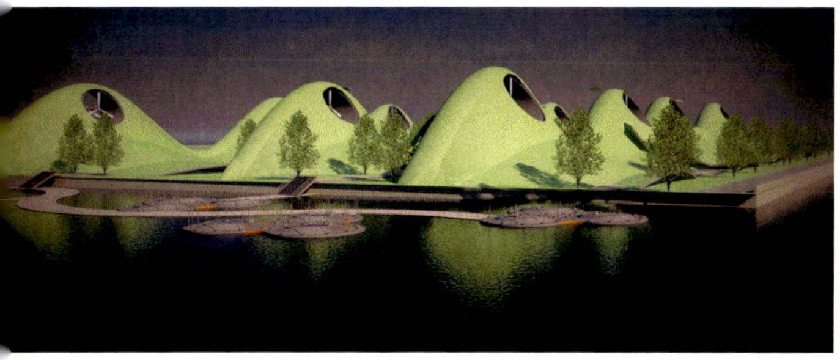

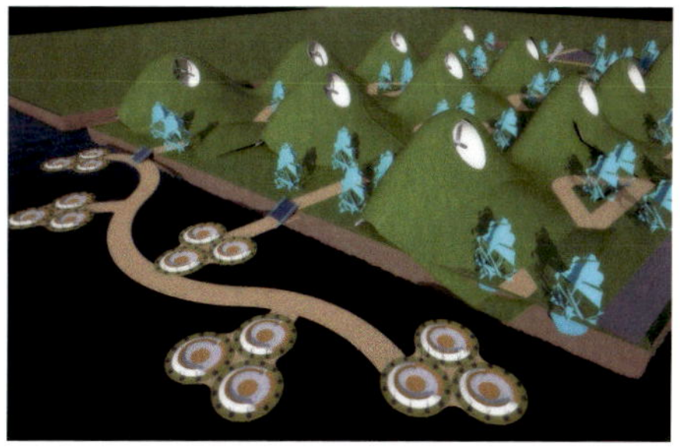

The *Kinectart Generator* is a combination of the sea-wind farm (wind-tidal energy offshore generators) and the windscape on the land (onshore wind generators).

Palace of the Prince

TEAM:
Jakub Med

TEAM LOCATION:
Prague, Czech Republic

ENERGY TECHNOLOGIES:
elastomeric piezoelectric toroids, wind-catching sails (leaves)

ANNUAL CAPACITY:
600 MWh

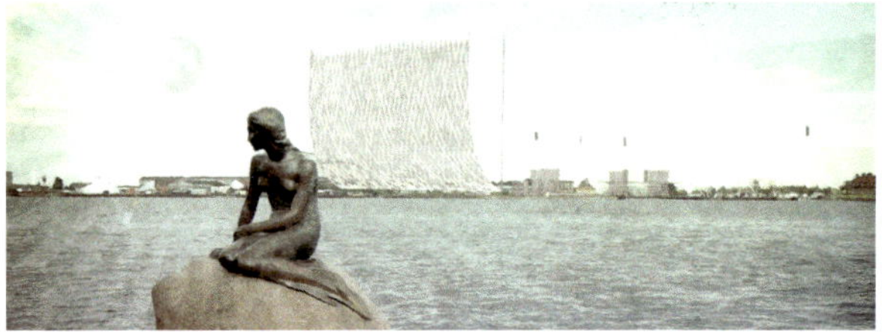

The Little Mermaid finally meets the prince. The construction and LEDs represent his crown.

The Little Mermaid finally meets the prince—his palace is situated on the opposite bank from her statue across the harbor. Her fate will perhaps not be so sad, as she will now be able to gaze upon the castle from her rocky perch.

The form of the *Palace of the Prince* and the public space below is variable. The sculpture changes its form by the impact of the wind. Leaves of sail fabric (Dacron®) are bent by the wind. During the night, LED diodes keep the candles burning in the palace.

The entire site is covered by areas with adaptable programs. Walls, ceilings, and floors can all be moved up and down by the hydraulic system.

The technology used in the design is a clean, energy-generating solution based on piezoelectric energy harvesting. The construction of each leaf consists of a surface and a bending shaft. The surface captures the wind energy by its area, and the shaft (containing elastomeric piezoelectric toroids and rigid plates) generates electric energy when it is bent.

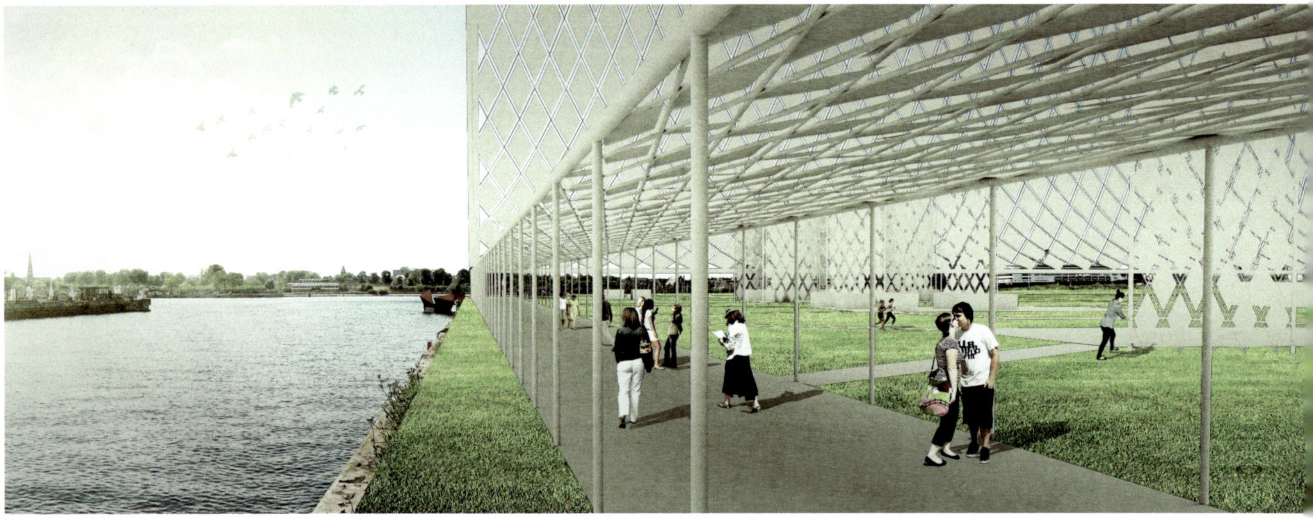

Different spaces connected into a grid are accessible from the promenade designed into the site.

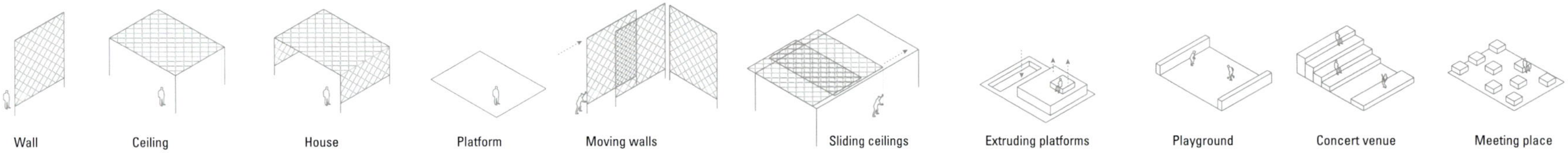

Changeability of uses

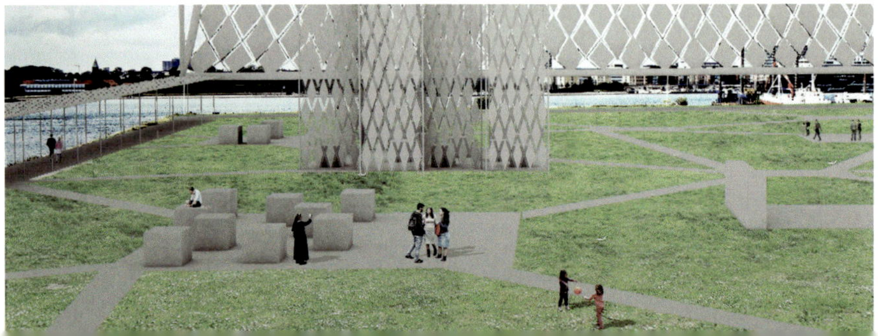

Cooperative urbanism. The site is designed as a space that will be used by the public with its usage and form determined by the visitors.

Tornado Tower

TEAM:
David Rieke, Stefanie Prümer

TEAM LOCATION:
Münster, Germany

ENERGY TECHNOLOGIES:
atmospheric vortex engine

ANNUAL CAPACITY:
21,000 MWh

Tornados produce incredibly strong energy forces. A medium tornado swirls with an energy as much as ten million kilowatts. This corresponds to the energy of 20,000 tons of TNT, or 1.5 times the force of the Hiroshima bomb. A tornado in the highest category can generate 10 to 100 times more energy. Even larger amounts of energy are converted in full-grown typhoons. Such cyclones generate daily energies of more than 400 billion kilowatt-hours, which could furnish, for example, the energy demand of Hong Kong for 90 years.

In the 1960s, the Frenchman Edgare Nazare designed the principle of an air vortex power plant. Its functionality is similar to that of the solar updraft tower.

Sunshine heats the air beneath a very wide greenhouse-like roofed collector structure surrounding the central base of a chimney tower. In a normal solar updraft tower, the chimney has to be very high (about 1,000 meters) to use the resulting convection by the chimney effect to drive the wind turbines and produce electricity. In an air vortex power plant, the chimney doesn't have to be as high. At the openings to the tower, the air is purposefully swirled so that a tornado effect arises. The tornado displaces the turbines in rotation and drives an electric generator.

Visitors to *Tornado Tower* can see into and learn about the technology, even climb to the 40-meter rim and look down from above while taking in the city views. Colored smoke is released into the vortices to make the tornado effects visible.

The *Tornado Tower* also incorporates piezoelectric elements, which are mounted on the glass-roofed collector and are rotated by the wind flow in vibration.

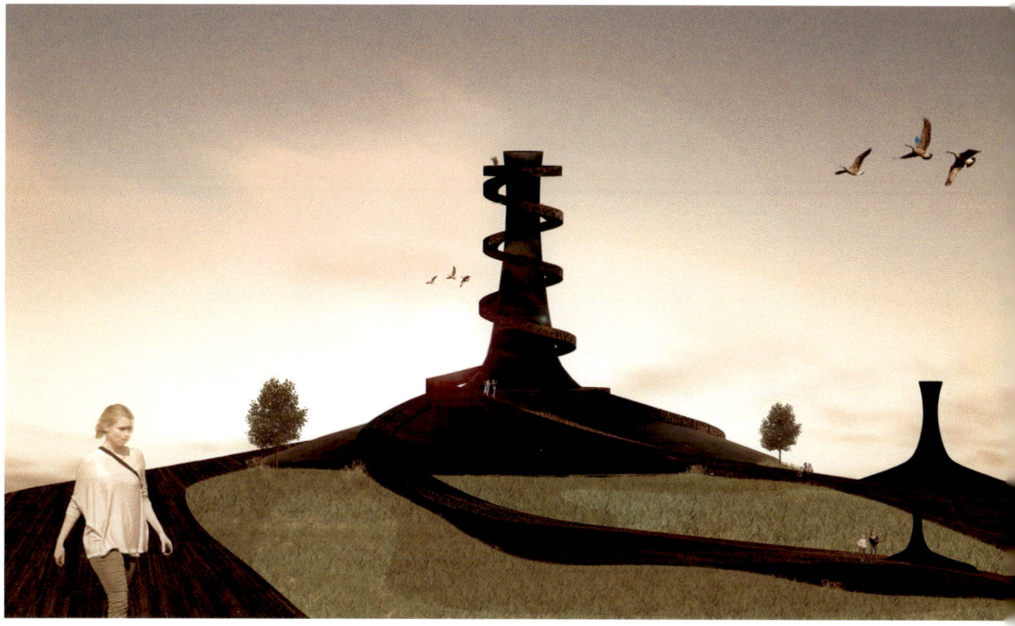

Tornado Tower allows visitors to hear, touch, and see the effects of vortex wind power.

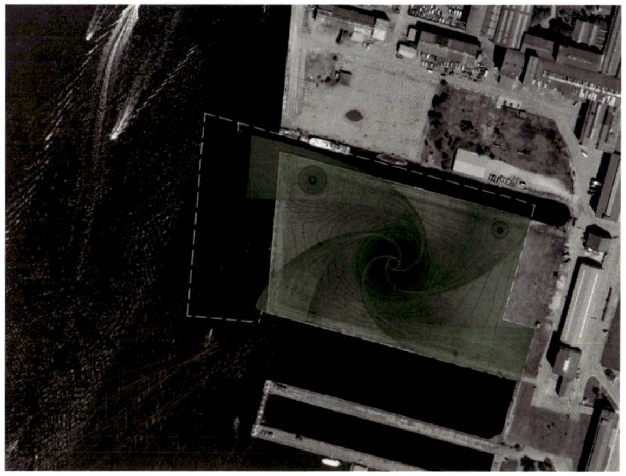

From the corners of the rectangular area, the four main routes run helically together in the middle of the site to feed the tornado wind.

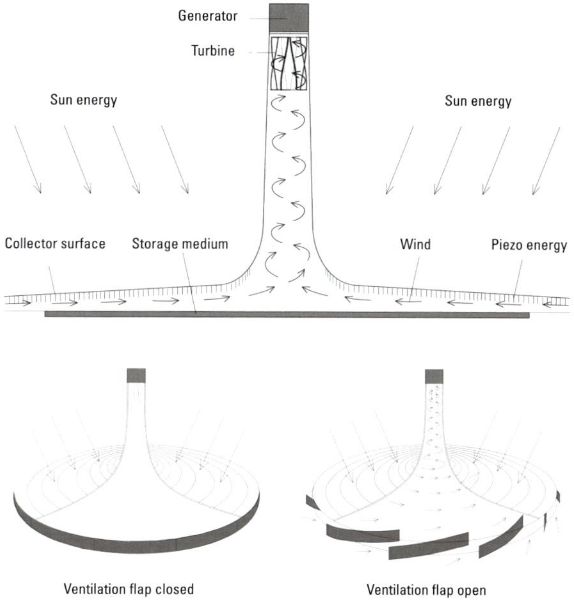

The Mermaid's Forest

TEAM:
Javier Fernandez Ponce

TEAM LOCATION:
Barcelona, Spain

ENERGY TECHNOLOGIES:
photovoltaic panels,
piezoelectric generators,
titanium dioxide air
purification

ANNUAL CAPACITY:
1,000 MWh

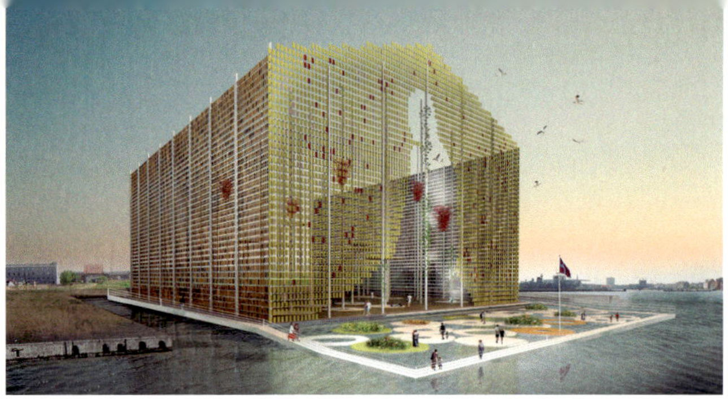

The Mermaid's Forest is a modular design, where the movement of panels and blades is translational with respect to wind, as opposed to rotating blades of current wind turbines.

The Mermaid's Forest is a 100-meter tall structure aligned within the plot perimeter limits of the design site. Its shape resembles a traditional shipyard warehouse, giving its interior a forest-like character where people and nature can meet. The warehouse becomes a new symbol for Copenhagen's industrial heritage, as well as one of sustainability that promotes biodiversity while generating utility-scale clean energy.

Three main technologies are integrated into the facade of this three-dimensional sculptural eco-warehouse, including a total of 5,532 photovoltaic panels (south facade), 3,456 wind power panels (west facade), and 5,544 air cleaning panels (north facade).

The smog-eater panels have a titanium dioxide coating and will clean the air via a superficial non-toxic chemical treatment. The generated free radicals attack and break down the NO_x in smog to a harmless nitrate.

The wind facade faces west toward the prevailing winds. The hinged movement of modular panels in the wind is transformed into electricity by way of piezoelectric generation rather than by rotating blades.

The south facade will produce PV energy in attractive and efficient polycrystalline solar panels that follow the same modular dimensions as the other facades.

Along the waterfront, a floating experimental garden will allow visitors to experience new types of aquaponic systems (aquaculture + hydroponics), which could potentially contribute to Copenhagen's future food supply.

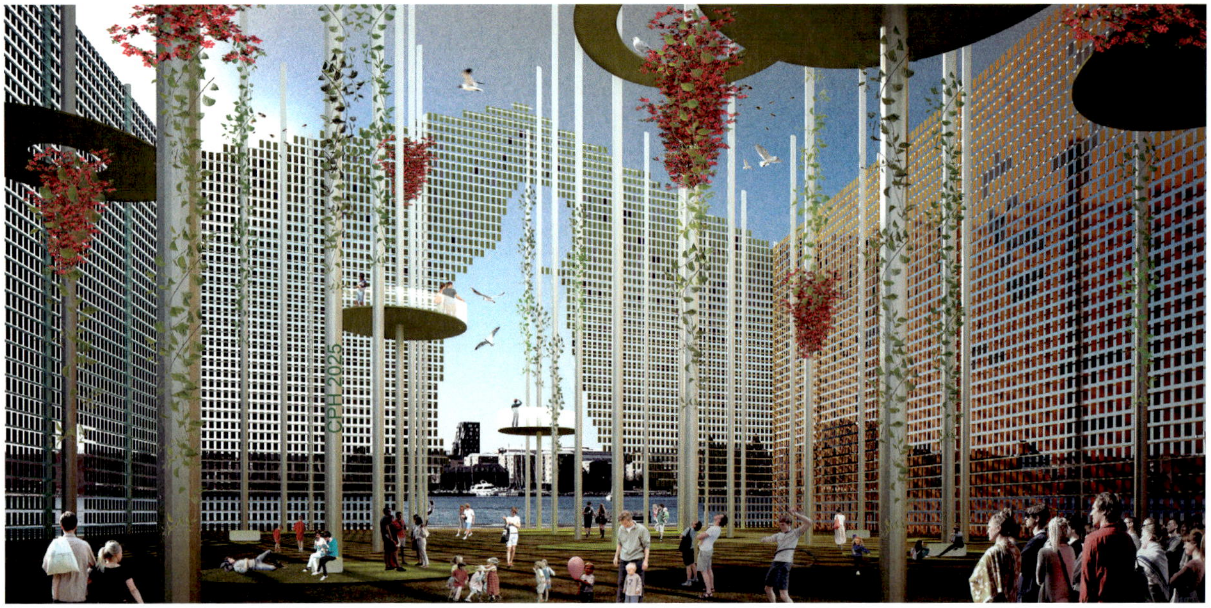

The artwork creates a new eco-destiny for all citizens. The design is a forest-like space surrounded by nature and ecosystems that will generate clean energy via its facades and allow filtered natural light into its interior. A portion of the PV energy can be used to elevate a series of viewing platforms in order to admire the surroundings.

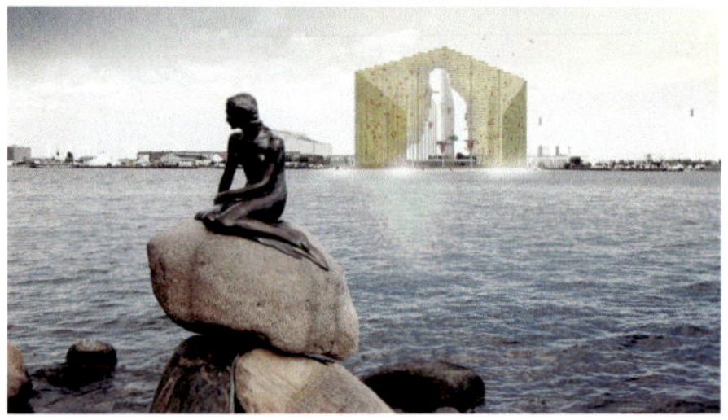

The site's view across the harbor to the Langelinie and the cherished statue of the Little Mermaid is the inspiration for the project.

Sight

Sight is installed on the front part of the design site, responding to the unique relationship the location has to its industrial past and the Little Mermaid.

Sight uses the image of the eye as an indicator of the sea level and as a reminder of climate change that will affect us all. The changing environment will affect the visibility of the "eye." As sea levels rise, the eye will gradually be less visible and will eventually disappear under the water. Until this time, the area will be a wetland—a station for birds and other coastal species. It will serve the citizens and visitors of Copenhagen as a place of rest, relaxation, and meditation.

An angled mirror is constructed on a steel frame. The reflective film is perforated in a form of flaps, equipped with piezoelectric modules. This makes the picture sensitive to weather effects and reduces wind impact by harnessing its energy. On the reverse of the plane is a large photovoltaic array.

Below the angled structure, an image of a child's eye is constructed pixel-by-pixel with a change of depth from zero to nearly two meters. In the shallow parts, where underwater currents accelerate, the bottom is covered with piezoelectric modules. In deeper parts, the bottom is covered with a local bright stone and local sand. The steep vertical surfaces dividing areas of different depths are reinforced with gabions.

TEAM:
Michal Pajakiewicz, Magdalena Rolka

TEAM LOCATION:
Moscow, Russia

ENERGY TECHNOLOGIES:
piezoelectric generators, photovoltaic thin-film

ANNUAL CAPACITY:
500 MWh

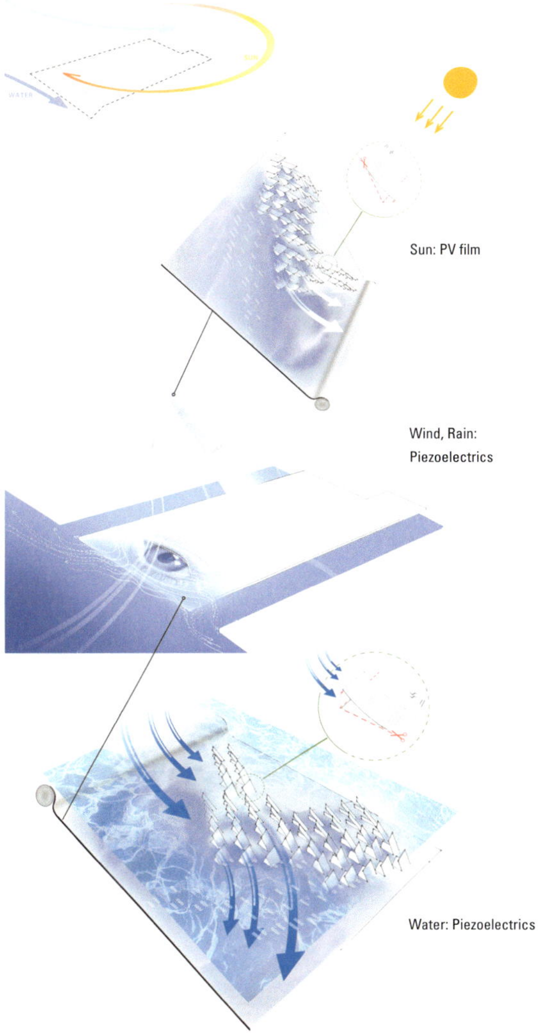

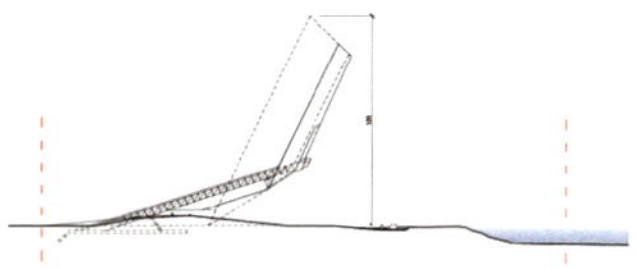

Section

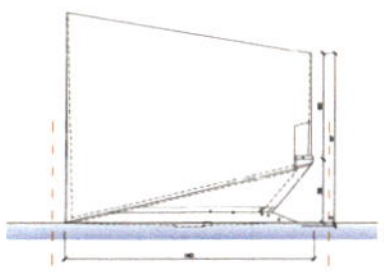

Front view

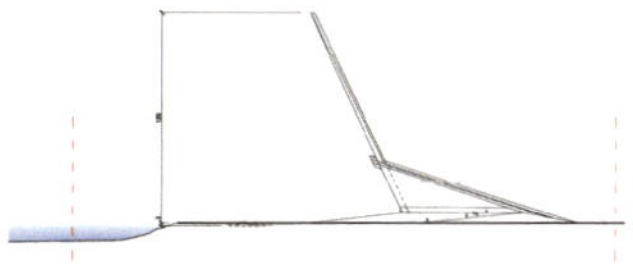

Side view

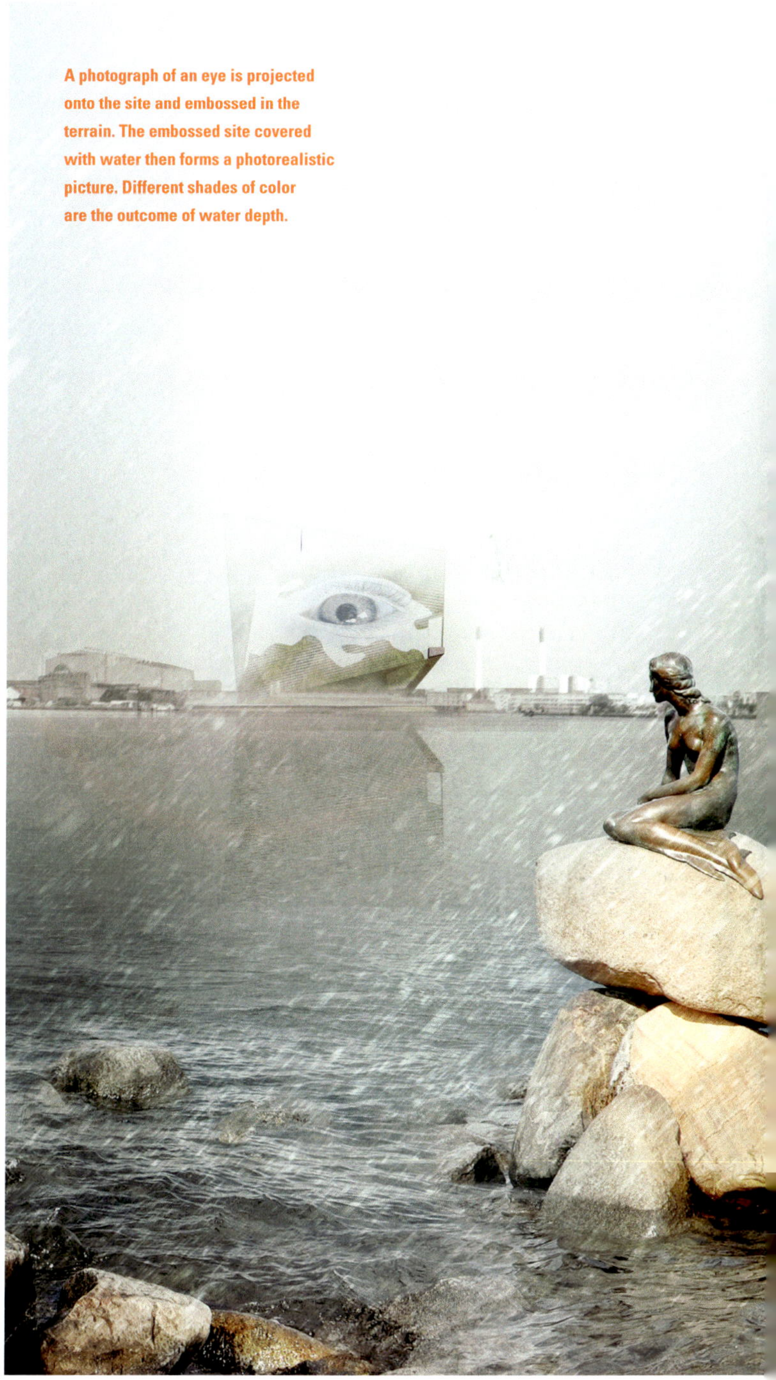

A photograph of an eye is projected onto the site and embossed in the terrain. The embossed site covered with water then forms a photorealistic picture. Different shades of color are the outcome of water depth.

Raised Farm

In *Raised Farm*, trees function as power stations, leaves are generating units, and roots are a transmission system. Special batteries are installed on roots, mimicking soil clinging to the roots of an actual tree.

Piezoelectric technology is the energy generation mechanism of the installation. The movement of wind through the leaves creates bending and collisions of piezoelectric generators. Although the energy generated by a single leaf is small, the total energy collected from all of the leaves in aggregate is considerable.

The roots of the trees in this energy farm are visible. They are designed to a specific shape—as vertical and horizontal arms to carry the lateral load of the structure, to transmit energy, and to house "TV-storages"—small (20 cm^3) LED illuminations. These modular pixels can be programmed to create one large image visible across the harbor.

Other aspects of the visitor experience will demonstrate the value of smart grids and energy demand curves, with the intention that they can further promote efficient electricity use at times that coincide with renewable production peaks.

TEAM:
Iman Amini, Alireza Hoobakht, Amin Amini, Mahdi Ghotbi, Mohsen Khanmohamadi, Mahdi Musavi

TEAM LOCATION:
Dallas (TX), USA

ENERGY TECHNOLOGIES:
piezoelectric generators

ANNUAL CAPACITY:
1,000 MWh

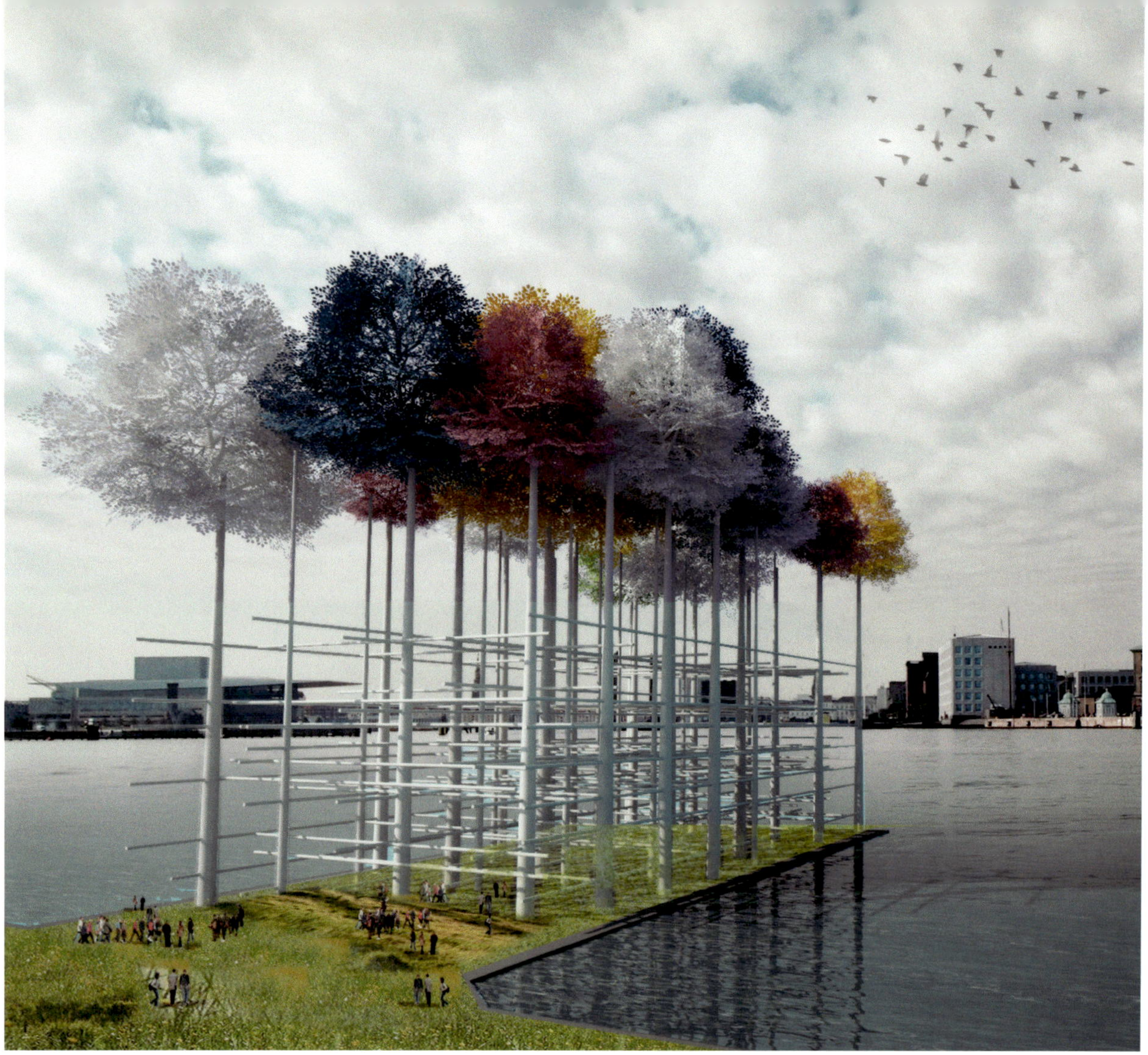

The energy of each leaf moves through the trunk of the tree to the roots. Energy is then transferred through the roots underground where they join and attach to electrical cables.

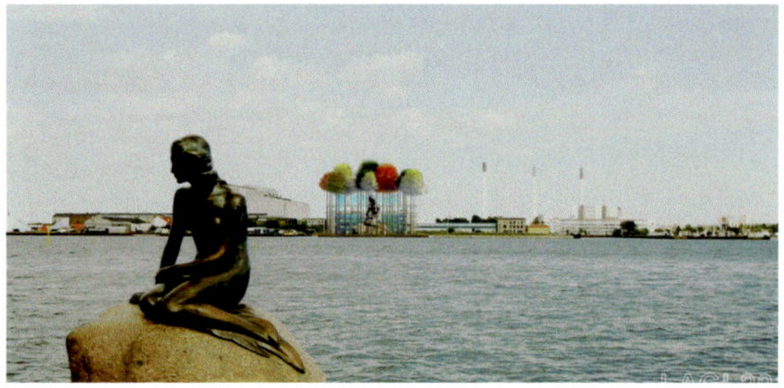

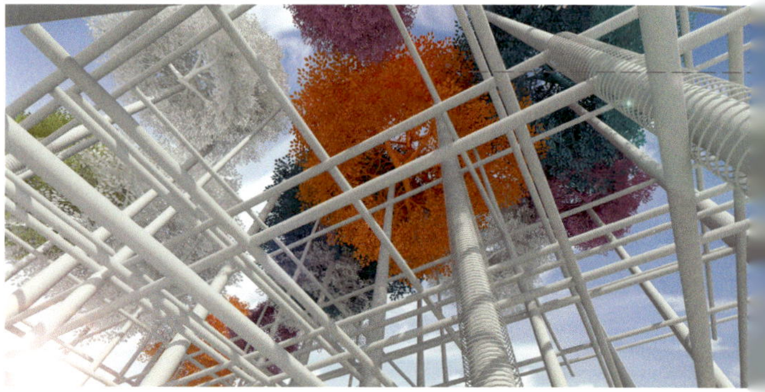

Silva

TEAM:
Clynt Ringrose

TEAM LOCATION:
Auckland, New Zealand

ENERGY TECHNOLOGIES:
vertical axis wind turbines,
Gravity Power™ storage

ANNUAL CAPACITY:
225 MWh

Silva is as much about the unbuilt as it is about the built. Its defining features communicate a narrative of environmentally sensitive development centered on the sustainable management and use of forest resources.

Vertical axis wind turbines (VAWT) are supported on curving structural timber supports. The turbines are arranged in a circular grid pattern based on John Dabiri's fish schooling theory in which close-spaced, counter-rotating VAWTs allow for greater energy density due to the performance-enhancing action of turbine vortices. This proposal uses similarly modified Windspire™ VAWTs but further explores the use of timber laminates for turbine blades and other non-structural elements.

The storage of surplus is a critical component of a resilient sustainable energy network. As an artistic demonstration of this, the installation utilizes a Gravity Power™ in-ground, closed-loop, modular-pumped storage system that is visible to visitors. With a storage shaft of 10 meters diameter and 1,000 meters in depth, this unit is estimated to provide 40 MW over 4 hours once charged.

A productive urban forest is split on an east-west axis by a timber boardwalk connecting the VAWT array to a transformer/education pavilion on the eastern edge of the site.

A timber and glass pavilion defines the main entry portal to the site on the eastern edge. This structure would house the transformer and other necessary electrical equipment behind glass and provide an educational benefit.

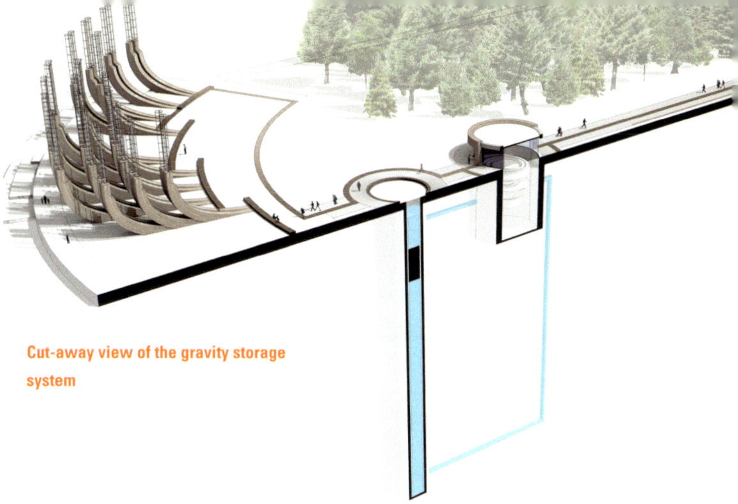

Cut-away view of the gravity storage system

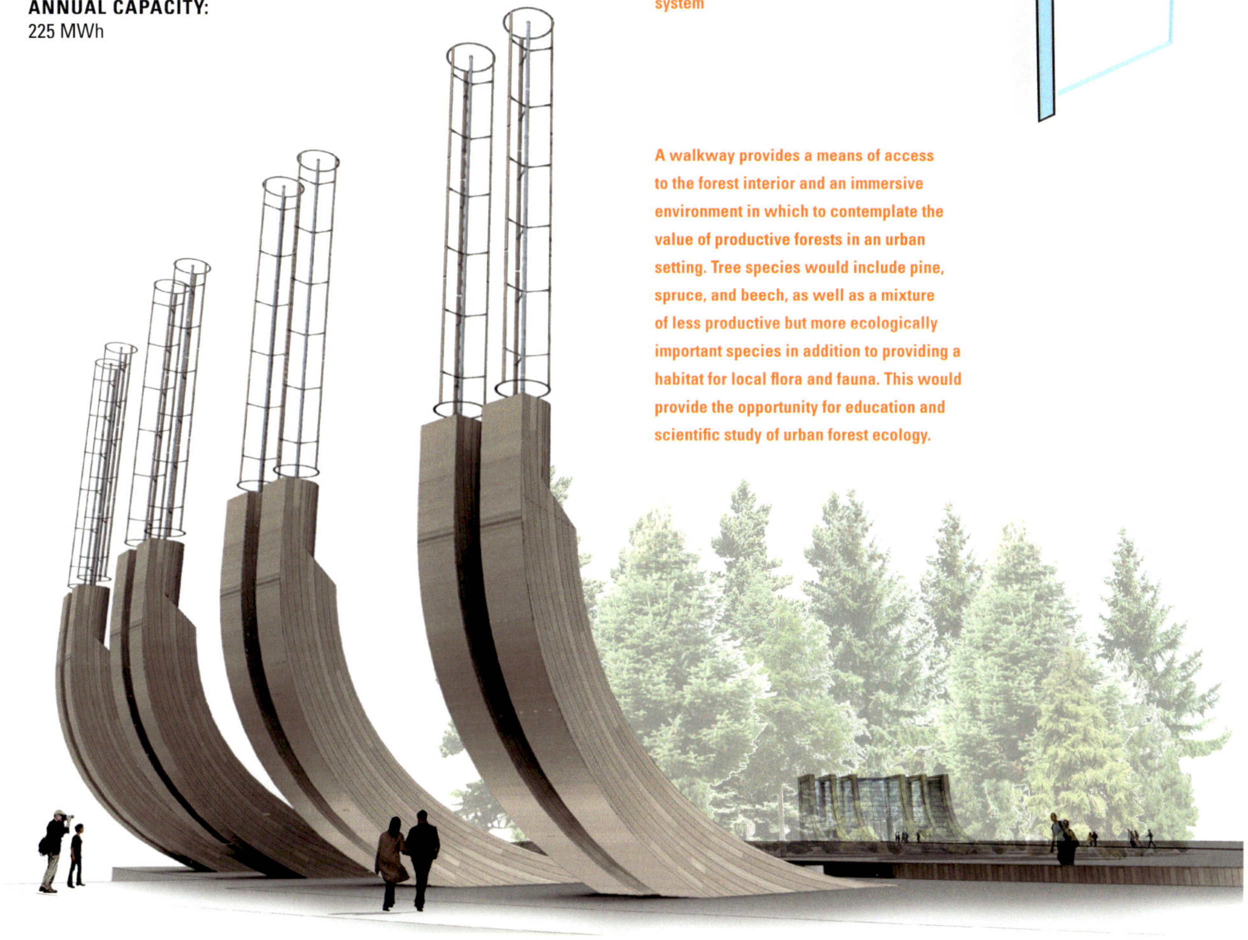

A walkway provides a means of access to the forest interior and an immersive environment in which to contemplate the value of productive forests in an urban setting. Tree species would include pine, spruce, and beech, as well as a mixture of less productive but more ecologically important species in addition to providing a habitat for local flora and fauna. This would provide the opportunity for education and scientific study of urban forest ecology.

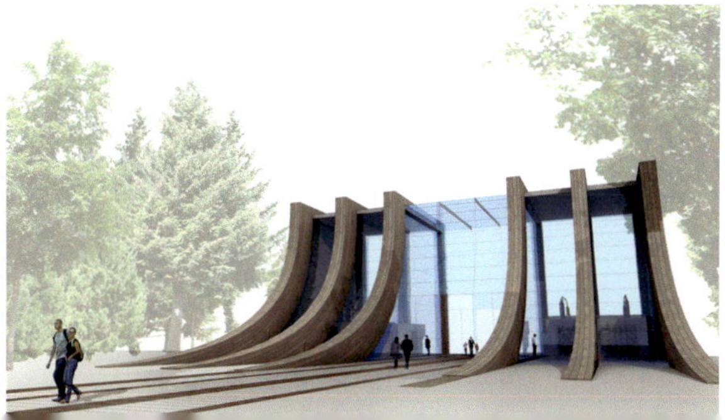

East view towards the transformer and education pavilion

Driven by the Wind:

Towards a Green Horizon

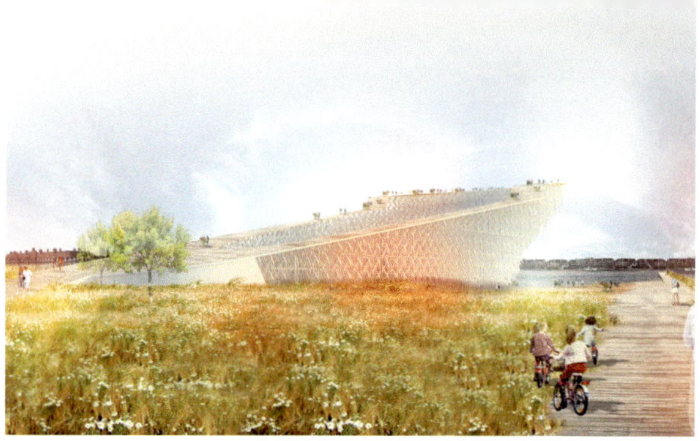

The design of the wind-ship represents the story of the site: the history of shipbuilding and the upcoming transition of the area.

The form of *Driven by the Wind* is based on the shape of the Selandia, the first large diesel ship in the world, that was built on the site of B&W at Refshaleøen. The design is a reference to the site's history and a symbol of the path to new horizons.

Its composition of vertical axis wind turbines creates a soft moiré pattern that offers a beautiful fairy tale background for the Little Mermaid. The dense pattern of turbines allows for a greater efficiency of power production as the turbine rotation is augmented by wind vortices created by the adjacent units. The wind-ship "sails" forward to a green horizon and a clean energy future.

An ecological park is incorporated into the design, with a riverbank rich in biodiversity. The rough vegetation and trees provide an attractive public space, and the ecological design contributes to the spatial experience of the area as a whole.

The ship uses a space frame structure, which supports 750 helical Darrieus-type vertical windmills that have their own inherent beauty. A combined staircase and ramp leads visitors to a recycled wood ship platform on the top of the space frame where they can experience unique views over the city from the 32-meter high bow of the ship.

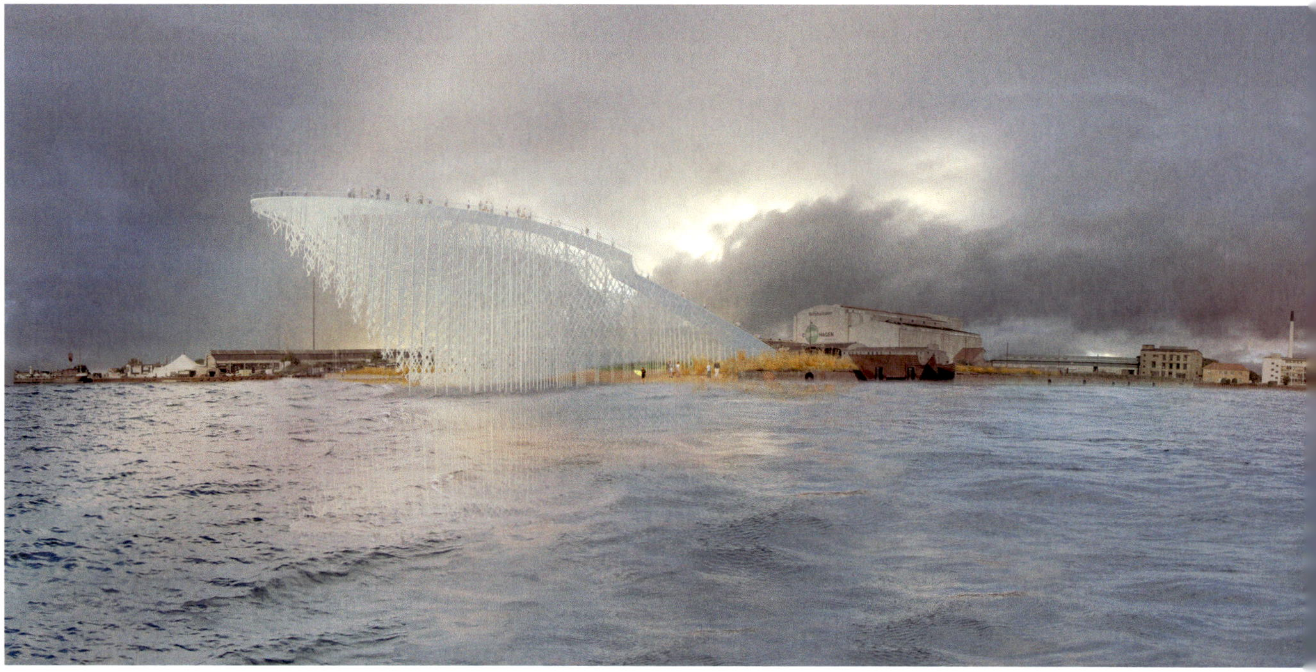

The ship slides over the gentle riverbank into the water, where a rich biodiversity will prosper.

TEAM:
Guido Zeck, Ingrid Ackermans, Peter Twisk

TEAM LOCATION:
Rotterdam, The Netherlands

ENERGY TECHNOLOGIES:
UGE VisionAIR3™ vertical axis wind turbines

ANNUAL CAPACITY:
862 MWh

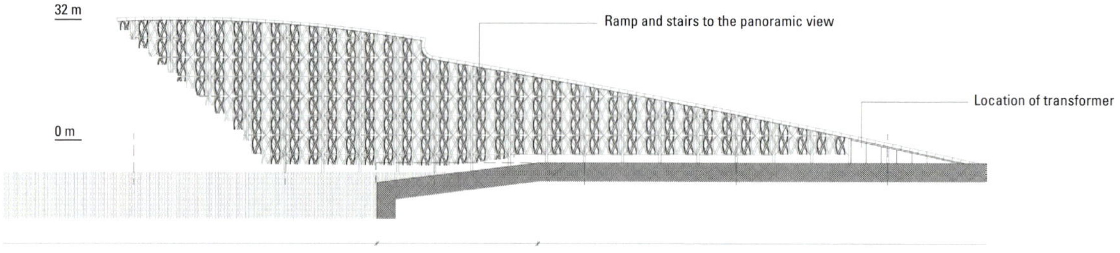

Elevation

Pixelated Wind

Pixelated Wind strives to ignite creative human energies by showcasing the invisible force that powers the city.

Beginning with landform manipulations, a series of mounds ripple from the center of the site and flow into Copenhagen's harbor. Windscreens are situated on top of the mounds, oriented to maximize wind energy harvesting. The screens are a surface of pixelations, created by small, circular wind turbines. The turbines can collectively manipulate wind to produce energy and, at the same time, allow for a memorable and unique interaction through embedded LED lights that change color based on wind speed.

The ever-changing luminescent pixelations make the energy of the wind visible. A diverse set of spatial experiences is created by the wind screens, which define rooms and walkways on the site. The large central space serves as an outdoor entertainment venue for concerts, festivals, speakers, and many other large outdoor events. At night the wind turbines will generate light in accordance to the direct wind speed hitting them.

The turbines include a series of high-performance cones that are used to concentrate the wind and increase its velocity as it passes through the rotors. The result is increased efficiency and noise reduction.

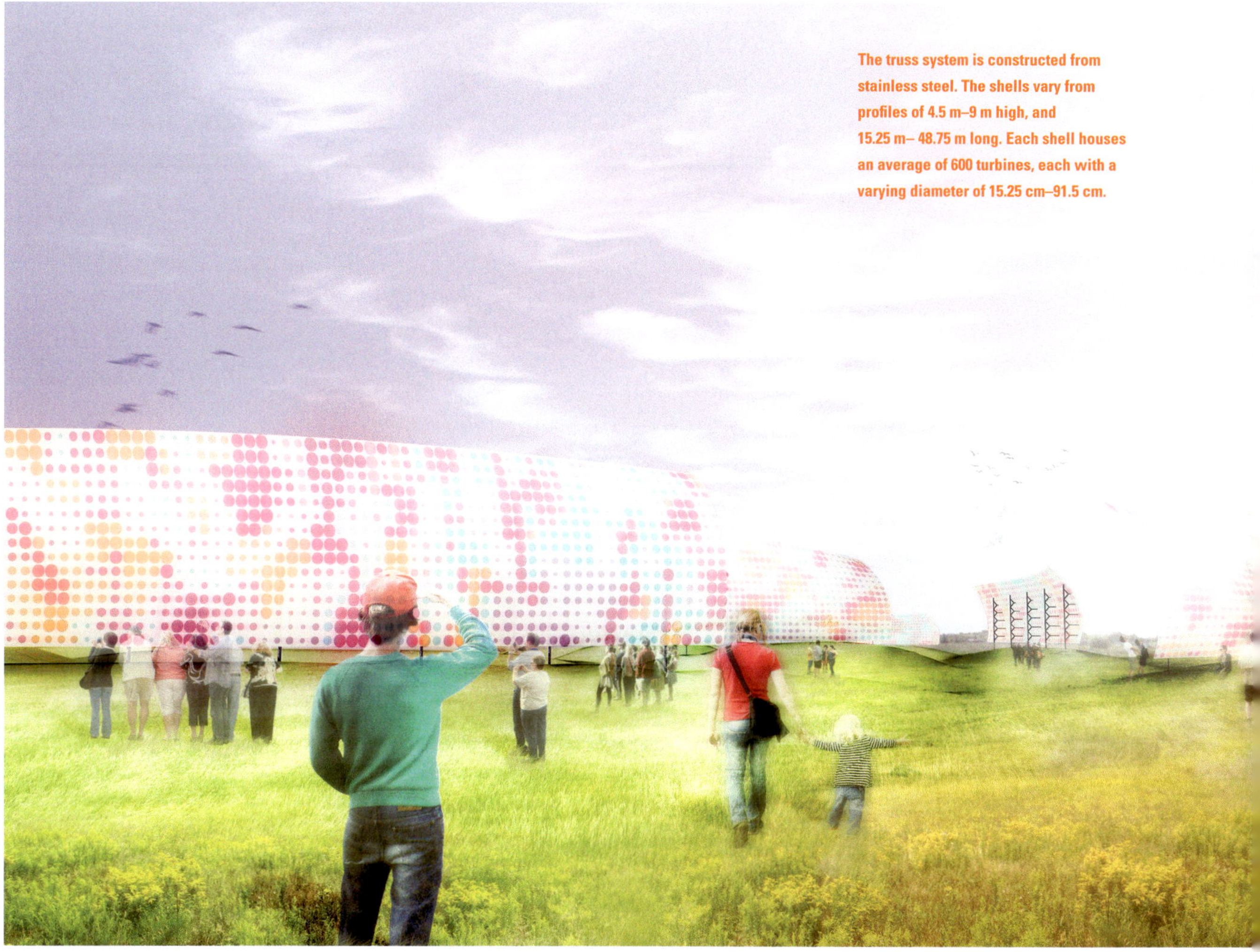

The truss system is constructed from stainless steel. The shells vary from profiles of 4.5 m–9 m high, and 15.25 m– 48.75 m long. Each shell houses an average of 600 turbines, each with a varying diameter of 15.25 cm–91.5 cm.

TEAM:
Ryan Helle, Delaney Rockwell,
Marcus Farr

TEAM LOCATION:
Boulder (CO), USA

ENERGY TECHNOLOGIES:
small compact acceleration
wind turbines (CWAT)

ANNUAL CAPACITY:
1,400 MWh

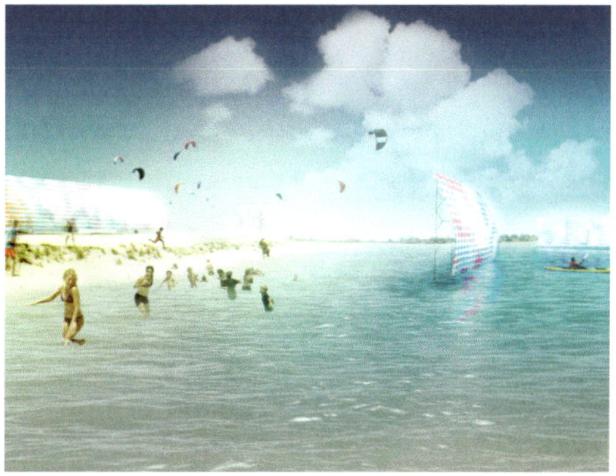

Research related to urban density suggests that the inclusion of sustainably integrated public design projects are necessary when dealing with the explosions of new populations in urban environments. This has an impact on how people live relative to economics, health, innovation, psychology, and geography, as well as the healthy demand for energy.

GREEN CITIES, DENMARK
LAGI 2014 PARTNERS

Green Cities, Denmark

Green Cities is a binding partnership of member municipalities in Denmark intended to provide inspiration for the development of climate and environmental policies. The member municipalities are geographically representative of the entire country and have track records of setting ambitious goals on their own.

For LAGI 2014, the Land Art Generator Initiative partnered with the Green Cities in Denmark, inviting teams to learn more about the four municipalities of Albertslund, Allerød, Herning, and Kolding, who all expressed interest in pursuing construction of energy-generating public artworks by working with the Land Art Generator Initiative & LAGI participants. Because Green Cities municipalities do their utmost to maintain, protect, and improve the environment while working towards creating a long-term sustainable local community, the partnership was a perfect fit.

Teams participating in the LAGI 2014 design competition for Copenhagen were invited to visit the Green Cities section of the LAGI design competition website, in order to learn more about the initiatives that are ongoing in the Green Cities, and to propose ideas for how their LAGI 2014 design ideas may be implemented in sites within the Green Cities municipalities.

The work of Green Cities is founded on the members' pooled experience, which creates a strong framework for the municipalities' work within the fields of climate and environment. In fact, Green Cities meets a specific need: at regular intervals the municipalities of Denmark hit an impenetrable wall of regulations and economic barriers. This effectively prevents the individual local authorities from realizing their sustainability goals on their own. Because citizens have high expectations for climate and environmental policy from their municipalities, Green Cities assists in working through these barriers to achieve real outcomes, and helps to advance political debate with regard to opportunities at the municipal level.

Green Cities has experience in everyday political decision-making processes, situating the organization as a perfect political platform from which to develop a coordinated climate and environment policy for all Danish municipalities. In doing so, the Green Cities initiative is creating a better environment for the benefit of the population. Green Cities has adopted the following shared goals:

A SUSTAINABLE SOCIETY

- Where environmental issues are concerned, the member municipalities of Green Cities are making determined efforts to set the most effective sustainable development goals in Denmark.
- Green Cities is taking a political lead and working to develop sustainable policies at the local level.
- Through national and international cooperation, Green Cities intends to further develop its environmental goals.
- Green Cities strives to improve the societal framework for sustainable development.
- Via its activities, Green Cities seeks to inspire and motivate other parties to intensify their efforts towards achieving sustainable development.
- The Green Cities partnership is based on involvement, cooperation, and commitment.

COMMON GOALS

Green Cities is committed to working towards achieving the following common goals:

1. Climate

- Before the end of 2015, CO_2 emissions to be reduced by 25% compared to 2006;
- Electricity and heating consumption to be CO_2 neutral by the end of 2025.

2. Groundwater

Green Cities is taking action to ensure that the groundwater is pure and plentiful. This includes:

- Protection of the areas surrounding wells and aquifers;
- Reduction of annual municipal water consumption with a goal of achieving:
  - an average consumption of 100 liters a day per person by 2015;
  - a 25% reduction in the consumption of water in municipal buildings by 2020 compared to 2010;
  - Reduction of water loss from water works to a maximum of 5% by 2020.

3. Organic food

Green Cities will promote organic food production at local and national levels by increasing production and consumption of organic food products. The individual municipalities are to achieve overall positive results in both of the following areas:

- A minimum of 75% of municipal food consumption to be organic by 2015;
- A minimum of 20% of agricultural land within the municipal boundaries to be farmed organically by 2025.

4. Natural environment

Green Cities will preserve and develop the natural environment for the benefit of fauna, flora, and the human population. Aims:

- To increase biological diversity by 2015;
- To make the natural environment, including the urban natural environment, more accessible;
- To encourage people to engage with and learn about the natural environment.

5. Traffic

Green Cities will encourage sustainable and accessible transportation, which must be maintained both within the towns and between town and country. Targets:

- Increase of bicycle traffic and public transport by 25% of the total volume of traffic in the member municipalities by 2020. Alternatively, vehicle traffic to be reduced by 25% compared to 2010;
- At least 50% of vehicular transportation to be powered by alternative types of fuel by 2025;
- Reduction of the quantity of microparticles in the urban atmosphere by 2015.

6. Waste

Green Cities will make an active and appreciable contribution to achieving more waste recovery. Targets:

- 30% reduction in the volume of combustible household waste by 2018 compared to 2010;
- 25% recycling of plastic materials and 20% recovery of metals from households by 2018 compared to potential development in these areas.

For further information visit www.greencities.dk.

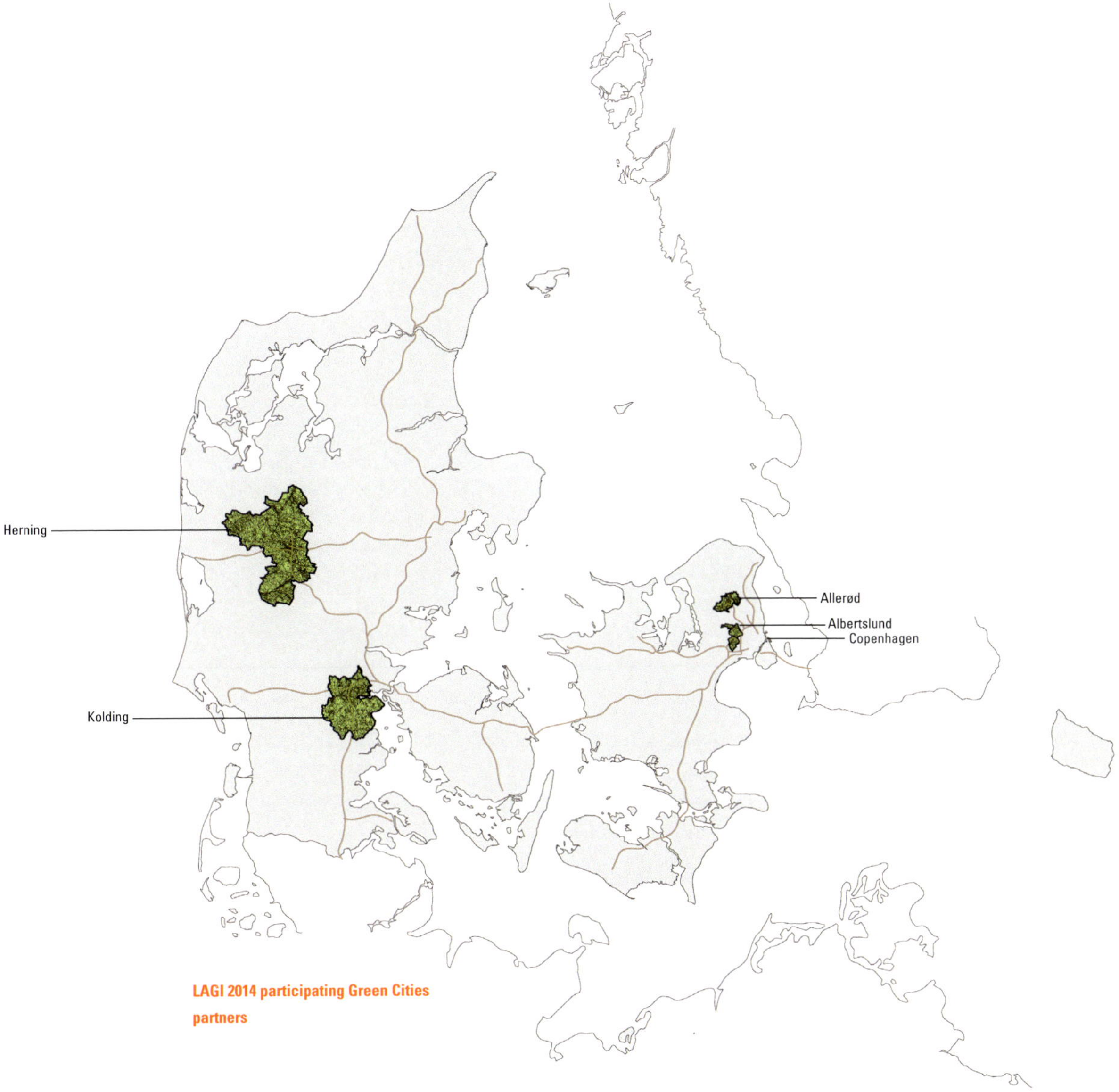

LAGI 2014 participating Green Cities partners

Albertslund, Denmark

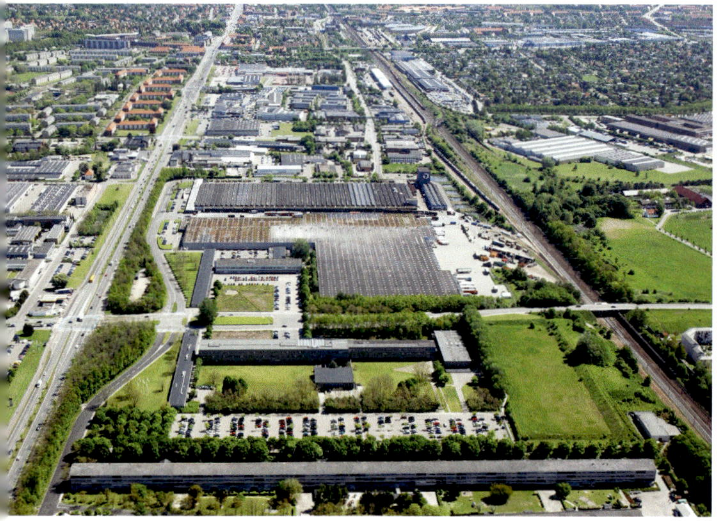

Roskildevej is a road that cuts through the city of Albertslund and is one of the old main roads to Copenhagen. The city of Albertslund is working to create better connections across Roskildevej in order to tie the city together.

Throughout the last 20 years sustainable development has been a high-profile political issue in the Albertslund Municipality. The municipality is focusing on sustainable planning and green growth through a variety of solutions. In 2009, Albertslund initiated the "triple helix" organization, Gate21, which has undertaken a wide range of climatic and environmental innovation projects based on partnerships between public, private, and research institutions.

The town serves as a living climate laboratory where new green technologies and solutions are invented in close relation to the city's development and people's lives. A great need for modernizing dwellings has led to The Albertslund Concept, a new method for energy renovation. The concept is suitable for many cities and has been implemented on a large scale in Albertslund, where many visitors come to experience these residential innovations. A renovated lighting system in Albertslund has led to new generations of the Albertslund Lamp—street and road lamps based on LED technology. Albertslund is also the initiator of DOLL, the Danish Outdoor Lighting Lab, which is Northern Europe's best test facility for energy efficient lighting, integration of sustainable energy sources, and other intelligent Smart City solutions.

Albertslund's motto is "Think global, act local." Since 1996, the Agenda 21 Center has initiated various environmental projects and supported citizens in sustainable actions and local initiatives. The municipality is environmentally certified, which means that environmental concerns and goals are part of daily life in all municipal workplaces.

The municipality of Albertslund has environmental goals, action plans, and green accounts to monitor its progress and the municipality has exhibited good results. For example, the climatic goal of 25% CO_2 reduction from 2006 to 2015 was reached ahead of schedule in 2013.

In 2007 Albertslund received the Nordic Council Nature and Environment Prize and in 2011 the municipality was appointed as a Nordic Energy Municipality, which "focuses on sustainable energy, green growth and energy-related climate work in the Nordic countries."

On these strong foundations, Albertslund continues to set new, ambitious goals for local sustainable development into the future.

Allerød, Denmark

Allerød City Council has high environmental ambitions and has been dedicated to ensuring sustainable development on a local scale since 2009. As a significant result, Allerød's CO_2 emissions related to activities within the municipality were reduced by 25% between 2006 and 2012.

Allerød uses the membership of Green Cities as a framework for working with sustainable development, and especially the environmental aspects. Areas of top priority include protecting groundwater, preserving nature, reducing CO_2 emissions, managing rainwater, and increasing waste recycling.

These priorities are embedded in the major municipal plans, and the management system is used to combine sustainability aims with all relevant municipal activities. When planning for urban development, the focus is on sustainable development, such as increased energy savings, use of bicycles, and biodiversity.

Because the City Council recognizes the importance of showing leadership in reducing CO_2 emissions, it has launched an energy renovation project with the aim of reducing CO_2 emissions from specific public buildings by 30%. Likewise, a new football stadium was built on sustainable principles, with heat pumps and solar cells, and a green roof to retain rainwater.

The municipality collaborates with a number of NGOs on different aspects of sustainability. For example, one collaborative project inspires private homeowners to reduce energy consumption and increase their use of sustainable energy sources. The municipality also leads a network of private companies that are committed to reduce their CO_2 emissions by 20%.

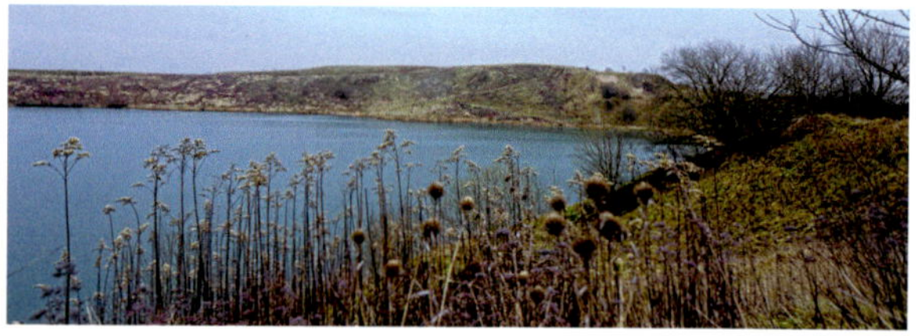

Future plans for Lynge Gravel Pit in Allerød include developing the area into a hot spot for recreational activity in the municipality. The western part of the pit has already been restored as a recreation site with paths for walking and running.

Herning, Denmark

Herning has a number of plans and strategies with green goals, including a Climate Plan, Nature Policy, and Sustainable Development Strategy, among others.

Herning is known for achieving high percentages of renewable energy in heat supply (84%) and electricity (50%), and has a tradition of collaboration with citizens and businesses with projects such as the "Climate Village Studstrup" and "New Circular Business Models." The municipality is proud of its ecological efforts such as urban stream reclamation and habitat restoration.

The municipality's ambition is to be among the municipalities with the lowest emission of greenhouse gases (measured in CO_2 equivalent) per citizen.

This will be achieved through:

- Energy savings by citizens, businesses, and municipalities;
- An increase of renewable energy to 45% by 2015 and 50% in 2025;
- Common internal and external ownership of the climate action;
- A focus on green industry;
- Ongoing monitoring and revision of the climate action plan every 4 years;
- Coordination with municipal and other local government planning.

Herning Municipality's climate targets are:

- 2007–2015: Emissions of CO_2 to be reduced by 25%;
- 2007–2030: Emissions of CO_2 to be reduced by 35%;
- Electricity and heat supply to be CO_2 neutral by 2025.

Herning Municipality's green commitments are:

- **Green Cities environmental cooperation between 6 municipalities**

  Targets and commitments in the climate, nature, groundwater, ecology, waste and transport, green purchasing, green accounting, environmental management, sustainability planning, civic and corporate involvement.

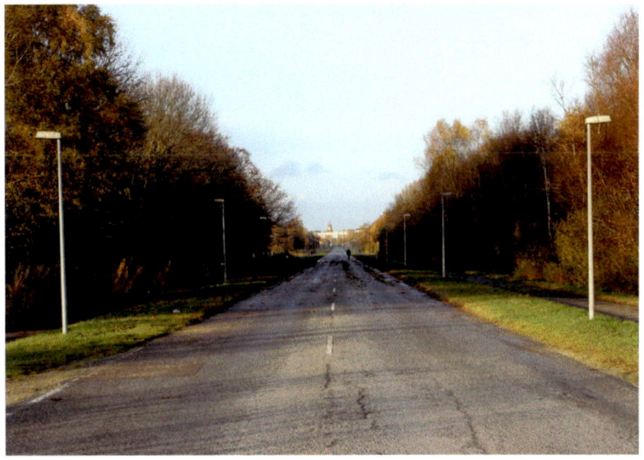

The "TimeLine" is a project to develop the closed Vejlevej road and use it for new purposes. The old road runs through the Knudsmosen (the Canute's bog) and in the future will be a unique setting for recreation, art, and nature.

- **Climate Municipality under the Danish Society for Nature Conservation**

  Goal: a 3.5% CO_2 reduction annually in municipal activities.

- **Covenant of Mayors**

  Goal: to meet and go beyond the European Union target of 20% CO_2 reduction by 2020.

- **Partnership for green procurement between 11 municipalities and regions**

  Setting of specific targets for municipal procurement of goods and services.

Kolding, Denmark

The Municipality of Kolding has a long history of working with ambitious climate and environmental projects. The first environmental plan in Kolding was realized in 1989. Around the same time, the municipality started substantial monitoring of energy use in municipal buildings. The focus on monitoring energy use has been maintained through the years and is now also addressing technical equipment as well as integrating sustainable energy and awareness campaigns.

There is an ongoing concern with both the environmental impacts of the municipality as an organization and the involvement of citizens and businesses. Therefore, the Kolding climate strategy initiated in 2008 embraces all target groups—businesses, organizations, and citizens—with several projects and activities involving each.

Examples of projects working with businesses:

- Carbon 20: companies are given assistance to reduce their CO_2 emissions;
- Green business growth: craftsmen and builders are trained to ensure higher quality while focusing on the sustainable renovation of houses;
- Green shops: shops are operated with care for the environment.

Examples of projects working with organizations:

- Green schools: focus on energy use in school buildings and on educating students to be environmentally friendly.

Examples of projects working with citizens:

- Low-energy villages: assistance to villages to become low consumers of energy in ways that match the local wishes and development of the respective villages;
- Electric vehicles: testing of electric vehicles by citizens, and establishment of a charging infrastructure;
- Green institutions: focus on energy usage in buildings, and teaching children to act in environmentally friendly ways.

BIOS, GLOSSARY, AND INDEX

BIOS

ALEXANDRA INSTITUTE

Alexandra Institute is an independent Danish research and technology organization whose primary focus is pervasive computing, i.e., Information and Communication Technologies (ICT) embedded into products and surroundings. They engage in research-based user-driven innovation projects with private corporations, public institutions, and citizens. www.alexandra.dk

ELSE MARIE BUKDAHL, D. PHIL. PROFESSOR

Else is the former president of the Royal Danish Academy of Fine Arts. She is a member of The Royal Danish and Norwegian Society of Sciences and Letters, an honored member of international *Who's Who*—selected as one of the leading educators of the world by the *International Biographical Centre, Cambridge*—and has been appointed *Officier des Palmes Académiques and Chevalier de l'Ordre des Arts et des Lettres, Paris*. She has a long list of publications spanning aesthetics, literature, visual art, and architecture, and has also written extensively about the intersection between art and the environment.

ROBERT FERRY

Robert is the founding co-director of the Land Art Generator Initiative and Studied Impact Design. His consulting work is focused on the design of net-zero and net-positive environments that achieve complete harmony with their local and global environments and with the people who use them. He has contributed to many high-profile green building projects around the world, and his design work has been featured in *Popular Science Magazine* and other publications.

His publications include *Regenerative Infrastructures, The Time is Now: Public Art of the Sustainable City, New Energies,* and *A Field Guide to Renewable Energy Technologies*.

CHRIS FREMANTLE

Chris is a researcher and producer based in Scotland. He is involved in the national development of artwork in public places as well as producing specific projects. He is also associated with art school practice-led research and knowledge exchange programs. He established ecoartscotland.net as a platform for research and practice addressing the cultural dimension of environmental crisis. Other aspects of his work can be found on www.chris.fremantle.org.

PETER KARNØE

Peter is a professor at Aalborg University Copenhagen, in the Centre for Design, Innovation, and Sustainable Transition. His research has focused on technological entrepreneurship as distributed socio-technical practices and he has proposed the concept of path creation to focus on emergent agency in innovation. Peter has studied the emergence of the Danish wind turbine industry. His current research explores valuation and market framing in relation to clean tech innovation and in disputes about re-setting electricity markets with wind power. He has published in various journals including *Research Policy, Journal of Management Studies,* and *Industrial Marketing Management.*

ELIZABETH MONOIAN

Elizabeth is the founder and director of Society for Cultural Exchange, a non-profit organization that is developing international exchanges between communities, academics, and artists. Under SCE she is the founding co-director of the Land Art Generator Initiative, which has brought together thousands of creatives from around the world. Elizabeth is an interdisciplinary artist and designer whose work has been screened and exhibited in venues throughout Europe, the Middle East, and the United States. She received an MFA from Carnegie Mellon University in 2000.

Her publications include *Regenerative Infrastructures, The Time is Now: Public Art of the Sustainable City, New Energies,* and *A Field Guide to Renewable Energy Technologies.*

TRINE PLAMBECH

Trine is Senior Green Urban Designer in the Energy & Sustainability Lab of the Alexandra Institute. She combines knowledge of people, information technology, and the environment in collaborative urban design projects. She has a special focus on energy and sustainability and is, among other things, involved in a smart city project where data of the city is used to generate knowledge and new solutions for urban environmental challenges. Trine is part of the Smart City Network initiated by the Ministry of Housing, Urban, and Rural Affairs in Denmark. Over the years, she has focused on user-centered design in business innovation. She has created insights into relevant human behavior issues and used her extensive knowledge about creativity and creative processes to facilitate innovation processes in a wide range of organizations together with end-users, clients, and colleagues.

Trine holds a MSc in Landscape Architecture from the University of Copenhagen, a BSc in Technological and Socio-Economic Planning from Roskilde University, and a MSc in Digital Design and Communication from the IT University of Copenhagen.

ANN T. ROSENTHAL

Ann's art installations address the local manifestation of global concerns, including climate change, food safety, and nuclear waste. Her work has been shown over the last decade at The Andy Warhol Museum and the Mattress Factory in Pittsburgh, Exit Art and the Hudson River Museum in New York, the Schuylkill Center for Environmental Education in Philadelphia, and Kunsthaus Kaufbeuren in Germany.

Her essays have been published most recently online in "Atomic Legacy Art" in the *Women Environmental Artists Directory Magazine* and the peer reviewed *Ecopsychology Journal*. She teaches environmental art and design courses in Pittsburgh, and has developed an online graduate course,

"Introduction to Eco/Community Art" for Plymouth State University in New Hampshire. She received her MFA from Carnegie Mellon University in 1999. Ann owns an industrial building in Pittsburgh where she directs LOCUS—a creative commons where art, community, and ecology meet.

LEA SCHICK

Lea is a PhD fellow at the IT University of Copenhagen, doing research in the field of smart grids and energy art. She is exploring how relations between humans, energy, and nature are imagined, made, and re-made in the performance and demonstration of green energy futures. She has a background in art and design and served as LAGI 2014 regional Project Manager.

PAUL SCHIFINO

Paul received his degree in Graphic Design from the Art Institute of Pittsburgh in 1979. His work has been published in design publications including *Communication Arts* (CA), *Graphis, Print, How,* and *ReadyMade* magazine. His art has been included in shows at The Andy Warhol Museum (AMP), the Mattress Factory (Gestures #4 and #14), TRAF Gallery (By Design), and the TRAF (Best of Pittsburgh Show).

He has served as president of the Pittsburgh chapter of the American Institute of Graphic Arts (AIGA), and is a former Advisory Board member of the American Shorts Reading Series, and the Art Institute of Pittsburgh. Paul currently serves on the board of the Society for Cultural Exchange.

ANNETTE SKOV

Annette is a Danish visual artist and writer working with interventions, books, video, photography, sound, performative walks, and secret tours. She travels around with two old caravans collecting stories in order to pass them on in different ways. In 2011 Annette founded Refshale Heathland on Refshaleøen, where she has her studio. Since the installation of 100 square meters of public heathland from Jutland, a wealth of life now dominates the grounds, including insects, birds, weeds, and other guests.

She is a member of Danish Visual Artists and the Danish Authors' Society, and is a Master of Arts, studying Art History and working at The National Gallery (SMK) in Copenhagen as an art educator. www.annetteskov.dk

LISELOTT STENFELDT

Liselott works as a Senior Interactions Designer and Architect MAA in the Interactive Spaces Lab of the Alexandra Institute. She manages projects in and for public surroundings. Her work focuses on technology integrated in physical surroundings (such as rooms and buildings in which we live and work, the landscapes, and the public and urban spaces in which we move). She examines future scenarios on how cities can be experiences and used in new ways. In all of her projects she aims to push the boundaries of research, design, architecture, engineering, and software (existing form and functions), to create something new. Liselott's main focus lies in how we can create awareness by making the invisible (such as energy usage and CO_2) visible around us, so as to better be able to act on it. She holds a Cand.Arch. from Aarhus School of Architecture in Denmark.

LAURA WATTS

Laura is an ethnographer and Associate Professor at the IT University of Copenhagen. Her research is concerned with the effect of landscape and location on how the future is imagined and made, and with different methods for writing the future. She has worked with the mobile telecoms industry, the renewable energy industry, and the public transport sector, and is currently working on the project "Alien Energy," in collaboration with the people and places of the marine energy industry in Orkney, Scotland. She is co-author of the artisan book "Data Stories," a set of ethnographic stories that engage with ideas of "Big Data." Much of her work is published on her website at www.sand14.com.

BRIT ROSS WINTHEREIK

Brit is Associate Professor at the IT University of Copenhagen, in the Technologies in Practice research group. Her research is concerned with information infrastructures, monitoring, partnerships, and accountability practices. Presently she is pursuing her research interests in relation to renewable and decentralized energy innovation. Brit has published widely on information technology in society. Her monograph *Monitoring Movements in Development Aid: Recursive Partnerships and Infrastructures,* co-authored with Casper Bruun Jensen (MIT Press, 2013), is a multi-sited ethnographic study of monitoring practices in development aid. Brit is a co-founder of the strategic research area Energy Futures, a co-founder of the Danish Association for Science and Technology Studies (DASTS), and she serves on the editorial team of the journal *Science and Technology Studies.*

ANNE SOPHIE WITZKE

Anne Sophie is a PhD fellow at Aarhus University. In her research, she investigates how new understandings of air and atmosphere are explored by new media artists through pervasive technologies, cloud computing, and environmental sensors.

Her curatorial work within art, sustainability, climate change, and technology includes her role as project manager and co-curator of the exhibition *RETHINK—Contemporary Art & Climate Change.*

GLOSSARY

AEROLASTIC FLUTTER

A self-feeding regular vibration that occurs when wind passes by an object and stimulates the object's natural frequency, creating a positive feedback loop. This often destructive effect is avoided in airplane wing and bridge engineering, but it can be intentionally utilized for wind energy power generation often in combination with piezoelectronics.

AIRFOIL (AEROFOIL)

The shape of a wing in section as designed to provide lift and drag forces by creating positive and negative air pressure on opposite sides of the shape. Turbine blades are also described by their airfoil shape.

ALGAE BIOFUEL

Algae can be grown and harvested (algaculture) as a feed stock for the production of alternatives to fossil fuels. Naturally occurring oils within algae (lipids) can be used directly (similar to straight vegetable oil), or they can be refined to burn more cleanly. Different production methods can result in biodiesel, biobutanol, biogasoline, methane, ethanol, or even jet fuel. The uptake of CO_2 by the algae during cultivation offsets the CO_2 that is emitted during the combustion of the algae-generated fuel. Algae can produce up to 300 times more oil per acre than conventional crops such as jatropha, palm, rapeseed, or soybean, and it can be cultivated in locations where these types of crops are not viable.

ALTERNATING CURRENT (AC)

An electrical system in which the flow of electric charge periodically reverses direction (as opposed to direct current in which the flow is constant). See "direct current."

AMORPHOUS SILICON (A-SI)

The functioning semi-conductor material within a type of photovoltaic system (thin-film) that is less expensive and more versatile in its application than crystalline silicon types. Conversion efficiency is generally less than crystalline silicone PV. See "monocrystalline silicon."

BIOGAS AND BIOMASS

Biogas is created through the breakdown of any organic material (biomass) in an oxygen-poor environment. The resulting gas byproduct is mostly methane and carbon dioxide. Biogas is similar in composition to conventional natural gas and as such can be compressed or fed into a municipal gas grid. It can be used for many different purposes including cooking, heating, lighting, transportation, and electricity production. It can be either tapped from the underground activity in a landfill site, or it can be produced in specially constructed anaerobic digester tanks. Farms with such tanks can process manure into biogas, reducing the amount of nitrous dioxide and methane that would otherwise enter the atmosphere. These two gases have a far greater atmospheric warming effect than does carbon dioxide (nitrous dioxide = 310 times greater, and methane = 21 times greater). Biomass is considered a sustainable energy resource because it is a product of organic processes, which naturally regenerate at a rapid cycle (as opposed to fossil fuel energy sources which take millions of years to form naturally). Biomass can be combusted directly as a solid fuel or converted to liquid or gas biofuels.

CARBON DIOXIDE (CO_2)

A naturally occurring chemical compound critical to life on earth, carbon dioxide also functions as a greenhouse gas (GHG) in the Earth's atmosphere (contributing to anthropogenic climate change and global warming). The emission of CO_2 through fossil fuel combustion by humans has, since modern industrialization, created an increase of 35% in the parts per million (ppm) concentration of the gas in the Earth's atmosphere. Since 1960, its concentration has risen from 320ppm to 402ppm (as of 2014) and further increases threaten rapid shifts upward in global temperature and sea levels. In order to avoid a temperature rise beyond 2° Celsius, between 2/3 and 4/5 of the known reserves of fossil fuel will need to remain unused until such time that proven methods of carbon capture and storage (CCS) can allow their safe combustion (no method of CCS has yet been proven suitable for long-term CO_2 storage). Increased atmospheric concentrations of CO_2 also have a secondary effect on the chemical composition of the oceans, as surface-level carbon dioxide dissolves forming other carbon compounds and leading to acidification.

CARBON FIBER REINFORCED RESIN POLES

A material that displays great strength in both tension and compression but that is ductile along its length. Similar to the material used in pole vaulting poles or tent structures.

CERAMIC MULTILAYER ACTUATORS

Piezoelectic ceramic discs that are fired together to create a stacked device that increases displacement, resulting in greater energy generation. See "piezoelectric generator."

COMPACT WIND ACCELERATION TURBINE (CWAT)

This type of horizontal axis wind turbine uses a cone or series of cones to concentrate the wind, increase the velocity of the wind as it passes through the rotor's swept area, and thus increase the efficiency of the overall system. They are also known as "ducted turbines" or "lens wind turbines." CWAT is a new acronym that encompasses the class of machines known as DAWTs in the 1970s and 1980s.

CONCENTRATED SOLAR POWER

Describes a variety of systems that use mirrors or lenses to concentrate the power of the sun in order to create heat energy that can then be converted into electricity.

CONVECTION LOOP

In the dynamics of fluids or gas, the tendency of higher pressure and lower pressure to equalize causes warm to migrate towards cool, thus creating a flow of gas or liquid. In closed systems with heat input in one area, a continuous loop is created as warm material flows to cool areas.

COPPER INDIUM GALLIUM SELENIDE (CIGS)

A semiconductor material alternative to silicon used in thin-film photovoltaic.

DATA MONITORING

Real-time statistics of how much electricity is being produced. Monitoring can be either on site or remotely accessed and is displayed in an easy-to-understand graphical interface that often simulates analog dials and meters.

DIRECT CURRENT (DC)

An electrical system in which the flow of electric charge is constant (as opposed to alternating current in which the flow periodically reverses direction). See "alternating current."

DYE-SENSITIZED SOLAR CELL (DSSC)

The DSSC solar cell is alternatively known as the Grätzel cell. They have the characteristic of being semi-transparent, flexible, and they are very durable. They also function comparatively better than other PV technologies in low light levels and indirect light.

ELECTROACTIVE POLYMERS (EAP)

Electroactive polymers are very complex and large molecules that alter their physical form when exposed to an electrical field. These polymers can be used to create materials that expand and contract in ways similar to that of animal muscle. There is some similarity between EAPs and piezoelectric materials, but EAPs can withstand far greater deformation and forces of strain. EPAM is an acronym for Electroactive Polymer Artificial Muscle and most closely relates to a product that can be licensed from SRI International.

ELECTRODE

The electrical conductor that makes contact with a semiconductor or other non-metallic material. Electrodes can be labeled either anode or cathode depending on which direction the electrical charge is flowing.

FRESNEL LENS

A magnifying lens that takes the sectional geometry of a simple convex lens and flattens it by slicing it in concentric circles and shifting the sliced segment profiles to create a flattened, corrugated surface. The optical effect of the lens is very similar to that of the original convex lens. Fresnel lenses are used in many concentrated photovoltaic systems to focus the sun's energy onto specially engineered photovoltaic cells.

FUEL CELL

Any mechanical system that converts the energy stored within a fuel source (e.g., hydrogen, methanol) into electricity through an oxidation process. Fuel cells require the replenishment of the fuel source (reactant) to maintain electrical output. Fuel cell technology has the potential to replace the internal combustion engine for the conversion of fuel into energy for use in transportation and machinery.

HELIOTROPIC (HELIOSTATIC)

The ability to follow the location of the sun in the sky and maintain an object's consistent relationship to it throughout the diurnal and seasonal shift. In solar energy technology, heliostatic mechanisms can maintain a solar cell perpendicular to the sunlight for ideal absorption and conversion, or mirrors can maintain an angle-of-incidence relationship to the sun so as to consistently reflect sunlight to a central collector.

HYDRAULIC TURBINE

A rotary engine that is driven by the force of passing water.

HYDROELECTRIC STORAGE

Excess capacity electricity is used to pump water temporarily into an upstream reservoir. The water can then later be released when there is demand for electricity. The force of gravity on the water drives generators just like at a conventional hydroelectric dam (which instead relies on natural precipitation cycles to provide its water source).

KILOWATT (KW)

Equal to 1,000 watts. See "watt."

KILOWATT-HOUR (kWh)

Equal to 1,000 watt-hours. See "watt-hour."

LED

Light-emitting diode, a semiconductor light source. An OLED is a LED that is fabricated using organic materials.

LINEAR ALTERNATOR

A linear motor used as a power generator for alternating current. Linear motors do not rely on torque and rotation but rather on simple linear motion.

MEGAWATT (MW)

Equal to 1,000,000 watts. See "watt."

MEGAWATT-HOUR (MWh)

Equal to 1,000,000 watt-hours. See "watt-hour."

MICROALGAE (OR MICROPHYTES)

Microscopic algae that can be found in both freshwater and marine ecosystems. Together these tiny simple species produce about half of the atmospheric oxygen by processing carbon dioxide gas. There are between 150,000 and 650,000 species of microalgae that have not yet been described and each has a unique chemical structure. Species that have been studied have been shown to produce useful compounds, such as enzymes, fatty acids, polymers, peptides, etc. See "algae biofuel."

MINTO WHEEL (ISKE MOTOR)

From an 1881 patent by Anthony and Albert Iske and later developed by Wally Minto in the 1970s, the Minto wheel uses a series of sealed containers arranged in a wheel formation (similar to the cars on a Ferris wheel) around a central hub. The containers are connected with a tube to their opposite on the other side of the wheel, and they are filled with a low-boiling-point liquid such as propane or isobutene before being sealed. As each container is heated slightly (one method is to have the containers pass through a bath of warm or hot water), the liquid expands and moves to the opposite container. The resulting weight imbalance causes a constant spinning of the entire mechanism, the energy of which can be used to run a motor or to generate electricity.

MONOCRYSTALLINE SILICON

Silicon (Si) is a semiconductor material that displays the photovoltaic effect. It was the first material to be employed in solar cells and is still the most prevalent. It can be applied for use in either a crystalline (wafer) form, or in a non-crystalline (amorphous) form. There are two types of crystalline silicon: monocrystalline and polycrystalline (aka multicrystalline). Monocrystalline is very expensive to manufacture (because it requires cutting slices from cylindrical ingots of silicon crystals that are grown with the Czochralski process) but it is the most efficient crystalline silicon technology. Its conversion efficiency is around 23%.

OMNI-DIRECTIONAL PHOTOVOLTAIC

Able to convert sunlight into electricity at any angle in relation to the sun. Most PV technologies either require or work best at an angle perpendicular to the sun's position in the sky.

ORGANIC PHOTOVOLTAIC (OPV)

Organic PV can be manufactured in solutions that can be painted or rolled onto proper substrate materials. They can be produced at very low cost in comparison with other PV technologies because they can take advantage of roll-to-roll production techniques in which the organic photovoltaic system is "printed" onto a long continuous sheet of substrate material. Current OPV technology has a conversion efficiency of only around 8%. But its low cost of production (and its good performance in lower level and indirect light) makes it an increasingly attractive option in the marketplace.

PARABOLIC TROUGH

A type of concentrated solar power that uses a long mirrored surface with the cross-sectional shape of a parabola. Sunlight that hits the mirror surface (at an angle parallel to the central axis of the parabola) is directed to the focal point of the parabola, thus providing energy to a heat transfer fluid that runs continuously along its length. The heated transfer fluid can be used to generate the steam required for turbine generators.

PHOTOVOLTAIC (PV)

The photovoltaic effect, first recognized by A. E. Becquerel in 1839, is the ability of a material to produce direct current electricity when exposed to solar radiation. It is related to the photoelectric effect, which is the ejection of an electron from a material substance (usually a more highly conductive metal as opposed to a semiconductor material) by electromagnetic radiation incident on that substance. However, in the photovoltaic effect, the electrons remain within the material (by the nature of the semiconductor material) creating positive and negative bands which can be harnessed by an electrical circuit.

PIEZOELECTRIC FIBER COMPOSITES (PFC)

An array of piezoelectric ceramic fibers arranged on a flexible substrate and protected by a resin film. PFCs provide higher performance, greater flexibility, and higher strength than monolithic piezoelectric materials.

PIEZOELECTRIC GENERATOR

A device that generates electrical power from pressure force. Common application of a piezoelectric device is as the ignition source for gas range and grill "push starters." Piezoelectric discs or torriods can be stacked within a flexible pole to generate electrical current from the bending of the pole.

POLYCRYSTALLINE SILICON

Polycrystalline is easier to manufacture than monocrystalline silicon and can be cut into square-shaped slices (monocrystalline typically has the beveled corners), but has lower conversion efficiency in its PV effect at around 18% (compared to 23% for monocrystalline). It is comprised of small crystals or crystallites. See "monocrystalline silicon" for more information.

RECIPROCATING

Operating with a repetitive back-and-forth or up-and-down linear motion.

SEEBECK THERMO ELECTRIC GENERATORS

In 1821 Thomas Johann Seebeck discovered that different metals respond uniquely to temperature differences to create a magnetic field. Modern thermoelectric generators that make use of this physical property of metals (their Seebeck coefficient) are able to generate voltage when the "hot side" is exposed to heat and the "cold side" is sufficiently insulated to maintain a differential temperature.

SEMITRANSPARENT PHOTOVOLTAICS

Solar cells that are encased in a transparent material in such a way that allows light to pass through partially. The pattern of the photovoltaic material placement can be dense or sparse, ordered dense or sparse, ordered or irregular.

SINGLE-WALLED CARBON NANOTUBES (CNT)

Carbon atoms arranged in a cylindrical form via a fullerene structure (in this linear case, a hexagonal arrangements of atoms that form a tube). These nanostructures (1/50,000th a human hair) display interesting properties of high strength and electrical conductivity. Spatial confinement of water inside carbon nanotubes alters its boiling point temperature.

SOLAR FABRIC

Flexible photovoltaic material integrated into canvas.

SOLAR THERMAL

Solar radiation used to heat a medium such as water or air.

SOLAR UPDRAFT (SOLAR CHIMNEY)

Combines the chimney/stack effect and greenhouse effect with wind turbines located at the base of a very tall tower. The tower is surrounded by a large greenhouse which serves to create superheated air at the ground level. With a sufficiently tall chimney structure, the air temperature at the top of the tower will be cool enough to provide a strong convection movement of air from the greenhouse area, into the bottom of the chimney and out of the top of the chimney (warmer air is higher pressure and moves to the lower pressure system at the top of the tower). At the outside perimeter of the greenhouse, new surface air is constantly taken into the system to be heated. As air passes from the greenhouse area into the base of the tower, it powers wind turbines located there.

STIRLING HEAT ENGINE

Device that converts heat into mechanical energy with high efficiency. This mechanical energy can then be used to power an electrical generator.

SYNERGETICS CONCEPT

As conceived by R. Buckminster Fuller, Synergetics is the study of self-organizational geometries inherent in nature and their transformative qualities. The tetrahedron is the fundamental unit of measure for Fuller's non-Euclidean and broad-reaching geometric theory of the universe.

TENSEGRITY

Individual structural members (usually metal bars) working in compression are suspended away from each other by means of a continuous tension net (usually comprised of metal cables). The term was coined by Buckminster Fuller as a portmanteau of "tensional" and "integrity." The structural system has been used in many works of public art, including those by Kenneth Snelson.

THERMOACOUSTIC ENGINE

A device that uses heat differentials to generate sound waves or, in reverse operation, uses sound wave energy to transfer heat.

THIN-FILM

As applied to photovoltaics, any of a variety of non-crystalline solar cell technologies that can be applied in very thin layers, thus reducing material costs. Sometimes referred to as "second-generation photovoltaic cells."

THIRD-GENERATION PHOTOVOLTAIC CELLS

Term applied to technological innovations that have allowed conversion efficiencies of solar cells to increase greatly via tandem or multi-junction structures that have varying bandgaps (physical property of the material that pertains to the flow of electrons through it or from it under conditions of excitation such as that provided by solar radiation). In single junction solar cells, there is energy lost when photoexcitation exceeds the limits of the particular semiconductor's bandgap.

TURBINE ROTOR

The moving part of a turbine engine which consists of a drum or a shaft with blades attached to it.

UTILITY-SCALE

Significant enough power generation so as to warrant the distribution of the energy to the utility grid (as opposed to "on-site" power generation for local use).

VENTURI EFFECT

When a fluid body, such as air, is in motion and is constricted in its path, such as by a funnel, the velocity of the fluid will increase and its static pressure will be reduced. The effect can be felt in urban areas when standing between two tall buildings, which together act as a funnel. This principle is incorporated into the design of compact wind acceleration turbines (CWAT).

VERTICAL AXIS WIND TURBINE

Any wind turbine in which the rotational axis is vertical in orientation (perpendicular to the ground plane). Types of vertical axis wind turbines include Savonius, Darrieus (eggbeater), and Giromills.

WASTE TO ENERGY (WtE)

The use of non-recyclable waste for combustion (incineration) to generate electricity, or in a small number of cases for processing into methane or similar fuel.

There are some emerging WtE technologies which do not require incineration:

- Gasification (produces hydrogen, synthetic fuels)
- Thermal depolymerization (produces synthetic crude oil)
- Pyrolysis (produces combustible tar, bio-oil, and biochars)
- Plasma arc gasification, PGP (produces syngas)

Non-thermal technologies:

- Anaerobic digestion (biogas rich in methane)
- Fermentation production (ethanol, lactic acid, hydrogen)
- Mechanical biological treatment (MBT)

Limestone scrubbers can greatly reduce the emission of harmful chemicals from WtE incineration, and while there is CO_2 released, the effect of this is less than the more toxic greenhouse gases that are produced by landfill off-gassing of methane, even if much of that methane is captured.

WATER-SOURCE HEAT SINK LOOP

A geothermal or ground-source heat pump operating in a cooling capacity. Water is passed via coiled tubing through the earth (which maintains a constant temperature between 10° and 16° Celsius depending on latitude) and then used for cooling interior space via a heat exchanger.

WATT (W)

Unit of measure of electrical power equivalent to 1/746 horsepower.
W = Volts x Amperes

WATT-HOUR

A measure of electrical energy equivalent to one watt of power used or produced for a one-hour duration.

WELLS TURBINE

Developed by Alan Wells in the late 1970s, the Wells Turbine incorporates symmetrical airfoil blades that provide continuous rotation regardless of the direction of air movement through the device. It is most typically used within oscillating water column wave power devices where the up and down movement creates compression of air into a chamber followed by the release of pressure from the chamber.

WIND MICROTURBINE

A small wind turbine with less than 3,000 watt peak capacity. Pico wind turbines are even smaller with peak capacity of less than 500 watts.

WIND ROSE

A graphic representation of the wind at any one location as measured over a long period of time. The wind rose diagram shows the frequency of wind speed and cardinal direction from which the wind is blowing.

WIND TURBINE

A rotary engine driven by the force of passing wind that can convert rotational force into electrical power.

INDEX

ACKNOWLEDGEMENTS

The LAGI 2014 design competition follows two editions held for Dubai/Abu Dhabi (2010) and New York City's Freshkills Park (2012). While developing LAGI 2014, we had the opportunity to visit renewable energy sector initiatives throughout Denmark, including projects such as the Nordic Folk Center for Renewable Energy, the forward thinking Energy Academy on Samsø Island (the island produces 140% of electrical needs from wind energy), and Wave Star (located at the Danish Wave Energy Centre test site in Hanstholm)—which is operating at 60% efficiency and rounding the corner to 80%. With all of the ambitious initiatives within and around Copenhagen, it is clear to us that this city is on target to meet its objective to become the first carbon neutral capital by 2025.

The backdrop of these initiatives created the perfect canvas for the LAGI project and we are grateful to so many institutions and individuals that helped along the way to create a successful 2014 design competition. Following are those who first come to mind, but there are many others behind the scenes without whom LAGI 2014 would not have been a success.

We were privileged to work with a Copenhagen based team whose vision and dedication helped to bring LAGI to their home. A very special thanks to Lea Schick, LAGI 2014 Project Manager. Lea generously took time out of her PhD to work with us to bring LAGI 2014 to Denmark. Thanks to IT University of Copenhagen for hosting this edition of the LAGI competitions, especially Energy Futures, the department at IT University that recognized a need to bring LAGI to Copenhagen. We are also very thankful to Region Hovedstaden and to the Culture and Leisure Committee of the City of Copenhagen for their co-financing and support of the competition, and to Dreyers Fund for providing vital financial support for this publication.

Our project partners and the institutions that provided us with critical support: Refshaleøen Holding, Danish Design Centre, Alexandra Institute, Information Studies at Aarhus University, and Green Cities.

Our media partner Shawati', the bilingual (Arabic/English) arts and culture publication based in Abu Dhabi, whose gorgeous design and thoughtful journalism is setting a new standard of excellence in socially-conscious print media.

Municipalities in Denmark, such as the Green Cities that have participated in the LAGI Green Cities partnership, are expressing interest in construction. We hope to see more than one "land art generator" built in this wonderful country over the coming years—beautiful contributions to Denmark's stated renewable energy goals.

A special thanks to Paul Schifino of Schifino Design who has designed LAGI materials along the way, including the Art+Energy Flash Cards and this very book that you are reading. As a board member of Society for Cultural Exchange, Paul has offered his advice and wisdom all along the way and we deeply cherish his friendship.

Deborah Hosking, another SCE board member, for everything including her beautiful video work for LAGI 2014.

The shortlist committee and the jurors who so graciously gave their time to the 2014 LAGI competition.

The cultural institutions that provided us with exhibit space, such as the Design Society (Danish Design Centre) and Dome of Visions.

The Danish politicians who have generously and wholeheartedly supported LAGI 2014, including Ida Auken, Martin Lidegaard, and Connie Hedegaard.

Our dear friend, Ann Rosenthal, whose copyediting skills have made this volume infinitely more readable.

All of the individuals who have helped us throughout the LAGI 2014 journey, including Gry Krogager Lund (who assumed the LAGI 2014 project management role in Denmark in January 2014), Anne Sophie Witzke, Trine Plambech, Ida Egedal Henriksson, Claudia Munkeboe, Claus Hovmøller Jensen, Natalie Mossin, Simone Kongsbak, Maria Hørmann, Marie Louise Gørvild, Gry Worre Hallberg, Peter Juul Nielsen, Karrah Beck, and our parents John and Jennifer Ferry, and Anita Monoian.

And especially to all who spent their dedicated time conceptualizing and designing for the 2014 LAGI design competition—with the belief that we can all take responsibility in working towards a more beautiful and sustainable energy future.

Robert Ferry and Elizabeth Monoian
Land Art Generator Initiative

Front cover: *Solar Hourglass*, Santiago Muros Cortés
Title page: *eMotions*, Antonio Maccà, Flavio Masi
Back cover: *Beyond the Wave*, Jaesik Lim, Ahyoung Lee, Sunpil Choi, Dohyoung Kim, Hoeyoung Jung, Jaeyeol Kim, Hansaem Kim (Heerim Architects & Planners)

Prestel Verlag, Munich
A member of Verlagsgruppe Random House GmbH

Prestel Verlag
Neumarkter Strasse 28
81673 Munich
Tel. +49 (0)89 4136-0
Fax +49 (0)89 4136-2335

www.prestel.de

Prestel Publishing Ltd.
14-17 Wells Street
London W1T 3PD
Tel. +44 (0)20 7323-5004
Fax +44 (0)20 7323-0271

Prestel Publishing
900 Broadway, Suite 603
New York, NY 10003
Tel. +1 (212) 995-2720
Fax +1 (212) 995-2733

www.prestel.com

Prestel books are available worldwide. Please contact your nearest bookseller or one of the above addresses for information concerning your local distributor.

Library of Congress Control Number is available; British Library Cataloguing-in-Publication Data: a catalogue record for this book is available from the British Library; Deutsche Nationalbibliothek holds a record of this publication in the Deutsche Nationalbibliografie; detailed bibliographical data can be found under: http://www.dnb.de.

Editorial direction: Robert Ferry, Elizabeth Monoian
Project management Prestel: Gabriele Ebbecke
Production Prestel: Andrea Cobré
Copyediting: Ann Rosenthal, David Shallis
Design: Paul Schifino, schifinodesign.com
Origination: royalmedia, Munich
Printing and binding: GCC Cuno, Calbe

Printed in Germany

ISBN 978-3-7913-5369-2

Verlagsgruppe Random House FSC® N001967
The FSC®-certified paper *Condat Perigord* was supplied by Papier-Union

FSC
www.fsc.org
MIX
Paper from responsible sources
FSC® C043106

Funding for this publication came from:
Dreyers Fond and Region Hovedstaden

dreyersfond

REGION H
The Capital Region of Denmark

The Land Art Generator Initiative is a project of Society for Cultural Exchange

www.landartgenerator.org